Environmental Analytical Chemistry

Environmental Analytical Chemistry

Editors

Stefan Tsakovski
Tony Venelinov

Basel • Beijing • Wuhan • Barcelona • Belgrade • Novi Sad • Cluj • Manchester

Editors

Stefan Tsakovski
Analytical Chemistry; Faculty
of Chemistry and Pharmacy
Sofia University "St. Kliment
Ohridski"
Sofia
Bulgaria

Tony Venelinov
Water Supply, Sewerage,
Water and Wastewater
Treatment; Faculty of
Hydraulic Engineering
University of Architecture,
Civil Engineering and
Geodesy
Sofia
Bulgaria

Editorial Office
MDPI
St. Alban-Anlage 66
4052 Basel, Switzerland

This is a reprint of articles from the Special Issue published online in the open access journal *Molecules* (ISSN 1420-3049) (available at: www.mdpi.com/journal/molecules/special_issues/Environmenta_Analytical).

For citation purposes, cite each article independently as indicated on the article page online and as indicated below:

Lastname, A.A.; Lastname, B.B. Article Title. *Journal Name* **Year**, *Volume Number*, Page Range.

ISBN 978-3-7258-0176-3 (Hbk)
ISBN 978-3-7258-0175-6 (PDF)
doi.org/10.3390/books978-3-7258-0175-6

© 2024 by the authors. Articles in this book are Open Access and distributed under the Creative Commons Attribution (CC BY) license. The book as a whole is distributed by MDPI under the terms and conditions of the Creative Commons Attribution-NonCommercial-NoDerivs (CC BY-NC-ND) license.

Contents

About the Editors vii

Stefan Tsakovski and Tony Venelinov
Environmental Analytical Chemistry
Reprinted from: *Molecules* 2024, 29, 450, doi:10.3390/molecules29020450 1

Andrea Mara, Ilaria Langasco, Sara Deidda, Marco Caredda, Paola Meloni, Mario Deroma, et al.
ICP-MS Determination of 23 Elements of Potential Health Concern in Liquids of e-Cigarettes. Method Development, Validation, and Application to 37 Real Samples
Reprinted from: *Molecules* 2021, 26, 6680, doi:10.3390/molecules26216680 6

Jungmin Jo, Ji-Yi Lee, Kyoung-Soon Jang, Atsushi Matsuki, Amgalan Natsagdorj and Yun-Gyong Ahn
Development of Quantitative Chemical Ionization Using Gas Chromatography/Mass Spectrometry and Gas Chromatography/Tandem Mass Spectrometry for Ambient Nitro- and Oxy-PAHs and Its Applications
Reprinted from: *Molecules* 2023, 28, 775, doi:10.3390/molecules28020775 24

Justyna Szulc, Małgorzata Okrasa, Małgorzata Ryngajłło, Katarzyna Pielech-Przybylska and Beata Gutarowska
Markers of Chemical and Microbiological Contamination of the Air in the Sport Centers
Reprinted from: *Molecules* 2023, 28, 3560, doi:10.3390/molecules28083560 36

Nimelan Veerasamy, Sarata Kumar Sahoo, Rajamanickam Murugan, Sharayu Kasar, Kazumasa Inoue, Masahiro Fukushi and Thennaarassan Natarajan
ICP-MS Measurement of Trace and Rare Earth Elements in Beach Placer-Deposit Soils of Odisha, East Coast of India, to Estimate Natural Enhancement of Elements in the Environment
Reprinted from: *Molecules* 2021, 6, 7510, doi:10.3390/molecules26247510 58

Edyta Boros-Lajszner, Jadwiga Wyszkowska and Jan Kucharski
Evaluation and Assessment of Trivalent and Hexavalent Chromium on *Avena sativa* and Soil Enzymes
Reprinted from: *Molecules* 2023, 28, 4693, doi:10.3390/molecules28124693 75

Galina Yotova, Mariana Hristova, Monika Padareva, Vasil Simeonov, Nikolai Dinev and Stefan Tsakovski
Multivariate Exploratory Analysis of the Bulgarian Soil Quality Monitoring Network
Reprinted from: *Molecules* 2023, 28, 6091, doi:10.3390/molecules28166091 93

Tony Venelinov, Veronika Mihaylova, Rositsa Peycheva, Miroslav Todorov, Galina Yotova, Boyan Todorov, et al.
Sediment Assessment of the Pchelina Reservoir, Bulgaria
Reprinted from: *Molecules* 2021, 26, 7517, doi:10.3390/molecules26247517 106

Mohammed Othman Aljahdali and Abdullahi Bala Alhassan
Rare Earth Elements and Bioavailability in Northern and Southern Central Red Sea Mangroves, Saudi Arabia
Reprinted from: *Molecules* 2022, 27, 4335, doi:10.3390/molecules27144335 123

Błażej Kudłak, Natalia Jatkowska, Wen Liu, Michael J. Williams, Damia Barcelo and Helgi B. Schiöth
Enhanced Toxicity of Bisphenols Together with UV Filters in Water: Identification of Synergy and Antagonism in Three-Component Mixtures
Reprinted from: *Molecules* **2022**, *27*, 3260, doi:10.3390/molecules27103260 138

Aleksander Kravos, Andreja Žgajnar Gotvajn, Urška Lavrenčič Štangar, Borislav N. Malinović and Helena Prosen
Combined Analytical Study on Chemical Transformations and Detoxification of Model Phenolic Pollutants during Various Advanced Oxidation Treatment Processes
Reprinted from: *Molecules* **2022**, *27*, 1935, doi:10.3390/molecules27061935 154

Kullapon Kesonkan, Chonnipa Yeerum, Kanokwan Kiwfo, Kate Grudpan and Monnapat Vongboot
Green Downscaling of Solvent Extractive Determination Employing Coconut Oil as Natural Solvent with Smartphone Colorimetric Detection: Demonstrating the Concept via Cu(II) Assay Using 1,5-Diphenylcarbazide
Reprinted from: *Molecules* **2022**, *27*, 8622, doi:10.3390/molecules27238622 174

Binta Hadi Jume, Niloofar Valizadeh Dana, Marjan Rastin, Ehsan Parandi, Negisa Darajeh and Shahabaldin Rezania
Sulfur-Doped Binary Layered Metal Oxides Incorporated on Pomegranate Peel-Derived Activated Carbon for Removal of Heavy Metal Ions
Reprinted from: *Molecules* **2022**, *27*, 8841, doi:10.3390/molecules27248841 185

Klaudia Stando, Ewa Korzeniewska, Ewa Felis, Monika Harnisz and Sylwia Bajkacz
Uptake of Pharmaceutical Pollutants and Their Metabolites from Soil Fertilized with Manure to Parsley Tissues
Reprinted from: *Molecules* **2022**, *27*, 4378, doi:10.3390/molecules27144378 200

About the Editors

Stefan Tsakovski

Stefan Tsakovski received his PhD degree from Sofia University "St. Kliment Ohriski" in 1999 in the field of Analytical Chemistry (Chemometrics). He is currently working as a full-time professor and head of the Chemometrics and Environmetrics Unit at the Department of Analytical Chemistry at Sofia University, "St. Kliment Ohridski". His research interests include many areas related to the data mining of information derived from chemical experiments, epidemiological studies, and environmental monitoring.

Tony Venelinov

Tony Venelinov received a PhD degree from Sofia University "St. Kliment Ohriski" in 2005 in the field of Analytical Chemistry. Upon completion, he was employed as a contract agent at the EC, DG–IRC, Institute for Reference Materials and Measurements. He is currently working as an associate professor at the Department of Water Supply, Sewerage, Water, and Wastewater Treatment at UACEG, Sofia. His current scientific interests include surface and wastewater quality analysis.

Editorial

Environmental Analytical Chemistry

Stefan Tsakovski [1] and Tony Venelinov [2,*]

[1] Chair of Analytical Chemistry, Faculty of Chemistry and Pharmacy, Sofia University St. Kliment Ohridski, 1 J. Bourchier Blvd., 1164 Sofia, Bulgaria; stsakovski@chem.uni-sofia.bg
[2] Chair of Water Supply, Sewerage, Water and Wastewater Treatment, Faculty of Hydraulic Engineering, University of Architecture, Civil Engineering and Geodesy, 1 Hr. Smirnenski Blvd., 1046 Sofia, Bulgaria
* Correspondence: tvenelinov_fhe@uacg.bg

1. Introduction

Environmental analytical chemistry has evolved into a well-established interdisciplinary field (analytical chemistry, pollution chemistry, chemical engineering, etc.), and it is currently in high demand. Investigations of the Earth's resources often exceed environmental capacity, causing various problems and changes. The management of these processes requires monitoring and analysis to gain information about organic, inorganic, and radioactive pollutants in air, water, soil, and biota. The development of new or enhancement of existing analytical methods is vital to achieve such goals. Examples include acquiring representative samples, improving sample preparation, lowering the quantification limits and measurement uncertainty, implementing appropriate methods and procedures for pollution risk assessment, revealing sources and pathways of exposure, as well as trends and the spatial distribution of analysed pollutants, among others.

The Special Issue "Environmental Analytical Chemistry" was introduced on 16 August 2021 and closed on 26 November 2023, with the submission deadline being on 20 May 2023. During this time, 18 papers were submitted seeking publication in *Molecules*.

2. An Overview of Published Articles

The diversity of studies related to the environment was highlighted by the papers received for publication in the SI "Environmental Analytical Chemistry" in the period from 13 October 2021 to 20 June 2023. These included the somewhat traditional topics "air", "soil", "sediment", "water", and "plant", and emerging ones such as "e-cigarettes" ("e-liquids").

Andrea Mara et al. (contribution 1) addressed the present lack of interest in the determination of toxic elements in electronic cigarette liquids (e-liquids). The analytical challenges for such investigations are strong matrix effects, which reflect on the existence of reliable, accurate, and validated analytical methods. In their study, the team developed and validated an ICP-MS method for the quantification of 23 elements in 37 e-liquids of different flavours, including sample pre-treatment and the optimisation of the ICP-MS conditions. Luckily, the results showed that all samples exhibited a very low amount of the investigated elements with a sum of their average concentration of ca. 0.6 mg/kg. Toxic elements in tobacco and tonic flavours (the highest and the lowest concentration of elements, respectively) were always below a few tens of μg/kg and very often below the quantification limits.

In their study, Jungmin Jo et al. (contribution 2) paid attention to some derivatives of PAHs—nitro-PAHs (NPAHs) and oxy-PAHs (OPAHs). They presented a validated method for the quantification of 18 NPAH and OPAH congeners in the atmosphere using gas chromatography coupled with chemical ionisation mass spectrometry. The application of negative chemical ionisation (NCI/MS) or positive chemical ionisation tandem mass spectrometry (PCI-MS/MS) achieved high sensitivity and selectivity for the quantification of individual NPAH and OPAH congeners without sample preparations. According to the results, the contribution of individual NPAHs and OPAHs to the total concentration differed according to the regional emission characteristics.

Markers of chemical and microbiological contamination in fitness centres were investigated by Justyna Szulc and her colleagues (contribution 3). Their study aimed to assess the particulate matter, CO_2, formaldehyde, volatile organic compound (VOC) concentration, the number and the biodiversity of microorganisms, and the presence of SARS-CoV-2 in the air, using various analytical methods. Their results showed that >99.6% of the particles are found in the $PM_{2.5}$ fraction. Different substances in various concentrations (CO_2, formaldehyde, 84 VOCs phenol, D-limonene, toluene, and 2-ethyl-1-hexanol), 422 genera of bacteria, 408 genera of fungi, and the SARS-CoV-2 virus were detected in the gym.

Inductively coupled plasma mass spectrometry (ICP-MS) was used by Nimelan Veerasamy et al. (contribution 4) to measure the concentration of trace and rare earth elements (REEs) in soils from Odisha, on the east coast of India. This analytical method was validated by the use of certified reference materials. The presented estimation of enrichment factor (EF) and geoaccumulation index (Igeo) showed that Cr, Mn, Fe, Co, Zn, Y, Zr, Cd, and U were significantly enriched, and Th was extremely enriched.

How soil contamination with Cr(III) and Cr(VI) in the presence of Na_2EDTA affects *Avena sativa* L. biomass was evaluated by Edyta Boros-Lajszner et al. (contribution 5). They assessed the remediation capacity of *Avena sativa* L. based on its tolerance index, translocation factor, and chromium accumulation, and they investigated how these chromium species affect the soil enzyme activity and physicochemical properties of soil. It was shown that the negative effect of chromium decreased the biomass of *Avena sativa* L. (aboveground parts and roots). The tolerance indices (TIs) showed that Avena sativa L. tolerates Cr(III) contamination better than Cr(VI) contamination and is of little use for the phytoextraction of chromium from the soil.

Galina Yotova et al. (contribution 6) used various indicators—primary nutrients (C, N, P), acidity (pH), physical clay content, and potentially toxic elements (PTEs: Cu, Zn, Cd, Pb, Ni, Cr, As, and Hg)—combined with chemometric and geostatistical methods to assess Bulgarian soil quality. The use of principal component analysis identified the contribution of each latent factor (the mountain soil factor, the geogenic factor, the ore deposit factor, the low nutrition factor, and the mercury-specific factor) to the overall soil quality. The spatial distribution of the soil quality patterns throughout the whole Bulgarian territory was visualised via mapping and was used to outline regions where additional measures for the monitoring of the phytoavailability of PTEs were required, -

Sediment cores were used by Tony Venelinov et al. (contribution 7) to study the temporal dynamics of anthropogenic impacts on the Pchelina Reservoir. They used ^{137}Cs activity to identify the layers corresponding to the 1986 Chernobyl accident and for the calculation of the average sedimentation rate. A Mann–Kendall test was used to reveal time trends in the elements' depth profiles (Ti, Mn, Fe, Zn, Cr, Ni, Cu, Mo, Sn, Sb, Pb, Co, Cd, Ce, Tl, Bi, Gd, La, Th, and U_{nat}) within the sediments in the sampling sites. The performed principal component analysis revealed two groups of chemical elements that were linked to anthropogenic impacts. The results obtained showed that the moderately contaminated, according to the Igeo, Pchelina Reservoir surface sediment samples have low ecotoxicity.

The bioavailability and fractionation of rare earth elements were assessed by Mohammed Othman Aljahdali and Abdullahi Bala Alhassan in mangrove ecosystems (contribution 8) using multi-elemental ratios, Igeo, bio-concentration factor (BCF), and the influence of sediment grain-size types. The results obtained showed BCF values of less than one for all the REEs determined, calling for the periodic monitoring of REE concentrations in mangroves to keep track of the sources of metal contamination and develop conservation and control strategies for these important ecosystems.

Błażej Kudłak et al. (contribution 9) utilised bioluminescent bacteria (Microtox assay) to monitor contaminants of emerging concern (CEC) mixtures at environmentally relevant doses and performed the first systematic study involving three sunscreen components (oxybenzone, OXYB; 4-methylbenzylidene-camphor, 4MBC and 2-ethylhexyl 4-methoxycinnamate, EMC) and three bisphenols (BPs—BPA, BPS or BPF) (contribution 9). A breast cell line and cell viability assay were used to determine the possible effect of these mixtures on human cells. The results from toxicity modelling with concentration addition and independent action approaches showed that mixtures containing any pair of three BPs (e.g., BPA + BPS, BPA + BPF and BPS + BPF), together with one sunscreen component (OXYB, 4MBC or EMC), interacted at environmentally relevant concentrations and had strong synergy or over additive effects. Mixtures containing any pair of OXYB, 4MBC, and EMC, and one BP had a strong propensity towards concentration-dependent underestimation. The UV filters (4MBC, EMC, and OXYB) were shown to be antagonistic toward each other.

Aleksander Kravos et al. (contribution 10) utilised advanced oxidation processes (AOPs) to understand the complex degradation processes of phenol, 2,4-dichlorophenol, and pentachlorophenol from a chemical and ecotoxicological point of view. They used instrumental analyses (HPLC–DAD, GC–MS, UHPLC–MS/MS, and ion chromatography) along with ecotoxicological assessment (Daphnia magna) to study the efficiency of ozonation, photocatalytic oxidation with immobilised nitrogen-doped TiO_2 thin films, and electrooxidation on boron-doped diamond (BDD) and mixed metal oxide (MMO) anodes. Monitoring the removal of target phenols, dechlorination, transformation products, and ecotoxicological impact, our results showed that ozonation was by far the most suitable for degradation. It showed rapid detoxification, contrary to photocatalysis, which was found to be slow and accumulated aromatic by-products.

Kullapon Kesonkan (contribution 11) proposed coconut oil as a natural solvent for the green downscaling solvent extractive determination of Cu(II) using 1,5-Diphenylcarbazide (DPC). Cu(II)-DPC complexes in an aqueous solution were transferred into the coconut oil phase, inducing a colour change, which enabled image processing on a smartphone. The developed new approach of green chemical analysis was applied to water samples in the range of 0–1 mg/mL Cu(II).

A novel biomass adsorbent based on activated carbon incorporated with sulphur-based binary metal oxide layered nanoparticles (SML-AC), including sulphur (S_2), Mn, and Sn oxide, was synthesised by Binta Hadi Jume et al. via the solvothermal method (contribution 12). The functional groups, surface morphology, and elemental composition of the newly synthesised SML-AC were studied using FTIR, FESEM, EDX, and BET, which were applied as efficient adsorbents to remove Pb^{2+}, Cd^{2+}, Cr^{3+}, and V^{5+} from the oil-rich region. The optimal pH, dosage, and time were found to provide a satisfactory adsorption capacity based on the Langmuir and Freundlich models.

Klaudia Stando et al. (contribution 13) evaluated the uptake of 14 veterinary pharmaceuticals by parsley from soil fertilised with manure. Pharmaceutical content (enrofloxacin, tylosin, sulfamethoxazole, and doxycycline) was determined in roots and leaves using liquid chromatography coupled with tandem mass spectrometry. Additionally, a solid–liquid extraction procedure combined with solid-phase extraction was developed, providing good recoveries for leaves and roots. The results obtained showed that enrofloxacin was present at the highest concentrations, doxycycline accumulated mainly in the roots, tylosin in the leaves, and sulfamethoxazole was found in both tissues.

3. Conclusions

Humanity irreversibly impacts the environment by polluting its natural vital resources—air, water, and soil. Therefore, environmental protection should become humanity's ultimate goal. A few directions should continue to be followed: (i) the development of new and the enhancement of existing analytical methods for the analysis of pollutants (organic, inorganic, and radioactive); (ii) the use of chemometric and statistical approaches for risk assessment; (iii) the determination of point and non-point pollution sources; (iv) the clarification of exposure pathways; and (v) the determination of trends and spatial distribution of analysed pollutants.

Following the initial Special Issue's success, a new Special Issue "Environmental Analytical Chemistry II" was introduced on 31 July 2023, with the submission deadline set to 31 May 2024 (to be closed on 27 November 2024). The Guest Editors would like to invite you to contribute to this Special Issue, so that your valuable unpublished research can find a worldwide audience among readers of *Molecules*.

Author Contributions: Conceptualization, S.T. and T.V.; writing—original draft preparation, S.T. and T.V.; writing—review and editing, S.T. and T.V. All authors have read and agreed to the published version of the manuscript.

Acknowledgments: The Special Issue editors attribute the contributions of all authors, reviewers, and technical assistants to the success of the SI "Environmental Analytical Chemistry".

Conflicts of Interest: The authors declare no conflicts of interest.

List of Contributions

1. Mara, A.; Langasco, I.; Deidda, S.; Caredda, M.; Meloni, P.; Deroma, M.; Pilo, M.I.; Spano, N.; Sanna, G. ICP-MS Determination of 23 Elements of Potential Health Concern in Liquids of e-Cigarettes. Method Development, Validation, and Application to 37 Real Samples. *Molecules* 2021, *26*, 6680. https://doi.org/10.3390/molecules26216680.
2. Jo, J.; Lee, J.-Y.; Jang, K.-S.; Matsuki, A.; Natsagdorj, A.; Ahn, Y.-G. Development of Quantitative Chemical Ionization Using Gas Chromatography/Mass Spectrometry and Gas Chromatography/Tandem Mass Spectrometry for Ambient Nitro- and Oxy-PAHs and Its Applications. *Molecules* 2023, *28*, 775. https://doi.org/10.3390/molecules28020775.
3. Szulc, J.; Okrasa, M.; Ryngajłło, M.; Pielech-Przybylska, K.; Gutarowska, B. Markers of Chemical and Microbiological Contamination of the Air in the Sport Centers. *Molecules* 2023, *28*, 3560. https://doi.org/10.3390/molecules28083560.
4. Veerasamy, N.; Sahoo, S.K.; Murugan, R.; Kasar, S.; Inoue, K.; Fukushi, M.; Natarajan, T. ICP-MS Measurement of Trace and Rare Earth Elements in Beach Placer-Deposit Soils of Odisha, East Coast of India, to Estimate Natural Enhancement of Elements in the Environment. *Molecules* 2021, *26*, 7510. https://doi.org/10.3390/molecules26247510.
5. Boros-Lajszner, E.; Wyszkowska, J.; Kucharski, J. Evaluation and Assessment of Trivalent and Hexavalent Chromium on Avena sativa and Soil Enzymes. *Molecules* 2023, *28*, 4693. https://doi.org/10.3390/molecules28124693.
6. Yotova, G.; Hristova, M.; Padareva, M.; Simeonov, V.; Dinev, N.; Tsakovski, S. Multivariate Exploratory Analysis of the Bulgarian Soil Quality Monitoring Network. *Molecules* 2023, *28*, 6091. https://doi.org/10.3390/molecules28166091.
7. Venelinov, T.; Mihaylova, V.; Peycheva, R.; Todorov, M.; Yotova, G.; Todorov, B.; Lyubomirova, V.; Tsakovski, S. Sediment Assessment of the Pchelina Reservoir, Bulgaria. *Molecules* 2021, *26*, 7517. https://doi.org/10.3390/molecules26247517.
8. Aljahdali, M.O.; Alhassan, A.B. Rare Earth Elements and Bioavailability in Northern and Southern Central Red Sea Mangroves, Saudi Arabia. *Molecules* 2022, *27*, 4335. https://doi.org/10.3390/molecules27144335.
9. Kudłak, B.; Jatkowska, N.; Liu, W.; Williams, M.J.; Barcelo, D.; Schiöth, H.B. Enhanced Toxicity of Bisphenols Together with UV Filters in Water: Identification of Synergy and Antagonism in Three-Component Mixtures. *Molecules* 2022, *27*, 3260. https://doi.org/10.3390/molecules27103260.
10. Kravos, A.; Žgajnar Gotvajn, A.; Lavrenčič Štangar, U.; Malinović, B.N.; Prosen, H. Combined Analytical Study on Chemical Transformations and Detoxification of Model Phenolic Pollutants during Various Advanced Oxidation Treatment Processes. *Molecules* 2022, *27*, 1935. https://doi.org/10.3390/molecules27061935.
11. Kesonkan, K.; Yeerum, C.; Kiwfo, K.; Grudpan, K.; Vongboot, M. Green Downscaling of Solvent Extractive Determination Employing Coconut Oil as Natural Solvent with Smartphone Colorimetric Detection: Demonstrating the Concept via Cu(II) Assay

Using 1,5-Diphenylcarbazide. *Molecules* **2022**, *27*, 8622. https://doi.org/10.3390/molecules27238622.

12. Jume, B.H.; Valizadeh Dana, N.; Rastin, M.; Parandi, E.; Darajeh, N.; Rezania, S. Sulfur-Doped Binary Layered Metal Oxides Incorporated on Pomegranate Peel-Derived Activated Carbon for Removal of Heavy Metal Ions. *Molecules* **2022**, *27*, 8841. https://doi.org/10.3390/molecules27248841.
13. Stando, K.; Korzeniewska, E.; Felis, E.; Harnisz, M.; Bajkacz, S. Uptake of Pharmaceutical Pollutants and Their Metabolites from Soil Fertilized with Manure to Parsley Tissues. *Molecules* **2022**, *27*, 4378. https://doi.org/10.3390/molecules27144378.

Disclaimer/Publisher's Note: The statements, opinions and data contained in all publications are solely those of the individual author(s) and contributor(s) and not of MDPI and/or the editor(s). MDPI and/or the editor(s) disclaim responsibility for any injury to people or property resulting from any ideas, methods, instructions or products referred to in the content.

Article

ICP-MS Determination of 23 Elements of Potential Health Concern in Liquids of e-Cigarettes. Method Development, Validation, and Application to 37 Real Samples

Andrea Mara [1], Ilaria Langasco [1], Sara Deidda [1], Marco Caredda [2], Paola Meloni [1], Mario Deroma [3], Maria I. Pilo [1], Nadia Spano [1] and Gavino Sanna [1,*]

[1] Dipartimento di Chimica e Farmacia, Università degli Studi di Sassari, Via Vienna 2, 07100 Sassari, Italy; a.mara@studenti.uniss.it (A.M.); ilangasco@uniss.it (I.L.); saradeidda96@tiscali.it (S.D.); paola94meloni@gmail.com (P.M.); mpilo@uniss.it (M.I.P.); nspano@uniss.it (N.S.)
[2] AGRIS Sardegna, Loc. Bonassai, S.S. 291 Km 18.6, 07100 Sassari, Italy; mcaredda@agrisricerca.it
[3] Dipartimento di Agraria, Università degli Studi di Sassari, Viale Italia 39/a, 07100 Sassari, Italy; mderoma@uniss.it
* Correspondence: sanna@uniss.it; Tel.: +39-079-229-500

Citation: Mara, A.; Langasco, I.; Deidda, S.; Caredda, M.; Meloni, P.; Deroma, M.; Pilo, M.I.; Spano, N.; Sanna, G. ICP-MS Determination of 23 Elements of Potential Health Concern in Liquids of e-Cigarettes. Method Development, Validation, and Application to 37 Real Samples. *Molecules* 2021, 26, 6680. https://doi.org/10.3390/molecules26216680

Academic Editor: Pawel Pohl

Received: 13 October 2021
Accepted: 29 October 2021
Published: 4 November 2021

Publisher's Note: MDPI stays neutral with regard to jurisdictional claims in published maps and institutional affiliations.

Copyright: © 2021 by the authors. Licensee MDPI, Basel, Switzerland. This article is an open access article distributed under the terms and conditions of the Creative Commons Attribution (CC BY) license (https://creativecommons.org/licenses/by/4.0/).

Abstract: The lack of interest in the determination of toxic elements in liquids for electronic cigarettes (e-liquids) has so far been reflected in the scarce number of accurate and validated analytical methods devoted to this aim. Since the strong matrix effects observed for e-liquids constitute an exciting analytical challenge, the main goal of this study was to develop and validate an ICP-MS method aimed to quantify 23 elements in 37 e-liquids of different flavors. Great attention has been paid to the critical phases of sample pre-treatment, as well as to the optimization of the ICP-MS conditions for each element and of the quantification. All samples exhibited a very low amount of the elements under investigation. Indeed, the sum of their average concentration was of ca. 0.6 mg kg^{-1}. Toxic elements were always below a few tens of a μg per kg^{-1} and, very often, their amount was below the relevant quantification limits. Tobacco and tonic flavors showed the highest and the lowest concentration of elements, respectively. The most abundant elements came frequently from propylene glycol and vegetal glycerin, as confirmed by PCA. A proper choice of these substances could further decrease the elemental concentration in e-liquids, which are probably barely involved as potential sources of toxic elements inhaled by vapers.

Keywords: e-cigarettes; e-liquids; toxic elements; trace elements; ICP-MS

1. Introduction

Electronic cigarettes (e-cigarettes) were introduced on the market in the mid-2000s and are considered a healthier alternative to traditional smoking. There are different types of devices, components, and e-liquids [1,2] to meet all consumer needs. However, all of them share the same components as well as principle of functioning. It consists of a power source, such as a rechargeable lithium battery, and a cartridge (or a tank) containing the liquid (henceforward called e-liquid). An electrical resistance (i.e., the atomizer), activated by a puff or by a button, is the heating element, which promotes the nebulization of the e-liquid and the formation of the aerosol inhaled by the vaper. An electronic system may be able to report the information related to the nebulization process (volts, watts, ohms, puffs) from a smartphone, LEDs, or an integrated screen. The flavored e-liquid is mainly formed by low amounts of water and two EU food additives, namely propylene glycol (E1520, according to the EU classification, (PG)) [3] and vegetal glycerin (E422, (VG)) [4]. Both generally recognized as safe (GRAS) by the U.S. Food and Drug Administration. In addition, it may contain variable amounts of nicotine, never exceeding the concentration of 20 mg cm^{-3} in EU countries [5].

Although they are considered healthier than traditional cigarettes, owing to their lack of toxic products formed by combustion [6–9], the health risks related to the use of e-cigarettes have also been widely studied [10–13]. The apparatuses mainly involved are the respiratory [14,15], the cardiovascular [16,17], the nervous [18], and the reproductive systems [19]. Nicotine [20] is likely one of the most toxic components assimilated by means of tobacco smoke. It is quickly assimilated [21] and adversely affects respiratory, cardiovascular, renal, and reproductive systems [22]. However, the presence of nicotine in e-liquids is optional, and it may be dosed to accomplish any nicotine replacement therapy needs. In a close analogy to what happens with traditional tobacco smoke, attention has been paid to the formation of potentially harmful compounds from the thermal degradation of the main ingredients of e-liquids, such as the flavors, PG and VG [23–25]. Hence, chromatographic methods devoted to determining species such as nitrosamines, VOCs, carbonyls, polycyclic aromatic hydrocarbons, and aldehydes have been developed for this purpose [26–29]. Nevertheless, and only with some exceptions [30], the absence of any combustion process and the low temperatures measured in the nebulization process of e-liquids greatly reduce the concentration and the number of harmful species in comparison to what is observed in the smoke of traditional cigarettes [7]. Moreover, the concentration of toxic metals and metalloids contained in the aerosol of e-cigarettes seems lower than that measured in traditional smoke [31–33]. Although, Badea et al. measured detectable amounts of rare earths in the serum of e-cigarette smokers [34]. Moreover, in this case, many analytical methods for the determination of elements in aerosols and—more rarely—in e-liquids were recently developed [35]. Only in a few cases, elemental determination is carried out on both liquids and aerosols [25,36–38]. In these contributions, the concentration of the elements measured was higher in aerosols rather than in e-liquids, with the only exception being the study performed by Beauval et al. [25], where the amount found in both matrices was comparable. The elemental amounts found in these studies are hardly comparable among them, due to the great differences in e-cigarettes, e-liquids, and sampling techniques and puffing protocols in the aerosol determination [39,40].

Moreover, the analytical technique used for elemental determination is a possible reason for the differences observed among the literature studies. Inductively coupled plasma (ICP) methods (e.g., ICP-mass spectrometry (ICP-MS) [7,25,33,36–46] or ICP-optical emission spectroscopy (ICP-OES) [47–50]) are the instrumental techniques mainly used for this purpose, although sometimes X-ray fluorescence methods have also been used [51]. As far as the toxic elements are concerned, those most frequently measured are As [7,25,36,37,39,41–45,49,51], Pb [7,25,33,36,37,41,43–45,47,51], Cd [7,25,33,36,37,41–45,51], and Cr [7,25,33,36,39,41,43–45,47,48,51], while sometimes also Sn [36,39,44,45,47–49], Sb [25,36,39,41,49], Hg [25,41,44,45], and Tl [25,41] were quantified.

Although it may seem obvious that determining the amounts of elements that are potentially health-threatening contained in e-liquids might be important in defining the overall level of exposure of vapers, until now the limits posed (or suggested) by countries [52,53] or international organizations [54] on the concentration of toxic elements are quite rare. This also reflects the scarce number of literature contributions addressed to this aim. From a purely analytical viewpoint, it is surprising that the possibility of a severe bias in the ICP-MS determination of these analytes, caused by a matrix almost entirely formed by organic species such as e-liquids, was considered only by one research group [41]. In addition, only two research groups [25,36,41] attempted to quantify a reasonable number among the toxic elements and oligoelements while, to the best of our knowledge, toxic elements such as Ba, Bi, and U were never quantified in e-liquids.

Therefore, the principal aim of this contribution was to develop and validate an ICP-MS method able to determine the total amount of the 23 elements of potential health concern, i.e., Al, As, B, Ba, Be, Bi, Cd, Co, Cr, Cu, Fe, Hg, Li, Mn, Mo, Ni, Pb, Se, Sb, Sn, Tl, U, and Zn, in a reliable sampling of e-liquids produced in Sardinia, Italy. Particular attention has been taken to optimize critical phases such as sample pre-treatment and quantification method. Samples were analyzed simply after a proper dilution or after microwave-assisted mineralization, whereas external calibration (in the absence or in the presence of the matrix) and internal calibration were kept into consideration for quantification. Since the need for studies aimed to evaluate the role of the flavor components in the contamination by trace elements in e-liquids was evidenced in previous publications [25], the optimized methods were tested on 37 different flavors and on the constituents of all e-liquids (i.e., PG, VG, water, and nicotine).

2. Results and Discussion
2.1. Method Assessment
2.1.1. Sample Pre-Treatment

The technique of sample pre-treatment most frequently described in the literature for e-liquids [25,36,37,41] is a simple dilution in an HNO_3 aqueous solution, exploiting their complete solubility in such a solvent. However, the e-liquids may represent in principle a strongly interfering matrix in ICP-MS measurement, due to their very high amount (always over 95% w/w) of organic matter. Consequently, low levels of dilution of e-liquids also potentially allow for quantifying the trace elements, but inevitably increase the noise and the interference of the carbon-based polyatomic ions (especially on the quantification of elements such as Cr and V [41]). On the other hand, high levels of dilution reduce bias due to matrix effects, but likely permit the quantification only of the most abundant elements. In order to allow a comparison with literature data, the first approach of sample pre-treatment chosen in this study was the dilution of e-liquids, according to Beauval et al. [25,41]. The same method was also used for pre-treating the pure organic constituents of the e-liquids (i.e., PG, VG, and nicotine). Hence, ca. 0.3 g of sample, exactly weighted on an analytical balance (±0.0001 g uncertainty), was diluted to a final volume of 15 cm^3 with an aqueous solution containing a 2% (v/v) HNO_3 solution and 0.1% (v/v) of Triton^{TM} X-100. In addition, by weighting 1.5 g and 0.15 g of the sample (finale volume 15 cm^3), 1:10 (w/v) and 1:100 (w/v) dilutions were also tested to optimize the method. All samples were filtered through a 0.22 μm nylon filter prior to ICP-MS analysis. Unfortunately, a high number of ionic counts for blanks were observed for several analytes (i.e., for Cr, V, and Zn) when the 1:10 and 1:50 dilutions were used, likely due to a very strong interference by polyatomic ions. Conversely, using a 1:100 dilution, this interference was reduced, but it was no more possible to quantify the trace analytes. In addition, the high amount of organic species conveyed to plasma causes serious instrumental issues, such as a thick deposit of soot in the RF coil, in the torch, and in the cones. For these reasons, after some preliminary tests, the pre-treatment for dilution was abandoned.

Another option for the decomposition of the organic matrix in the e-liquids is an acid/oxidant attack assisted by microwaves. An amount of 0.3 g of the sample, exactly weighted on an analytical balance, was treated with 0.5 cm^3 of HNO_3 and 6 cm^3 of water inside a 15 cm^3 internal volume polytetrafluoroethylene (PTFE) vessel. Due to the high amounts of polyols contained in this matrix, the operative conditions used were the best compromise among the maximization of the oxidizing power and the minimization of the risk of unpredictable and potentially violent reactions inside the vessels. Figure 1 shows the trends of the microwave power and of the internal temperature and pressure of the vessel.

The whole mineralization cycle lasts 70 min. After this time, the vessels were opened at room temperature and the mineralized samples were diluted up to 15 cm^3 and filtered through a 0.22 μm nylon filter.

2.1.2. ICP-MS Method

Table 1 reports the instrumental parameters and the elemental settings used for the ICP-MS determination of 23 toxic elements and oligoelements in e-liquids.

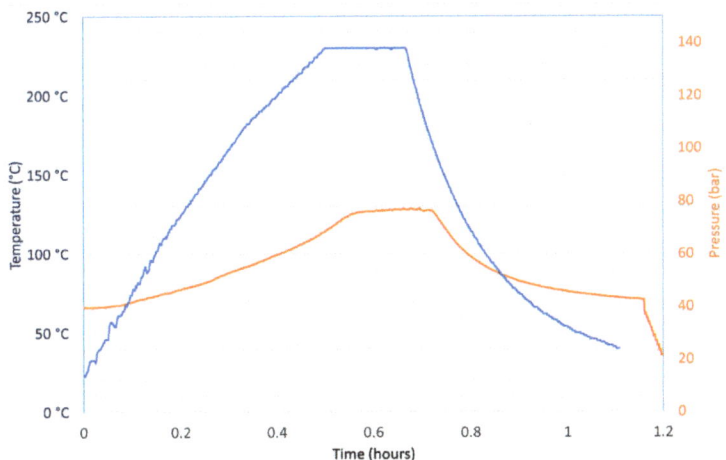

Figure 1. Trends of temperature and pressure inside vessels along a typical mineralization cycle of 0.3 g of an e-liquid dissolved in 0.5 cm^3 of HNO$_3$ and 6 cm^3 of water.

Table 1. Instrumental parameters and elemental settings used for the ICP-MS determination of 23 toxic elements and oligoelements in e-liquids.

RF power generator (W)	1300	KED ^{a} mode cell entrance voltage (V)	−8.0
Ar plasma flow (dm^3 min^{-1})	18.0	KED mode cell exit voltage (V)	−25.0
Ar auxiliary flow (dm^3 min^{-1})	1.20	Resolution (Da)	0.7
Ar nebulizer flow (dm^3 min^{-1})	0.91	Scan mode	Peak hopping
Nebulizer	Meinhard®, glass	Detector mode	Dual
Spray chamber	Cyclonic, glass	Dwell time (ms)	50
Skimmer and sampling cones	Nickel	Number of points per peak	3
Sampling depth (mm)	0	Acquisition time (s)	6
Deflector voltage (V)	−8.00	Acquisition dead time (ns)	35
Analog stage voltage (V)	−1750	KED gas	Helium, 99.999%
Pulse stage voltage (V)	+1350	Masses of optimization	^{7}Li, ^{89}Y and ^{205}Tl

Quantification Ion (% abundance)	Interferents	Analyzing Mode	He Flow (cm^3 min^{-1})	Correction Equation
^{27}Al$^+$ (100)	^{11}B^{16}O$^+$; ^{13}C^{14}N$^+$; ^{11}Be^{16}O$^+$; ^{26}Mg^{1}H$^+$; ^{12}C^{15}N$^+$; ^{54}Cr^{2+}; ^{54}Fe^{2+}	KED	3.5	none
^{75}As$^+$ (100)	^{40}Ar^{35}Cl$^+$; ^{59}Co^{16}O$^+$; ^{39}K^{36}Ar$^+$; ^{63}Cu^{12}C$^+$; ^{40}Ca^{35}Cl$^+$; ^{58}Ni^{16}O^{1}H$^+$	KED	3.0	none
^{11}B$^+$ (80.1)	none	Normal		none
^{138}Ba$^+$ (71.7)	^{40}Ar$_2$^{58}Ni$^+$; ^{138}La$^+$; ^{122}Sn^{16}O$^+$; ^{137}Ba^{1}H$^+$; ^{121}Sb^{16}O^{1}H$^+$	KED	4.0	−0.000901 × ^{139}La −0.002838 × ^{140}Ce
^{9}Be$^+$ (100)	none	Normal		none
^{209}Bi$^+$ (100)	none	Normal		none

Table 1. Cont.

Isotope	Interferences	Mode	KED (a)	Correction
^{111}Cd$^+$ (12.80)	^{95}Mo^{16}O$^+$; ^{97}Mo^{14}N$^+$; ^{79}Br^{16}O$_2^+$; ^{94}Zr^{16}O^{1}H$^+$; ^{71}Ga^{40}Ar$^+$	KED	4.0	none
^{59}Co$^+$ (100)	^{43}Ca^{16}O$^+$; ^{42}Ca^{16}O^{1}H$^+$; ^{24}Mg^{35}Cl$^+$; ^{40}Ar^{18}O^{1}H$^+$; ^{118}Sn^{2+}; ^{27}Al^{16}O$_2^+$; ^{58}Ni^{1}H$^+$; ^{24}Mg^{35}Cl$^+$	KED	3.5	none
^{52}Cr$^+$ (83.79)	^{40}Ar^{12}C$^+$; ^{36}Ar^{16}O$^+$; ^{1}H^{35}Cl^{16}O$^+$; ^{104}Pd^{2+}; ^{51}V^{1}H$^+$; ^{40}Ca^{12}C$^+$; ^{38}Ar^{14}N$^+$	KED	3.0	none
^{63}Cu$^+$ (69.17)	^{40}Ar^{23}Na$^+$; ^{31}P^{16}O$_2^+$; ^{47}Ti^{16}O$^+$; ^{28}Si^{35}Cl$^+$; ^{51}V^{12}C$^+$	KED	4.0	none
^{57}Fe$^+$ (2.12)	^{40}Ar^{16}O^{1}H$^+$; ^{40}Ca^{16}O^{1}H$^+$; ^{40}K^{16}O^{1}H$^+$	KED	3.0	none
^{7}Li$^+$ (92.50)	none	Normal		none
^{202}Hg$^+$ (22.86)	^{186}W^{16}O$^+$	Normal		none
^{55}Mn$^+$ (100)	^{40}Ar^{14}N^{1}H$^+$; ^{37}Cl^{18}O$^+$; ^{39}K^{16}O$^+$	KED	3.0	none
^{98}Mo$^+$ (24.13)	^{98}Ru$^+$; ^{81}Br^{17}O$^+$; ^{40}K$_2^{18}$O$^+$; ^{58}Ni^{40}Ar$^+$; ^{63}Cu^{35}Cl$^+$	Normal		$-0.10961 \times ^{101}$Ru
^{60}Ni$^+$ (26.22)	^{44}Ca^{16}O$^+$; ^{43}Ca^{16}O^{1}H$^+$; ^{23}Na^{37}Cl$^+$; ^{25}Mg^{35}Cl$^+$; ^{28}Si^{16}O$_2^+$	KED	3.5	none
^{208}Pb$^+$ (52.40)	none	Normal		none
^{121}Sb$^+$ (57.21)	^{107}Ag^{14}N$^+$; ^{109}Ag^{12}C$^+$; ^{105}Pd^{16}O$^+$; ^{81}Br^{40}Ar$^+$; ^{120}Sn^{1}H$^+$	KED	3.5	none
^{82}Se$^+$ (8.73)	^{82}Kr$^+$; ^{81}Br^{1}H$^+$; ^{66}Zn^{16}O$^+$; ^{68}Zn^{14}N$^+$; ^{164}Dy^{2+}; ^{65}Cu^{16}O^{1}H$^+$	KED	3.5	$-0.00783 \times ^{83}$Kr
^{120}Sn$^+$ (32.58)	^{39}K^{81}Br$^+$; ^{80}Se^{40}Ar$^+$; ^{104}Pd^{16}O$^+$; ^{104}Ru^{16}O$^+$	KED	3.5	none
^{205}Tl$^+$ (70.26)	^{189}Os^{16}O$^+$	Normal		none
^{238}U$^+$ (99.3)	none	Normal		none
^{66}Zn$^+$ (27.90)	^{50}Ti^{16}O$^+$; ^{34}S^{16}O$_2^+$; ^{132}Ba^{2+}; ^{50}Cr^{16}O$^+$; ^{65}Cu^{1}H$^+$; ^{26}Mg^{40}Ar$^+$; ^{31}P^{35}Cl$^+$; ^{52}Cr^{14}N$^+$	KED	3.0	none

a Kinetic Energy Discrimination, KED.

Whereas the elemental settings used for the determination of Al, As, Cd, Cr, Cu, Fe, Hg, Mn, Mo, Ni, Pb, Se, Sb, Tl, and Zn were from methods previously assessed by this research group [55–57], slightly modified to optimize them towards this matrix, those used to quantify B, Ba, Be, Bi, Co, Li, Sn, and U were specifically devoted to e-liquids. In particular, the choice of quantification ion is extremely important for the overall reliability of the method. It should be the best compromise between the lowest LoD and the minimization of any possible interference. Be, Bi, and Co are monoisotopic, hence no alternative is possible. For elements showing a multiplicity of isotopes, the most abundant one provides generally the highest instrumental sensitivity, and this is a very appealing feature when very low concentrations must be measured. For this reason, ^{11}B, ^{138}Ba, ^{7}Li, ^{120}Sn, and ^{238}U, respectively, were chosen for the quantification of these elements in this study. The interference from molecular ions (mainly from oxides, carbides, nitrides, hydrides, or Ar-based species formed in the plasma, but also from polyatomic ions that originated from elements contained in high amounts in the matrix) is one of the most meaningful causes of bias in ICP-MS measurements [58]. Its presence/absence for each analyte has been established based on the behavior of the ionic signal measured at the increasing of He flows. A change of the slope of the decreasing trend of the ionic signal measured on a real sample at the increasing of the He flow accounted for the presence of an interference by molecular ions [59]. For explanatory purposes, Figure 2 reports the behavior of the ionic signal of the $^{52}Cr^+$ ion at variations of the He flow. This behavior is similar to those observed for the remaining elements determined in KED mode.

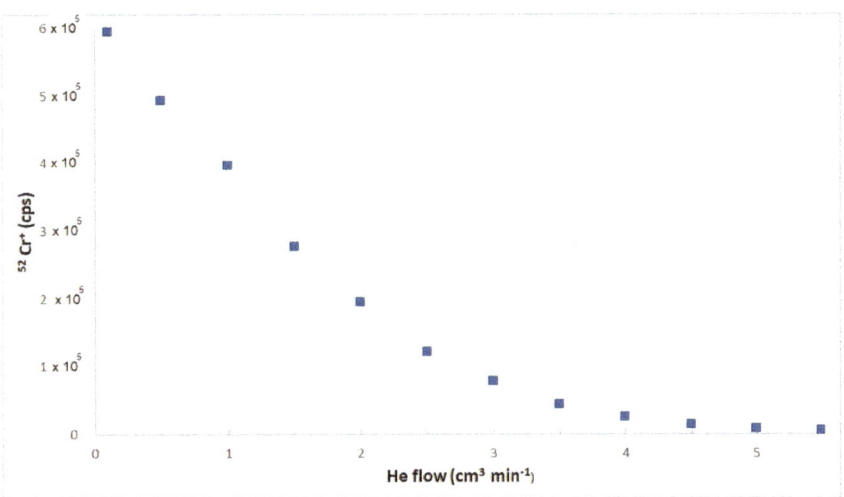

Figure 2. Dependence of the ionic signal of the $^{52}Cr^+$ ion at variations of the He flow.

It is evident that the first linear tract of the curve (He flow up to 2 cm^3 min^{-1}) gives an account for the removal of the molecular ionic interference, likely due to the $^{40}Ar^{12}C^+$ ion, whereas the second linear tract of the curve (roughly parallel to the x-axis, He flows higher than 4 cm^3 min^{-1}) accounts for a slight reduction of signal of the elemental ion as a function of the increase of the He flow. The abscissa of the intersection among both linear plots provided the optimized He flow aimed to remove the interference, which has been observed for 14 elements out 23. Hence, while the number of ionic counts of B, Be, Bi, Li, Hg, Mo, Pb, Tl, and U was measured in normal mode, the bias from polyatomic ions has been minimized for the remaining elements, using He flows between 3 cm^3 min^{-1} and 4 cm^3 min^{-1}.

2.1.3. Quantification, Quality Assurance and Quality Control

Both external and internal calibration approaches have been used in this study. External calibration (using either analyte standard solutions dissolved in water or in a synthetic matrix, henceforward called SM, formed by 54% of PG, 43% of VG, and 3% of water, respectively) has always been used for all pre-treatments chosen (dilution or microwave-assisted mineralization), but the ascertainment of a severe matrix effect in the quantification of almost all elements suggested the use of the internal calibration, accomplished by means of multiple additions of standard solutions. Figure 3 reports the linear calibration plots obtained using both external and internal calibration for the determination of As.

Constant and proportional bias due to matrix effects are well evident by the comparison of the behaviors of the two external calibration plots (obtained either on a 2% (v/v) HNO$_3$ solution in water, line 1, or on a 2% (v/v) HNO$_3$ solution in SM, line 2) and of the three internal calibration plots, obtained on tobacco, tonic, and fruity flavors, respectively. A similar behavior is also evident for the remaining elements quantified. It is well known that the internal calibration method has evident disadvantages with respect to the external calibration method. Firstly, it is cumbersome and quite time-consuming, since it requires the preparation of calibration curves for each individual sample and the relevant blanks [60]. In addition, it is characterized by a very high uncertainty, due to the widening of the confidence range of the internal calibration plot [61]. Conversely, its careful execution can almost suppress any matrix interferences, allowing the obtainment of a basically bias-free measurement [62]. A 10 µg dm^{-3} solution of Rh was used as an internal standard to compensate for any possible signal instability, while a washing cycle of at least 80 s was interposed between two consecutive samples to eliminate any possible memory effect. All data have been blank-corrected. In order to constantly monitor the overall level of accuracy of the method, one reagent blank every five samples was analyzed, whereas a standard solution containing 0.1 µg kg^{-1} of Be, Bi, Cd, Co, Tl, and U, 1 µg kg^{-1} of Li, Pb, Sb, and Se, 10 µg kg^{-1} of As, Al, Cu, and Mn, 50 µg kg^{-1} of B, Ba, Cr, and Hg, and 100 µg kg^{-1} of Fe, Mo, Ni, Sn, and Zn in SM was analyzed every ten samples. Each sample was analyzed in duplicate, and each piece of analytical data is the average of four replicated ICP-MS measurements.

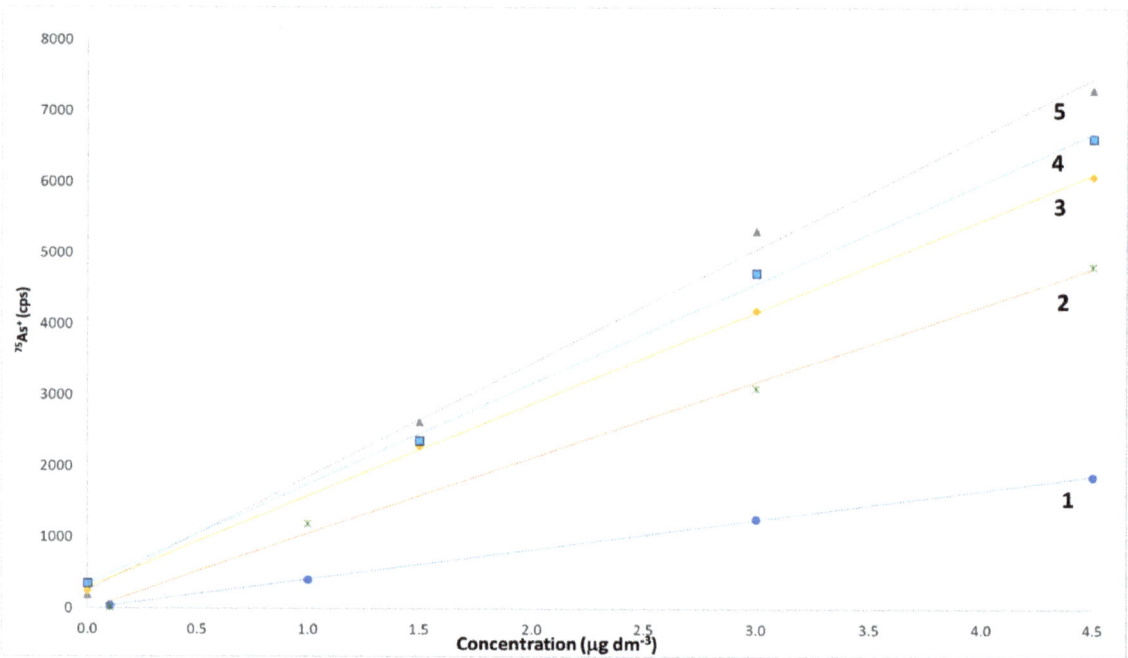

Figure 3. Linear calibration plots obtained using both external and internal calibration for the determination of As. Line 1, external calibration, As concentration in the range between 0.1 and 4.5 µg dm^{-3} in 2% (v/v) HNO$_3$ in water; line 2, external calibration, As concentration in the range between 0.1 and 4.5 µg dm^{-3} in 2% (v/v) HNO$_3$ in synthetic matrix; line 3, internal calibration for a tobacco flavor sample; line 4, internal calibration for a tonic flavor sample; line 5, internal calibration for a fruity flavor sample. In calibration lines 3–5, the amounts of As added to samples are of 113 pg, 226 pg, and 339 pg, respectively.

2.2. Validation

Validation was performed in terms of Limit of Detection (LoD), Limit of Quantification (LoQ), and precision. Table 2 reports the relevant figures for all the parameters evaluated.

Table 2. Validation parameters for the ICP-MS determination of the total amount of 23 toxic elements and oligoelements in e-liquids.

Element	LoD (µg kg^{-1})	LoQ (µg kg^{-1})	Repeatability (CV%)	Element	LoD (µg kg^{-1})	LoQ (µg kg^{-1})	Repeatability (CV%)
Al	26	84	10	Li	0.37	1.2	40
As	0.51	1.7	40	Mn	1.6	5.1	40
B	37	120	60	Mo	0.45	1.5	70
Ba	15	50	30	Ni	2.3	7.4	60
Be	0.057	0.19	70	Pb	0.80	2.7	40
Bi	0.089	0.29	80	Sb	1.1	3.7	50
Cd	0.12	0.39	90	Se	4.6	15	100
Co	0.089	0.29	60	Sn	0.24	0.78	30
Cr	4.2	14	70	Tl	0.055	0.18	50
Cu	5.2	17	100	U	0.21	0.69	30
Fe	53	180	90	Zn	62	200	30
Hg	4.5	15	90				

LoD and LoQ have been calculated according to Currie [63]. The standard deviation needed for the calculation of the LoD (and, hence, also for the LoQ) was calculated on the repeated measurement of 15 blanks. At last, for the elements previously measured in e-liquids, the data obtained are comparable with those reported in the literature [25,36,41,42]. Precision has been measured in terms of repeatability, twice quantifying the same real sample along the whole analytical process in the same analytical session by means of an internal calibration method (multiple standard addition). Quite high CV% values have been obtained for all parameters, ranging between 10% (for Al) and 100% (for Cu and Se). Since Se has always been found below its LoD for all samples of e-liquids, the precision was evaluated for this element by spiking the samples with the amount of Se standard solution needed to reach a final concentration in the samples slightly over the LoQ (i.e., 15 µg kg^{-1}). Regardless, it is well known that, for the peculiar features of the standard addition method, the precision parameters measured that work in this way are much worse than those measured by quantifying by external calibration [61]. Trueness is usually measured by (i) analyzing a certified reference material (CRM), (ii) by comparison with an independent and validated analytical method, or, ultimately, (iii) with recovery tests, performed by spiking the sample with repeated amounts of pure analyte. Unfortunately, none of these methods can be used in this case. To the best of our knowledge, no CRM of e-liquids is commercially available, and the heavy matrix effect observed in this study, well substantiated in Figure 3, discourages using any apparently "similar" matrix. In addition, no independent and validated analytical method capable of measuring the concentration of 23 elements at concentrations no higher than a few µg dm^{-3} is currently available in our labs. Finally, the close similarities among the application of both the standard addition method and the recovery tests based on spiked samples suggest also discarding this option. In the attempt to obtain a tentative evaluation of the trueness of the method, a "synthetic e-liquid" was prepared, where known amounts of each analyte were added to the SM solution to reach a concentration close to the average value measured for the real samples. For the analytes that were not frequently quantified, their final concentration in the synthetic e-liquid was the LoQ. Multiple spiking tests, performed on this sample, allowed the obtainment of recoveries between 81% (for Cr) and 118% (for Zn), exhibiting hence a negligible level of bias if compared to the very low level of concentration measured.

2.3. Concentration of 23 Elements in 37 e-Liquids

The data relative to the determination of the amount of 23 elements in 37 e-liquids belonging to three main flavors (i.e., fruity, tobacco, and tonic) are summarized, in terms of both mean and ranges, in Table 3.

Table 3. Mean amounts, ranges (both in µg kg^{-1}), and percentage of quantified samples (C > LoQ) in the determination of 23 toxic elements and oligoelements in 37 different e-liquids.

Elements	All Flavors (n = 37) (Mean; Range; % Samples > LoQ)			Fruity Flavors (n = 9) (Mean; Range; % Samples > LoQ)			Tobacco Flavors (n = 16) (Mean; Range; % Samples > LoQ)			Tonic Flavors (n = 12) (Mean; Range; % Samples > LoQ)		
Al	<33;	<26–160	8%	<35;	<26–110	11%	<37;	<26–160	13%	<26;	<26; <26	0%
As	5;	0.6–11	100%	6;	0.6–8	100%	5;	1.5–11	100%	6;	0.8–10	100%
B	<61;	<37–140	54%	<50;	<37–100	33%	<60;	<37–100	56%	<70;	<37–140	67%
Ba	<27;	<15–130	62%	<55;	<15–130	78%	<20;	<15–45	63%	<18;	<15–30	50%
Be	<0.06; <0.057–0.12		14%	<0.07; <0.057–0.12		33%	<0.058; <0.057–0.06		13%	<0.057; <0.057–<0.057		0%
Bi	<0.11; <0.089–0.3		27%	<0.09; <0.089–0.1		11%	<0.14; <0.089–0.3		56%	<0.089; <0.089–<0.089		0%
Cd	<0.14; <0.12–1		5%	<0.12; <0.12–<0.12		0%	<0.18; <0.12–1		6%	<0.12; <0.12–0.13		8%
Co	<0.34; <0.089–0.9		84%	<0.3; <0.089–0.9		78%	<0.4; 0.1–0.8		100%	<0.3; <0.089–0.6		67%
Cr	34;	20–40	100%	38;	30–40	100%	34;	30–40	100%	32;	20–40	100%
Cu	<7.1;	<5.2–20	41%	<10;	<5.2–20	56%	<7;	<5.2–14	50%	<5.4;	<5.2–7	17%
Fe	<308;	<53–3000	65%	<66;	<53–100	56%	<570;	<53–3000	88%	<140;	<53–1000	42%
Hg	<5;	<4.5–14	8%	<5;	<4.5–10	11%	<5;	<4.5–14	6%	<5;	<4.5–5	8%
Li	1.9;	0.7–9	100%	1.5;	0.8–2	100%	2.5;	0.9–9	100%	1.5;	0.7–2.2	100%
Mn	<10;	<1.6–80	73%	<2;	<1.6–4	67%	<18;	<1.6–80	69%	<5;	<1.6–20	83%
Mo	<0.8;	<0.45–3	57%	<0.6;	<0.45–1	44%	<1;	<0.45–3	81%	<0.7;	<0.45–2	33%
Ni	<3.5;	<2.3–14	46%	<3.4;	<2.3–7	67%	<3.8;	<2.3–14	44%	<3.2;	<2.3–6	33%
Pb	<1.0;	<0.8–3	19%	<1.4;	<0.8–3	44%	<0.84;	<0.8–1.3	13%	<0.81;	<0.8–1	8%
Sb	<2.9;	<1.1–10	73%	<1.8;	<1.1–4	78%	<3.1;	<1.1–7	75%	<3.4;	<1.1–10	67%
Se	<4.6;	<4.6–<4.6	0%	<4.6;	<4.6–<4.6	0%	<4.6;	<4.6–<4.6	0%	<4.6;	<4.6–<4.6	0%
Sn	<0.6;	<0.24–4	65%	<1;	<0.24–4	67%	<0.5;	<0.24–1.6	81%	<0.37; <0.24–1.5		42%
Tl	<0.07; <0.055–0.16		46%	<0.07; <0.055–0.16		22%	<0.06; <0.055–0.15		38%	<0.08; <0.055–0.15		75%
U	<0.29; <0.21–0.7		41%	<0.29; <0.21–0.6		44%	<0.31; <0.21–0.6		44%	<0.26; <0.21–0.7		33%
Zn	<109;	<62–300	76%	<150;	<62–300	89%	<90;	<62–220	63%	<100;	<62–170	83%

Each sample has been analyzed twice. In italics: data below the LoD; in underlined: data below the LoQ. All average data have been rounded after calculation. The average data prefixed with the sign "<" have been calculated based on at least one concentration that has been found below the corresponding LoD.

The data reported in Table 3 show that all e-liquids considered exhibit a very low amount of the evaluated analytes, being only ca. 0.6 mg kg^{-1}, the sum of their average concentrations measured in this study. These amounts are comparable with the literature data [25,36,41] and confirm that the amount of potentially toxic elements contained in e-devices is orders of magnitude less than in traditional cigarettes [44,45]. Among all elements considered, only Se has always been found below its LoD in all samples analyzed. This is not surprising, knowing the scarce sensitivity of Se in terms of counts per second as a function of its concentration [59]. In addition, the mineralization process implies a dilution of 50 times of the sample, hence the effective LoD for mineralized solutions is 92 ng dm^{-3}, i.e., quite close to the instrumental detection limit [64]. As far as the remaining elements are concerned, only six elements out of 36 (i.e., As, Co, Cr, Fe, Li, and Mn) exhibit a mean concentration higher than the LoQ. Only three elements, i.e., As, Cr, and Li, are always quantified (100% of the samples), but none of them seem to represent an effective health threat for vapers. Indeed, the highest amount measured for these elements is 11 µg kg^{-1} for As and 40 µg kg^{-1} for Cr. Only for the sake of comparison, the limits posed by the EU guidelines for these elements in water intended for human use are 10 µg kg^{-1} for As and 50 µg kg^{-1} for Cr, respectively [65], whereas the amount of daily intake for humans of water is, obviously, several orders of magnitude higher than that of e-liquids. Additionally, Li is less abundant than As and Cr, and its average concentration is only

1.9 µg kg^{-1}. The average concentration of the elements most frequently quantified are, in decreasing percentage, Co (average concentration <0.34 µg kg^{-1}, quantified in the 84% of samples), Zn (<109 µg kg^{-1}, 76%), Mn (<10 µg kg^{-1}, 73%), Sb (<2.9 µg kg^{-1}, 73%), and Fe (<308 µg kg^{-1}, 65%). Among the remaining toxic elements, only Sn and Ba have been quantified in over 25% of the samples, where the remaining elements, i.e., Be, Cd, Hg, Pb, Tl, and U, had seldom reached the relevant LoQs.

Table 4 reports a comparison between the data (mean and range) here obtained and those reported in the literature. To favor a reliable comparison among data obtained in different studies, only those obtained by using ICP-MS methods have been considered in this table.

Table 4. Average amounts and ranges (in µg kg^{-1}) of elements of health concern in e-liquids. Concentration was measured by means of ICP-MS methods.

Elements	Ref. [25] [a] (n = 6)	Ref. [33] [b] (n = 5)	Ref. [36] [c] (n = 56)	Ref. [37] [d] (n = 1)	Ref. [41] [a] (n = 27)	Ref. [42] [e] (n = 2)	Ref. [43] [f] (n = 3)	This Study (n = 37)
Al	12; 10–15		50.3; 46.22–59.6	7.7 ± 0.5	12.9; 8.82–30.7			≤33; <26–160
As	1.2; <1–1.5			0.08 ± 0.04	1.57; <1–3.42	<430	2.18; 0.83–3.04	5; 0.6–11
B								≤61; <37–140
Ba								≤27; <15–130
Be					<0.1			≤0.06; <0.057–0.12
Bi								≤0.11; <0.089–0.3
Cd	<0.4	43.5; 0.137–755	<0.1	<0.01	<0.4	<220	0.54; <0.25–1.28	≤0.14; <0.12–1
Co	0.15; <0.1–0.27				0.262; <0.1–0.884			<0.34; <0.089–0.9
Cr	5.2; 4.1–7.7	669; 41.5–16900	12; 12–14.26		7.16; 4.08–11.5			34; 20–40
Cu	23; <20–32		5.14; <1.0–16.1	<0.01	27.0; <20–30.6			≤7.1; <5.2–20
Fe			66.5; 48.74–130.9	4.1 ± 0.2				<308; <53–3000
Hg	<4				4.38; <4–4.54			≤5; <4.5–14
Li								1.9; 0.7–9
Mn	2.1; <1.6–3.3	1627; 11.8–31500	1.09; <1.0–2.74	0.159 ± 0.006	3.99; <1.6–8.42			<10; <1.6–80
Mo								≤0.8; <0.45–3
Ni	<16	7613; 13.7–72700	7.33; 5.30–47.4	0.161 ± 0.007	<16		3.43; 1.42–5.11	≤3.5; <2.3–14
Pb	<1	444; 3.17–4870	0.476; 0.243–1.05	<0.01	<1		12.28; <0.25–23.49	≤1.0; <0.8–3
Sb	1.6; 1.2–1.5		1.0; 1.0–1.219		7.21; 0.400–214			≤2.9; <1.1–10
Se								<4.6; <4.6–<4.6
Sn			1.53; 0.689–3.75					≤0.6; <0.24–4
Tl	<0.1				<0.1			≤0.07; <0.055–0.16
V	0.45; <0.4–0.64				0.602; <0.4–1.36			-
U								≤0.29; <0.21–0.7
Zn	<200		18.2; 11.94–28.2	0.51 ± 0.03	418; <200–510			≤109; <62–300

In italics: data below the LoD; in underlined: data below the LoQ. [a] Solvents: PG < 65%, VG < 35%, samples with (16 mg cm^{-3}) or without nicotine. [b] E-liquids were popular brands sold in the USA. An interlaboratory trial confirmed these data. [c] Solvents: PG, 70% VG, 30%. Data are relative to median, where the range is within the 25th and the 75th percentile. Data reported in this column are the sum of those reported in Table 2 of the paper and the amount of the relevant blanks reported in Table S1 of the supplementary material. Data are from only one measurement for each sample. Trueness has been evaluated through an interlaboratory trial and by analysis of a NIST SRM®® 1640a (Trace Elements in Natural Water). [d] E-liquid was 1:100 diluted with HNO$_3$ 1% in water before ICP-MS analysis. All trace metal analyses were performed as a contracted service. No details were provided on the validation of the ICP-MS method. [e] MarkTen®® Menthol and Classic e-liquids, both containing 1.5% nicotine. No details were provided on the validation of the ICP-MS method. [f] No details were provided on the nature and validation of the ICP-MS determination.

Among the literature contributions considered in Table 4, V is the only element not analyzed in this study, but it previously was quantified by Beauval et al. [25,41]. Conversely, the amounts of B, Ba, Bi, Li, Mo, and U in e-liquids were measured for the first time in this study. Another element never quantified in previous studies, i.e., Se, was found to be below its LoD in this research. The amounts of elements analyzed in this study are in good agreement with results reported using validated methods for the determination of several elements in a statistically significant number of samples [25,36,41]. The most important differences regard the most abundant (and the most interfered) elements, such as Al, Fe, and Zn, which were frequently present in higher concentrations in this study. On the other hand, the amounts of another "critical" element, such as Ni, are higher in the literature studies than in this research. For elements such as As, Cd, Ni, and Pb, data here reported are in fair agreement with those reported by Song et al. [43], even if the validation of the method used in that contribution was not reported in the paper. On the whole, and with only very rare exceptions, such as the quite high amounts of Sb in a few of the e-liquids measured by Beauval et al. [41], the amounts of toxic elements found in e-liquids in this, as well as in previous studies [25,36,37,41–43], are coherent with a negligible health risk associated to its intake by vapers. The data measured here are several orders of magnitude below the worrying amounts measured by Hess et al. [33] for Cd, Cu, Mn, Ni, and Pb in two of the most popular brands of e-liquids sold in the USA. On the other hand, these concentrations, albeit much higher than those measured in the cited literature studies, are within the amounts recommended by the AFNOR [52], i.e., 1 mg kg^{-1} for Cd and Hg, 3 mg kg^{-1} for As, 5 mg kg^{-1} for Sb, and 10 mg kg^{-1} for Pb.

Data reported also account for some differences among the three classes of flavors. Figure 4 shows the box-whisker plots for the concentrations of the most representative elements under exam in nine fruity, sixteen tobacco, and twelve tonic e-liquids. As a general behavior, tobacco flavors are the richest in the considered elements, whereas tonic flavors are the less abundant. In particular, tobacco flavors exhibit the highest amounts of Al, Bi, Cd, Fe, Hg, Li, Mn, Mo, and Ni, whereas the amounts of Al, Be, Bi, and Cd are always below the relevant LoDs in tonic flavors. Interestingly, fruity flavors show the highest amounts of some bivalent elements such as Ba, Be, Cu, Pb, Sn, and Zn.

2.4. Concentration of 23 Elements in PG, VG, Water, and Nicotine Used in the Composition of the e-Liquids

Table 5 reports the concentration of the 23 elements under exam in all the compounds used for the preparation of the e-liquids, with the only exclusion being of the concentrated flavor, always present in a 10% (w/w) amount in the final composition of the e-liquid.

Organic samples were analyzed using the same method developed for e-liquids, whereas water was analyzed using literature official methods [66,67]. The data reported in Table 5 give an account for an overall negligible contribution of toxic elements in the composition of the e-liquids. On the other hand, it is evident that the most abundant elements in e-liquids come from VG (i.e., As, Cr, Cu, Li, and Zn) and PG (i.e., B, Cr, Fe, and Zn). Furthermore, the water is quite rich in B, whereas nicotine (always absent in all the samples analyzed, see Section 3.1) is very rich in Al (5 mg kg^{-1}). Moreover, the amounts of Cr, Zn, and Sb are significant. Since the concentration of Cr is relatively elevated in both VG and PG, it is likely that the high amount of this element measured in the aerosols can depend not only on the metal constituents of the e-cigarette, but also on the aliquot of Cr present in the organic solvents in the e-liquids. Hence, a reduction of the elemental concentrations of the main trace elements contained in the e-liquids would be an effective tool to further reduce the amount of oligoelements in them.

2.5. Principal Components Analysis

In order to verify which elements were deriving either from the constituents or from the flavors, the whole dataset consisting of 37 samples of e-liquids and 22 elements (Se was always found below its LoD and thus excluded) was normalized with respect to the concentration of each element measured in the SM (i.e., in the absence of any contribution given by any flavor), obtaining a dataset expressed in terms of standard deviation referred to in the SM concentration. A principal components analysis (PCA) was carried out on the normalized dataset. Figure 5 shows the loading plot (a), with its zoomed view (b), and the score plot (c), with its zoomed view (d).

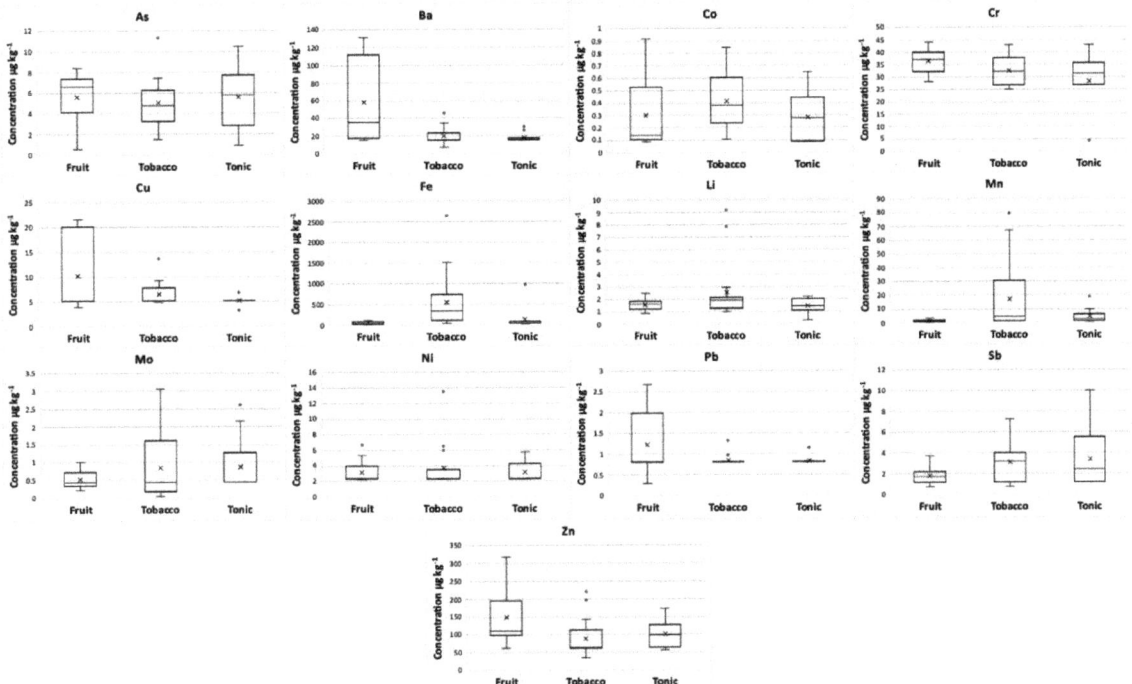

Figure 4. Box-whisker plots of the concentrations of As, Ba, Co, Cr, Cu, Fe, Li, Mn, Mo, Ni, Pb, Sb, and Zn in the three different flavor classes of e-liquids. The horizontal lines in the box represent the 25th percentile, the mean value, and the 75th percentile, respectively, and the interval between the ends of the whiskers represents the range. The "x" symbol is the median value, and the "∘" symbol represents the outlier amounts.

Table 5. Mean amounts (in µg kg^{-1} ± SD) of 23 toxic elements and oligoelements in VG, PG, water, and nicotine used for the preparation of the e-liquids. $n = 3$.

Element	VG	PG	Water	Nicotine
Al	<26	<26	<5	5000 ± 1000
As	7 ± 2	2.6 ± 0.6	0.020 ± 0.005	0.8 ± 0.2
B	<37	110 ± 10	470 ± 70	<37
Ba	<15	<15	0.18 ± 0.02	<15
Be	<0.057	<0.057	<0.01	<0.057
Bi	<0.089	<0.089	<0.005	<0.089
Cd	<0.12	<0.12	0.04 ± 0.02	<0.12
Co	0.2 ± 0.1	<0.089	<0.005	<0.089
Cr	56 ± 7	49 ± 6	<0.1	59 ± 8
Cu	22 ± 4	<5.2	<0.1	<5.2
Fe	0.6 ± 0.1	100 ± 20	<1	0.6 ± 0.1
Hg	<4.5	<4.5	<0.005	<4.5
Li	6 ± 1	<0.37	0.374 ± 0.001	≤1.2
Mn	<1.6	<1.6	2.8 ± 0.2	11 ± 2
Mo	<0.45	<0.45	0.8 ± 0.2	<0.45

Table 5. Cont.

Element	VG	PG	Water	Nicotine
Ni	<2.3	<2.3	2.5 ± 0.6	<2.3
Pb	<0.80	<0.80	1.6 ± 0.1	<0.80
Sb	<u><3.7</u>	<1.1	0.67 ± 0.01	120 ± 20
Se	<4.6	<4.6	0.035 ± 0.015	<4.6
Sn	<0.24	<0.24	<0.01	<0.24
Tl	<u><0.18</u>	<0.055	<0.001	<0.055
U	<0.21	<0.21	<0.003	<0.21
Zn	180 ± 40	41 ± 9	5 ± 1	160 ± 30

VG: vegetal glycerin; PG: propylene glycol; SD: standard deviation; in italics: amounts below the LoD; underlined: amounts below the LoQ.

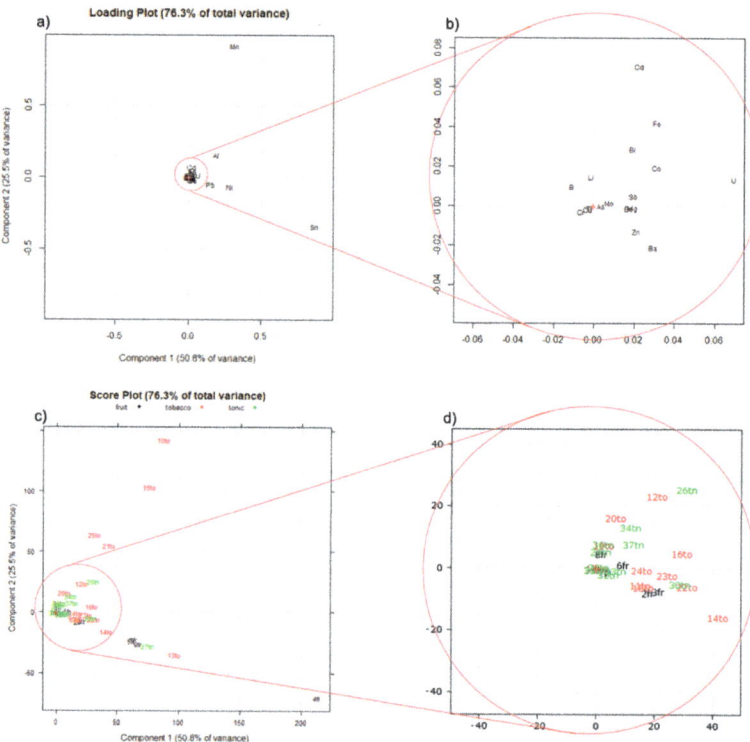

Figure 5. PCA performed on the same dataset processed and normalized, for each analyte, with respect to the relevant concentration measured in the synthetic matrix (SM). (**a**) Loading plot; (**b**) zoomed view of the loading plot; (**c**) score plot; (**d**) zoomed view of the score plot; to: tobacco flavored samples; fr: fruity-flavored samples; tn: tonic-flavored samples.

The two principal components explain 76% of the total variance. In the loading plot (Figure 5a,b), 17 out the 22 elements quantified are crowded around the center of the plot, giving no contribution to the explained variance, whereas elements such as Al, Pb, Ni, and mainly Mn and Sn are responsible for the variability expressed by the first two components. In the score plot, most of the samples are grouped in the origin of the two PC axes, with only a few samples highly differing from the others. To interpret the obtained PCA, it must be kept in mind that data were normalized with respect to the concentrations of the elements in the SM; this means that the samples grouping around the origin of the axes in the score plot (Figure 5c,d) show no variability with respect to the SM, whereas samples distant from the origin are those that mostly differ from the SM. This difference is due to a few elements (Mn, Sn and, to a minor extent, Al, Pb, and Ni are responsible for the variability seen in the plots). This means that the tobacco samples 10to, 15to, 21to, and 25to, described by high positive score values of PC2, are the richest in Al and, in particular, Mn, with respect to the SM. On the other hand, the samples 4fr, 13to, 27tn, 5fr, 9fr, and 1fr, described by high positive score values of PC1 and negative score values of PC2, are richer in Sn, Ni, and Pb, with respect to the SM.

Looking at the zoomed view of the plots (Figure 5b,d), it is interesting to observe that positive scores of PC1 are correlated by the increase of the amount of each analyte in the e-liquid, with respect to the amount present in the SM. The almost complete absence of elements and samples described by negative scores of PC1 is, in our opinion, a clear marker of the overall reliability of the data here obtained. Summarizing, Sn, Mn, and—in decreasing amounts—Ni, Al, and Pb, are elements that mainly originated from the flavors, whereas the remaining elements are derived mainly from the constant compounds of the e-liquids, i.e., PG, VG, and water. In addition, it is interesting to observe that tonic e-liquids are the category closest to the matrix concentration, whereas fruity and tobacco e-liquids are the most scattered samples.

3. Materials and Methods

3.1. Samples and Reagents

The 37 samples of e-liquids were provided by a local producer. Eight of them were fruity-flavored, twelve were tonic-flavored, and seventeen were tobacco-flavored. Table 6 reports the composition of each class of e-liquids.

Table 6. Composition of the three classes of e-liquids.

e-Liquid Flavor Class	PG (%)	VG (%)	Concentrated Flavor (%)	Water (%)
Fruity	50	40	8	2
Tobacco	50	40	6	4
Tonic	50	40	7	3

The same producer also provided the pure compounds (i.e., propylene glycol, glycerin, and water) that were added to the concentrated flavor to produce the samples analyzed. Albeit they were all nicotine-free, the producer also provided a sample of pure nicotine to evaluate its contribution to the overall amount of trace elements in the nicotine-added e-liquids.

HNO_3 69% (NORMATON for ultra-trace analysis, VWR, Milan, Italy) and type I water (MilliQ plus System, Millipore, Vimodrone, Italy) were used in all the phases of the study. Triton™ X-100 aqueous solution, 10%(w/v) was from Merck, Milan, Italy. ICP-MS elemental standards of Al, As, B, Ba, Be, Bi, Cd, Co, Cr, Cu, Fe, Hg, Li, Mn, Mo, Ni, Pb, Sb, Sn, Tl, U, and Zn (1000 mg dm^{-3} each) were purchased from LabKings (Hilversum, The Netherlands). The ICP-MS setup solution (a 1% v/v aqueous solution of HNO_3 containing 1 μg dm^{-3} each of Be, Ce, Fe, In, Li, Mg, Pb, and U), the ICP-MS KED setup solution (a 1% (v/v) aqueous solution of HNO_3 containing 10 μg dm^{-3} of Co and 1 μg dm^{-3} of Ce), and the internal standard solution (10 μg dm^{-3} of Rh in a 1% v/v aqueous solution of HNO_3) were all from Perkin Elmer, Milan, Italy. All the samples were filtered before the analysis using a polypropylene filter (pore diameter: 0.22 μm) from VWR, Milan, Italy.

3.2. Instrumentation

The samples were mineralized using a microwave SRC system (UltraWave™, Milestone, Sorisole, Italy). The elemental determination was performed using a NexION 300X ICP-MS spectrometer (Perkin Elmer, Milan, Italy) equipped with a nebulization system composed of a glass concentric nebulizer, a glass cyclonic spray chamber, an autosampler model S10, and a KED collision cell. All the elements were determined simultaneously, and the information about the analytes and operational conditions has already been reported in Table 1.

3.3. Statistical Analysis

Principal components analysis (PCA) was performed by means of the R-based software Chemometric Agile Tool (CAT) developed by the Italian group of chemometrics [68].

4. Conclusions

The perceived need in the literature for sensitive, reliable, and validated ICP-MS methods to quantify a wide range of both toxic elements and oligoelements in e-liquids is the main reason for accomplishing this study. Hence, an original method capable of determining the concentration of 23 elements in such a matrix has been developed and validated in terms of LoD, LoQ, precision, and trueness. The need to circumvent the very strong matrix effect has required the introduction of significant changes with respect to the methods described in the literature. Namely, the sample pre-treatment step has been accomplished by microwave-assisted mineralization. Moreover, the ICP-MS instrumental parameters have been optimized to enhance the sensitivity and reduce the spectral interference by the polyatomic ions. Finally, the quantification, performed by means of an internal calibration with three additions of standards, has allowed for the obtainment of accurate data. The method has been applied to 37 different e-liquids of three different classes of flavors (i.e., fruity, tobacco, and tonic). The average concentration of all analytes is aligned with the lowest amounts measured in the literature, confirming hence that, also in this case, the potential health risk related to the amount of toxic elements in e-liquids is orders of magnitude less than that measured for traditional cigarettes. Almost all toxic elements have been found below an amount lower than a few $\mu g\ kg^{-1}$ and, very often, they were below the relevant LoQ. The data obtained also give an account for the slight differences among the classes of flavors. Tobacco and tonic flavors exhibit the highest and the lowest amounts of elements, respectively. The results of the PCA showed that the multivariate approach allowed easy differentiation of the elements derived from constituents and those from added flavors. For this reason, this approach may be generalized in order to minimize the amount of each element in the final e-liquids. In conclusion, e-liquids seem to be scarcely involved as potential sources of toxic elements inhaled by vapers. Hence, further investigations are needed to ascertain the influence of the conditions of use for e-cigarettes on the dragging in the vapor phase for the elements of potential toxicity in humans.

Supplementary Materials: The following table is available online. Table S1. Elemental concentration of toxic elements and oligoelements in 37 different e-liquids.

Author Contributions: Conceptualization, A.M. and G.S.; methodology, A.M., M.D. and G.S.; validation, A.M., I.L. and G.S.; formal analysis, A.M. and M.C.; investigation, A.M., I.L., S.D. P.M. and M.D.; resources, G.S.; data curation, A.M., M.C. and G.S.; writing—original draft preparation, A.M., M.C. and G.S.; writing—review and editing, A.M., I.L., M.C., M.I.P., N.S. and G.S.; visualization, A.M., M.C. and G.S.; supervision, G.S.; project administration, G.S.; funding acquisition, G.S. All authors have read and agreed to the published version of the manuscript.

Funding: This research was funded by Università degli Studi di Sassari ("Fondo di Ateneo per la ricerca 2020").

Institutional Review Board Statement: Not applicable.

Informed Consent Statement: Not applicable.

Acknowledgments: The authors gratefully thank the New Flavours company, Sassari, Sardinia, Italy, for the generous supply of e-liquids and their common co-formulants (i.e., PG, VG, water, and nicotine). Moreover, the GAUSS UNISS, Grandi Attrezzature Università di Sassari is gratefully acknowledged for the access granted to their ICP-MS instrument.

Conflicts of Interest: The authors declare no conflict of interest.

Sample Availability: Samples of the compounds are not available from the authors but may be requested from the New Flavours company.

References

1. Williams, M.; Talbot, P. Variability Among Electronic Cigarettes in the Pressure Drop, Airflow Rate, and Aerosol Production. *Nicotine Tob. Res.* **2011**, *13*, 1276–1283. [CrossRef]
2. Breland, A.; Soule, E.; Lopez, A.; Ramôa, C.; El-Hellani, A.; Eissenberg, T. Electronic cigarettes: What are they and what do they do? *Ann. N. Y. Acad. Sci.* **2017**, *1394*, 5–30. [CrossRef]
3. European Commission. Food Additives. E1520. Propane-1,2-Diol (Propylene Glycol). Available online: https://webgate.ec.europa.eu/foods_system/main/index.cfm?event=substance.view&identifier=351 (accessed on 12 October 2021).
4. European Commission. Food Additives. E422. Glycerol. Available online: https://webgate.ec.europa.eu/foods_system/main/index.cfm?event=substance.view&identifier=166 (accessed on 12 October 2021).
5. European Directive 2014/40/EU of the European Parliament and of the Council of 3 April 2014 on the approximation of the laws, regulations and administrative provisions of the Member States concerning the manufacture, presentation and sale of tobacco and related products and repealing Directive 2001/37/EC Text with EEA relevance. *Off. J. Eur. Union. L* **2014**, *127/1*. Available online: https://eur-lex.europa.eu/legal-content/EN/TXT/HTML/?uri=CELEX:32014L0040&from=EN (accessed on 12 October 2021).
6. Talhout, R.; Schulz, T.; Florek, E.; Van Benthem, J.; Wester, P.; Opperhuizen, A. Hazardous Compounds in Tobacco Smoke. *Int. J. Environ. Res. Public Health* **2011**, *8*, 613–628. [CrossRef]
7. Goniewicz, M.L.; Knysak, J.; Gawron, M.; Kosmider, L.; Sobczak, A.; Kurek, J.; Prokopowicz, A.; Jabłońska-Czapla, M.; Rosik-Dulewska, C.; Havel, C.; et al. Levels of selected carcinogens and toxicants in vapour from electronic cigarettes. *Tob. Control.* **2013**, *23*, 133–139. [CrossRef]
8. Hecht, S.S.; Carmella, S.G.; Kotandeniya, D.; Pillsbury, M.E.; Chen, M.; Ransom, B.W.S.; Vogel, R.; Thompson, E.; Murphy, S.E.; Hatsukami, D.K. Evaluation of Toxicant and Carcinogen Metabolites in the Urine of E-Cigarette Users Versus Cigarette Smokers. *Nicotine Tob. Res.* **2015**, *17*, 704–709. [CrossRef] [PubMed]
9. Grana, R.; Benowitz, N.; Glantz, S.A. E-Cigarettes. *Circulation* **2014**, *129*, 1972–1986. [CrossRef] [PubMed]
10. Pisinger, C.; Døssing, M. A systematic review of health effects of electronic cigarettes. *Prev. Med.* **2014**, *69*, 248–260. [CrossRef]
11. Wang, G.; Liu, W.; Song, W. Toxicity assessment of electronic cigarettes. *Inhal. Toxicol.* **2019**, *31*, 259–273. [CrossRef]
12. Sapru, S.; Vardhan, M.; Li, Q.; Guo, Y.; Li, X.; Saxena, D. E-cigarettes use in the United States: Reasons for use, perceptions, and effects on health. *BMC Public Health* **2020**, *20*, 1518. [CrossRef]
13. Cao, Y.; Wu, D.; Ma, Y.; Ma, X.; Wang, S.; Li, F.; Li, M.; Zhang, T. Toxicity of electronic cigarettes: A general review of the origins, health hazards, and toxicity mechanisms. *Sci. Total. Environ.* **2021**, *772*, 145475. [CrossRef]
14. Chun, L.F.; Moazed, F.; Calfee, C.S.; Matthay, M.A.; Gotts, J.E. Pulmonary toxicity of e-cigarettes. *Am. J. Physiol. Lung Cell. Mol. Physiol.* **2017**, *313*, L193–L206. [CrossRef] [PubMed]
15. Leigh, N.J.; Tran, P.L.; O'Connor, R.J.; Goniewicz, M.L. Cytotoxic effects of heated tobacco products (HTP) on human bronchial epithelial cells. *Tob. Control.* **2018**, *27*, s26–s29. [CrossRef]
16. Putzhammer, R.; Doppler, C.; Jakschitz, T.; Heinz, K.; Förste, F.; Danzl, K.; Messner, B.; Bernhard, D. Vapours of US and EU Market Leader Electronic Cigarette Brands and Liquids Are Cytotoxic for Human Vascular Endothelial Cells. *PLoS ONE* **2016**, *11*, e0157337. [CrossRef]
17. Carnevale, R.; Sciarretta, S.; Violi, F.; Nocella, C.; Loffredo, L.; Perri, L.; Peruzzi, M.; Marullo, A.G.M.; De Falco, E.; Chimenti, I.; et al. Acute Impact of Tobacco vs. Electronic Cigarette Smoking on Oxidative Stress and Vascular Function. *Chest* **2016**, *150*, 606–612. [CrossRef]
18. Kuntic, M.; Oelze, M.; Steven, S.; Kröller-Schön, S.; Stamm, P.; Kalinovic, S.; Frenis, K.; Vujacic-Mirski, K.; Jimenez, M.T.B.; Kvandova, M.; et al. Short-term e-cigarette vapour exposure causes vascular oxidative stress and dysfunction: Evidence for a close connection to brain damage and a key role of the phagocytic NADPH oxidase (NOX-2). *Eur. Heart J.* **2020**, *41*, 2472–2483. [CrossRef]
19. Whittington, J.R.; Simmons, P.M.; Phillips, A.M.; Gammill, S.K.; Cen, R.; Magann, E.F.; Cardenas, V.M. The Use of Electronic Cigarettes in Pregnancy: A Review of the Literature. *Obstet. Gynecol. Surv.* **2018**, *73*, 544–549. [CrossRef]
20. Schweitzer, K.S.; Chen, S.X.; Law, S.; Van Demark, M.; Poirier, C.; Justice, M.J.; Hubbard, W.C.; Kim, E.S.; Lai, X.; Wang, M.; et al. Endothelial disruptive proinflammatory effects of nicotine and e-cigarette vapor exposures. *Am. J. Physiol. Lung Cell. Mol. Physiol.* **2015**, *309*, L175–L187. [CrossRef] [PubMed]
21. Vansickel, A.R.; Cobb, C.O.; Weaver, M.F.; Eissenberg, T.E. A Clinical Laboratory Model for Evaluating the Acute Effects of Electronic "Cigarettes": Nicotine Delivery Profile and Cardiovascular and Subjective Effects. *Cancer Epidemiol. Biomark. Prev.* **2010**, *19*, 1945–1953. [CrossRef]
22. Bodas, M.; Van Westphal, C.; Carpenter-Thompson, R.; Mohanty, D.K.; Vij, N. Nicotine exposure induces bronchial epithelial cell apoptosis and senescence via ROS mediated autophagy-impairment. *Free Radic. Biol. Med.* **2016**, *97*, 441–453. [CrossRef] [PubMed]

23. Bekki, K.; Uchiyama, S.; Ohta, K.; Inaba, Y.; Nakagome, H.; Kunugita, N. Carbonyl Compounds Generated from Electronic Cigarettes. *Int. J. Environ. Res. Public Health* **2014**, *11*, 11192–11200. [CrossRef] [PubMed]
24. Khlystov, A.; Samburova, V. Flavoring Compounds Dominate Toxic Aldehyde Production during E-Cigarette Vaping. *Environ. Sci. Technol.* **2016**, *50*, 13080–13085. [CrossRef]
25. Beauval, N.; Antherieu, S.; Soyez, M.; Gengler, N.; Grova, N.; Howsam, M.; Hardy, E.M.; Fischer, M.; Appenzeller, B.M.R.; Goossens, J.-F.; et al. Chemical Evaluation of Electronic Cigarettes: Multicomponent Analysis of Liquid Refills and their Corresponding Aerosols. *J. Anal. Toxicol.* **2017**, *41*, 670–678. [CrossRef]
26. Kim, H.-J.; Shin, H.-S. Determination of tobacco-specific nitrosamines in replacement liquids of electronic cigarettes by liquid chromatography–tandem mass spectrometry. *J. Chromatogr. A* **2013**, *1291*, 48–55. [CrossRef]
27. Famele, M.; Ferranti, C.; Abenavoli, C.; Palleschi, L.; Mancinelli, R.; Draisci, R. The Chemical Components of Electronic Cigarette Cartridges and Refill Fluids: Review of Analytical Methods. *Nicotine Tob. Res.* **2015**, *17*, 271–279. [CrossRef]
28. Bansal, V.; Kim, K.-H. Review on quantitation methods for hazardous pollutants released by e-cigarette (EC) smoking. *TrAC Trends Anal. Chem.* **2016**, *78*, 120–133. [CrossRef]
29. Ogunwale, M.A.; Li, M.; Raju, M.V.R.; Chen, Y.; Nantz, M.H.; Conklin, D.J.; Fu, X.-A. Aldehyde Detection in Electronic Cigarette Aerosols. *ACS Omega* **2017**, *2*, 1207–1214. [CrossRef] [PubMed]
30. Behar, R.Z.; Davis, B.; Wang, Y.; Bahl, V.; Lin, S.; Talbot, P. Identification of toxicants in cinnamon-flavored electronic cigarette refill fluids. *Toxicol. Vitr.* **2014**, *28*, 198–208. [CrossRef] [PubMed]
31. Pappas, R.S.; Polzin, G.M.; Zhang, L.; Watson, C.H.; Paschal, D.C.; Ashley, D.L. Cadmium, lead, and thallium in mainstream tobacco smoke particulate. *Food Chem. Toxicol.* **2006**, *44*, 714–723. [CrossRef]
32. Fresquez, M.R.; Pappas, R.S.; Watson, C.H. Establishment of Toxic Metal Reference Range in Tobacco from US Cigarettes. *J. Anal. Toxicol.* **2013**, *37*, 298–304. [CrossRef]
33. Hess, C.A.; Olmedo, P.; Navas-Acien, A.; Goessler, W.; Cohen, J.E.; Rule, A.M. E-cigarettes as a source of toxic and potentially carcinogenic metals. *Environ. Res.* **2017**, *152*, 221–225. [CrossRef]
34. Badea, M.; Luzardo, O.P.; González-Antuña, A.; Zumbado, M.; Rogozea, L.; Floroian, L.; Alexandrescu, D.; Moga, M.; Gaman, L.; Radoi, M.; et al. Body burden of toxic metals and rare earth elements in non-smokers, cigarette smokers and electronic cigarette users. *Environ. Res.* **2018**, *166*, 269–275. [CrossRef] [PubMed]
35. Zhao, D.; Aravindakshan, A.; Hilpert, M.; Olmedo, P.; Rule, A.M.; Navas-Acien, A.; Aherrera, A. Metal/Metalloid Levels in Electronic Cigarette Liquids, Aerosols, and Human Biosamples: A Systematic Review. *Environ. Health Perspect.* **2020**, *128*, 036001. [CrossRef] [PubMed]
36. Olmedo, P.; Goessler, W.; Tanda, S.; Grau-Perez, M.; Jarmul, S.; Aherrera, A.; Chen, R.; Hilpert, M.; Cohen, J.E.; Navas-Acien, A.; et al. Metal Concentrations in e-Cigarette Liquid and Aerosol Samples: The Contribution of Metallic Coils. *Environ. Health Perspect.* **2018**, *126*, 027010. [CrossRef]
37. Palazzolo, D.L.; Crow, A.P.; Nelson, J.M.; Johnson, R.A. Trace Metals Derived from Electronic Cigarette (ECIG) Generated Aerosol: Potential Problem of ECIG Devices That Contain Nickel. *Front. Physiol.* **2017**, *7*, 663. [CrossRef] [PubMed]
38. Zhao, J.; Nelson, J.; Dada, O.; Pyrgiotakis, G.; Kavouras, I.G.; Demokritou, P. Assessing electronic cigarette emissions: Linking physico-chemical properties to product brand, e-liquid flavoring additives, operational voltage and user puffing patterns. *Inhal. Toxicol.* **2018**, *30*, 78–88. [CrossRef] [PubMed]
39. Mikheev, V.B.; Brinkman, M.C.; Granville, C.A.; Gordon, S.M.; Clark, P.I. Real-Time Measurement of Electronic Cigarette Aerosol Size Distribution and Metals Content Analysis. *Nicotine Tob. Res.* **2016**, *18*, 1895–1902. [CrossRef] [PubMed]
40. Halstead, M.; Gray, N.; Gonzalez-Jimenez, N.; Fresquez, M.; Valentin-Blasini, L.; Watson, C.; Pappas, R.S. Analysis of Toxic Metals in Electronic Cigarette Aerosols Using a Novel Trap Design. *J. Anal. Toxicol.* **2020**, *44*, 149–155. [CrossRef] [PubMed]
41. Beauval, N.; Howsam, M.; Antherieu, S.; Allorge, D.; Soyez, M.; Garçon, G.; Goossens, J.F.; Lo-Guidice, J.M.; Garat, A. Trace elements in e-liquids—Development and validation of an ICP-MS method for the analysis of electronic cigarette refills. *Regul. Toxicol. Pharmacol.* **2016**, *79*, 144–148. [CrossRef]
42. Flora, J.W.; Meruva, N.; Huang, C.B.; Wilkinson, C.T.; Ballentine, R.; Smith, D.C.; Werley, M.S.; McKinney, W.J. Characterization of potential impurities and degradation products in electronic cigarette formulations and aerosols. *Regul. Toxicol. Pharmacol.* **2016**, *74*, 1–11. [CrossRef]
43. Song, J.-J.; Go, Y.Y.; Mun, J.Y.; Lee, S.; Im, G.J.; Kim, Y.Y.; Lee, J.H.; Chang, J. Effect of electronic cigarettes on human middle ear. *Int. J. Pediatr. Otorhinolaryngol.* **2018**, *109*, 67–71. [CrossRef] [PubMed]
44. Tayyarah, R.; Long, G.A. Comparison of select analytes in aerosol from e-cigarettes with smoke from conventional cigarettes and with ambient air. *Regul. Toxicol. Pharmacol.* **2014**, *70*, 704–710. [CrossRef] [PubMed]
45. Margham, J.; McAdam, K.; Forster, M.; Liu, C.; Wright, C.; Mariner, D.; Proctor, C. Chemical Composition of Aerosol from an E-Cigarette: A Quantitative Comparison with Cigarette Smoke. *Chem. Res. Toxicol.* **2016**, *29*, 1662–1678. [CrossRef]
46. Zhao, D.; Navas-Acien, A.; Ilievski, V.; Slavkovich, V.; Olmedo, P.; Adria-Mora, B.; Domingo-Relloso, A.; Aherrera, A.; Kleiman, N.J.; Rule, A.M.; et al. Metal concentrations in electronic cigarette aerosol: Effect of open-system and closed-system devices and power settings. *Environ. Res.* **2019**, *174*, 125–134. [CrossRef]
47. Williams, M.; Villarreal, A.; Bozhilov, K.; Lin, S.; Talbot, P. Metal and Silicate Particles Including Nanoparticles Are Present in Electronic Cigarette Cartomizer Fluid and Aerosol. *PLoS ONE* **2013**, *8*, e57987. [CrossRef]

48. Williams, M.; To, A.; Bozhilov, K.; Talbot, P. Strategies to Reduce Tin and Other Metals in Electronic Cigarette Aerosol. *PLoS ONE* **2015**, *10*, e0138933. [CrossRef]
49. Williams, M.; Bozhilov, K.; Ghai, S.; Talbot, P. Elements including metals in the atomizer and aerosol of disposable electronic cigarettes and electronic hookahs. *PLoS ONE* **2017**, *12*, e0175430. [CrossRef] [PubMed]
50. Williams, M.; Bozhilov, K.N.; Talbot, P. Analysis of the elements and metals in multiple generations of electronic cigarette atomizers. *Environ. Res.* **2019**, *175*, 156–166. [CrossRef] [PubMed]
51. Kamilari, E.; Farsalinos, K.; Poulas, K.; Kontoyannis, C.G.; Orkoula, M.G. Detection and quantitative determination of heavy metals in electronic cigarette refill liquids using Total Reflection X-ray Fluorescence Spectrometry. *Food Chem. Toxicol.* **2018**, *116*, 233–237. [CrossRef]
52. AFNOR. *XP D90-300-2—Electronic Cigarettes and E-Liquids-Part 2: Requirements and Test Methods for E-Liquids*; European Association for the Co-ordination of Consumers Representation in Standardisation; AISBL Press: Brussels, Belgium, 2015.
53. BSI. PAS 54115:2015—Vaping Products, including Electronic Cigarettes, e-Liquids, e-Shisha and Directly-Related Products. In *Manufacture, Importation, Testing and Labelling Guide*; British Standards Institution (BSI) Press: London, UK, 2015.
54. European Association for the Coordination of Consumer Representation in Standardization (ANEC). *E-Cigarettes and E-Liquids—Limits for Chemicals*; Basis for discussion. Position Paper ANEC-PT-2019-CEG-005; ANEC: Brussels, Belgium, 2019. Available online: https://www.anec.eu/images/Publications/position-papers/Chemicals/ANEC-PT-2019-CEG-005.pdf (accessed on 12 October 2021).
55. Langasco, I.; Caredda, M.; Sanna, G.; Panzanelli, A.; Pilo, M.I.; Spano, N.; Petretto, G.; Urgeghe, P.P. Chemical Characterization of Craft Filuferru Spirit from Sardinia, Italy. *Beverages* **2018**, *4*, 62. [CrossRef]
56. Spanu, A.; Valente, M.; Langasco, I.; Leardi, R.; Orlandoni, A.M.; Ciulu, M.; Deroma, M.A.; Spano, N.; Barracu, F.; Pilo, M.I.; et al. Effect of the irrigation method and genotype on the bioaccumulation of toxic and trace elements in rice. *Sci. Total. Environ.* **2020**, *748*, 142484. [CrossRef]
57. Langasco, I.; Barracu, F.; Deroma, M.A.; López-Sánchez, J.F.; Mara, A.; Meloni, P.; Pilo, M.I.; Estrugo, À.S.; Sanna, G.; Spano, N.; et al. Assessment and validation of ICP-MS and IC-ICP-MS methods for the determination of total, extracted and speciated As. Application to samples from a soil-rice system at varying the irrigation method. *J. Env. Manag.* **2021**. submitted.
58. May, T.W.; Wiedmeyer, R.H. A Table of Polyatomic Interferences in ICP-MS. *At. Spectrosc.* **1998**, *19*, 150–155.
59. Spanu, A.; Langasco, I.; Valente, M.; Deroma, M.A.; Spano, N.; Barracu, F.; Pilo, M.I.; Sanna, G. Tuning of the Amount of Se in Rice (*Oryza sativa*) Grain by Varying the Nature of the Irrigation Method: Development of an ICP-MS Analytical Protocol, Validation and Application to 26 Different Rice Genotypes. *Molecules* **2020**, *25*, 1861. [CrossRef] [PubMed]
60. Goncalves, D.A.; Jones, B.T.; Donati, G.L. The reversed-axis method to estimate precision in standard additions analysis. *Microchem. J.* **2016**, *124*, 155–158. [CrossRef]
61. Andersen, J.E.T. The standard addition method revisited. *TrAC Trends Anal. Chem.* **2017**, *89*, 21–33. [CrossRef]
62. Harris, D.C. *Quantitative Chemical Analysis*, 8th ed.; W.H. Freeman: New York, NY, USA, 2010.
63. Currie, L.A. Nomenclature in evaluation of analytical methods including detection and quantification capabilities. *Anal. Chim. Acta* **1999**, *391*, 105–126. [CrossRef]
64. Chemnitzer, R. Strategies for Achieving the Lowest Possible Detection Limits in ICP-MS. *Spectrosc. (St. Monica)* **2019**, *34*, 12–16. Available online: https://www.spectroscopyonline.com/view/strategies-achieving-lowest-possible-detection-limits-icp-ms (accessed on 12 October 2021).
65. Council of the European Union. Council Directive 98/83/EC of 3 November 1998 on the Quality of Water Intended for Human Consumption. *Off. J. Eur. Union. L* **1998**, *330*, 32–54. Available online: https://eur-lex.europa.eu/legal-content/EN/TXT/PDF/?uri=CELEX:31998L0083&from=IT (accessed on 12 October 2021).
66. Water quality—Application of inductively coupled plasma mass spectrometry (ICP-MS)—Part 1: General Guidelines. In *International Organization for Standardization: Method ISO 17294-1*; ISO: Geneva, Switzerland, 2004.
67. Water quality—Application of inductively coupled plasma mass spectrometry (ICP-MS)—Part 2, determination of selected elements including uranium isotopes. In *International Organization for Standardization: Method ISO 17294-2*; ISO: Geneva, Switzerland, 2016.
68. CAT (Chemometric Agile Tool). Available online: http://gruppochemiometria.it/index.php/software (accessed on 12 October 2021).

Article

Development of Quantitative Chemical Ionization Using Gas Chromatography/Mass Spectrometry and Gas Chromatography/Tandem Mass Spectrometry for Ambient Nitro- and Oxy-PAHs and Its Applications

Jungmin Jo [1], Ji-Yi Lee [1], Kyoung-Soon Jang [2], Atsushi Matsuki [3], Amgalan Natsagdorj [4] and Yun-Gyong Ahn [5,*]

[1] Department of Environmental Science and Engineering, Ewha Womans University, Seoul 03760, Republic of Korea
[2] Bio-Chemical Analysis Team, Korea Basic Science Institute, Cheongju 28119, Republic of Korea
[3] Institute of Nature and Environmental Technology, Kanazawa University, Kanazawa 920-1192, Japan
[4] Department of Chemistry, National University of Mongolia, Ulaanbaatar 14200, Mongolia
[5] Western Seoul Center, Korea Basic Science Institute, Seoul 03759, Republic of Korea
* Correspondence: ygahn@kbsi.re.kr

Abstract: The concentration of polycyclic aromatic hydrocarbons (PAHs) in the atmosphere has been continually monitored since their toxicity became known, whereas nitro-PAHs (NPAHs) and oxy-PAHs (OPAHs), which are derivatives of PAHs by primary emissions or secondary formations in the atmosphere, have gained attention more recently. In this study, a method for the quantification of 18 NPAH and OPAH congeners in the atmosphere based on combined applications of gas chromatography coupled with chemical ionization mass spectrometry is presented. A high sensitivity and selectivity for the quantification of individual NPAH and OPAH congeners without sample preparations from the extract of aerosol samples were achieved using negative chemical ionization (NCI/MS) or positive chemical ionization tandem mass spectrometry (PCI-MS/MS). This analytical method was validated and applied to the aerosol samples collected from three regions in Northeast Asia—namely, Noto, Seoul, and Ulaanbaatar—from 15 December 2020 to 17 January 2021. The ranges of the method detection limits (MDLs) of the NPAHs and OPAHs for the analytical method were from 0.272 to 3.494 pg/m^3 and 0.977 to 13.345 pg/m^3, respectively. Among the three regions, Ulaanbaatar had the highest total mean concentration of NPAHs and OPAHs at 313.803 ± 176.349 ng/m^3. The contribution of individual NPAHs and OPAHs in the total concentration differed according to the regional emission characteristics. As a result of the aerosol samples when the developed method was applied, the concentrations of NPAHs and OPAHs were quantified in the ranges of 0.016~3.659 ng/m^3 and 0.002~201.704 ng/m^3, respectively. It was concluded that the method could be utilized for the quantification of NPAHs and OPAHs over a wide concentration range.

Keywords: nitro-PAHs; oxy-PAHs; GC-NCI/MS; GC-PCI-MS/MS; aerosol samples

1. Introduction

Atmospheric particulate matter (PM), which is composed of a significant proportion of carbonaceous compounds, is emitted directly into the atmosphere. A secondary organic aerosol is also formed from the photo-oxidation of mixed anthropogenic volatile organic compounds [1]. Among the organic compounds related to PM, polycyclic aromatic hydrocarbons (PAHs) are the representative air toxic substances that are regulated because of their carcinogenic and mutagenic properties [2–4]. PAHs mainly originate from the incomplete combustion of coal, vehicle exhausts, and biomass burning; nitro-PAHs (NPAHs) and oxy-PAHs (OPAHs) can also arise by a secondary formation from the reactions of PAHs with atmospheric oxidants [5,6]. They are known to be much more toxic than the parent PAHs. Accordingly, increasing attention has been paid to research topics such as their formation mechanism, toxicity, source identification, and risk assessments [7–9]. For example, the value of the toxic equivalent factor for 6-nitrochrysene, which is an NPAH congener, is 10 times higher than that of Benzo[a]pyrene in the PAH group [10]. Thus, although the concentrations of NPAHs and OPAHs are generally lower than those of their parent PAHs in the atmosphere, increasing attention has been paid to these PAH derivatives. Furthermore, their monitoring in the atmosphere, as well as that of PAHs, is becoming more important in order to determine whether their formation is due to primary emissions or secondary sources [11–15]. However, quantitative data of PAH derivatives in aerosol samples are much fewer than those for parent PAHs because it is difficult to quantify them using the universal method applied to parent PAHs. The concentrations of PAH derivatives, especially NPAHs in aerosol samples, have been detected at 10~100 times lower than those of parent PAHs [16–18]. Thus, a sensitive analytical method with a high selectivity that considers their trace concentration level is required [19].

Several analytical methods have been employed for the determination of PAH derivatives [20]. In the case of NPAH measurements, gas chromatography (GC) combined with various detection methods such as electron capture detection [21,22], nitrogen selective detection [23,24], reductive electro-chemical detection [25,26], and negative and positive chemical ionization (NCI and PCI) mass spectrometry (MS) have been reported [27–30]. In the method of electron ionization mass spectrometry, the study of PAHs and nitro-PAHs using thermal desorption gas chromatography has been reported [31]. Among these methods, GC coupled with MS has been the most used [32,33]. Furthermore, the application of tandem mass spectrometry (MS/MS) by using a triple quadrupole mass spectrometer (QqQ-MS) has recently increased in terms of analytical sensitivity and specificity enhancement [34,35]. There are two common ionization methods, electron impact (EI) and chemical ionization (CI), in a GC/MS analysis. The EI method is commonly used to quantify parent PAHs because of its advantages such as a mass spectral library search to assist compound identification by providing a reference for mass spectra [36,37]. In the case of the CI method, negative chemical ionization (NCI) and positive chemical ionization (PCI) can be selected according to the chemical properties of the target analytes. As NPAHs and OPAHs exist at lower concentration levels in the atmosphere than parent PAHs, the CI method has been attempted more often than EI for a sensitive and selective detection [38]. Consequently, a rapid and accurate quantitative analytical method is required; this method, without a sample preparation, is preferred due to the difficulty in procuring samples and having to obtain a variety of chemical information from one specific sample alone. Nicol et al. developed a GC-MS/MS analysis of the EI method for OPAHs in an air concentration range of 30–170 pg/m^3 without a purification of the extract [39]. The detection limits of NPAHs and OPAHs reported so far are mostly results based on the signal-to-noise ratio; the method detection limit (MDL) has not yet been evaluated. In view of this, the aims of this study were: (1) to optimize the analytical conditions for 8 NPAH and 10 OPAH congeners using negative chemical ionization mass spectrometry (NCI/MS) or positive chemical ionization tandem mass spectrometry (PCI-MS/MS); (2) to validate the method with respect to the linearity, recovery, and MDL; and (3) to assess the practical applicability for the quantification of NPAHs and OPAHs over a wide concentration range in aerosol samples collected from Northeast Asian sites.

2. Results and Discussion
2.1. Congener-Specific Determination of PAH Derivatives
2.1.1. Optimal Ionization Mode Selection

The use of NCI can be the most selective detection method for specific classes of molecules containing electro-negative or acidic groups [40]. The mass spectra of 1-nitronaphthalene (1-NNAP) obtained by each ionization mode is shown in Figure 1. The most intense peak of the molecular ion for 1-NNAP is shown in the two types of chemical ionization modes. In the EI mode, the mass spectrum (Figure 1a) contained many fragmented ions and less sensitive molecular ions. In the PCI mode (Figure 1c), where the protonated molecular ion [M+H]⁺ was generated by methane gas, added ions such as $[M+C_2H_5]^+$ and $[M+C_3H_5]^+$ as well as a fragmented ion $[M-CH_3]^+$ from the loss of a methyl group were identified. On the other hand, the strongest molecular ion, [M−H]⁻, was produced in the NCI mode (Figure 1b). As a result, the NCI mode provided a high sensitivity and selectivity for the detection of 1-NNAP. When the intensities of the NPAH molecular ions were compared with the NCI and EI modes, their sensitivities were 3 to 15 times higher under the NCI condition (Table S1).

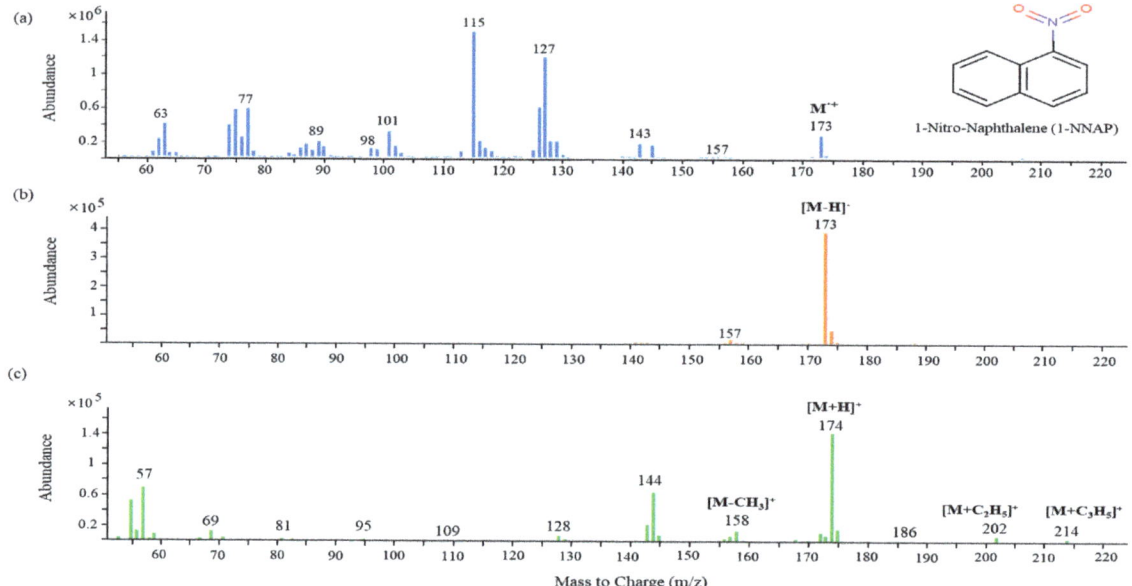

Figure 1. Mass spectra of 1–nitronaphthalene obtained from each ionization mode: (**a**) EI; (**b**) NCI; (**c**) PCI.

2.1.2. Ionization Efficiency

In a GC-MS/MS analysis, the molecular ions of the target analytes in the two types of chemical ionization mode are selected as the precursor ion that is then fragmented by a collision-induced dissociation (CID) in MS/MS to generate the product ions [41]. Despite the advantage of GC-MS/MS in that it is capable of a high-sensitivity analysis, low ionization efficiencies of the product ions from NPAH molecular ions were obtained because of the formation of strong and stable molecular ions in the case of the NCI condition. Consequently, the data acquisitions for the analysis of atmospheric NPAHs and OPAHs were performed by selected ion monitoring (SIM) under the NCI condition and selected reaction monitoring (SRM) under the PCI condition. Most NPAHs showed a better sensitivity in the SIM mode under the NCI condition, but the use of the PCI mode under the SRM condition was more appropriate in terms of the sensitivity and selectivity for the OPAHs, as shown in Figure 2. The sensitivities of xanthone (XT), phenalenone (PH), and 5,12-naphthacenequinone (Ncq) in the class of OPAHs and 6-nitrochrysene (6-NCHR) in the class of NPAHs particularly showed a significant difference between the two detection methods.

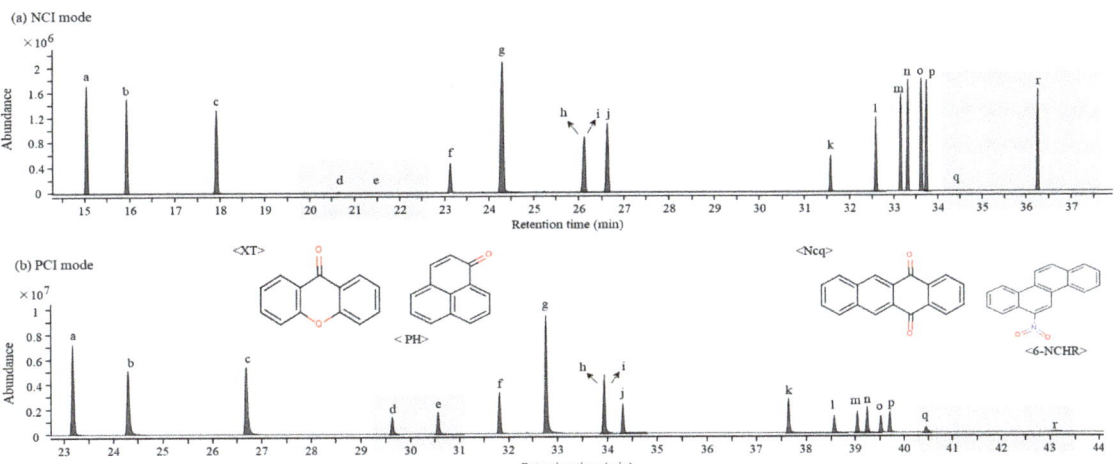

Figure 2. Comparison between SIM and SRM modes of a standard solution with NPAHs and OPAHs (4 µg/mL): (**a**) SIM chromatogram under NCI condition on DB-5MS UI capillary column (30 m × 0.25 mm × 0.25 µm); (**b**) SRM chromatogram under PCI condition on DB-5MS UI capillary column (60 m × 0.25 mm × 0.25 µm). The gray shading indicates four congeners, with a significant difference in sensitivity between the two detection methods. The peak identities were as follows: a—1-nitronaphthalene (1-NNAP); b—2-nitronaphthalene (2-NNAP); c—9-fluorenone (9-Flu); d—xanthone (XT); e—phenalenone (PH); f—anthraquinone (Anq); g—1,8-naphthalic anhydride (1,8-NA); h—2-nitrofluorene (2-NFLUO); i—2-methylanthraquinone (2-Maq); j—9-nitroanthracene (9-NANT); k—benzo[b]fluoren-11-one (BbFLU); l—menzoanthrone (BZA); m—3-nitrofluoranthene (3-NFL); n—4-nitropyrene (4-NPYR); o—benz[a]anthracene-1,2-quinone (BAQ); p—1-nitropyrene (1-NPYR); q—5,12-naphthacenequinone (Ncq); r—6-nitrochrysene (6-NCHR).

The results obtained by GC-NCI/MS and GC-PCI-MS/MS in comparing the sensitivities of four analytes at three different concentrations (Figure 3) showed that the detection method for individual OPAHs and NPAHs was determined through results that provided a clear change of response sensitivity in accordance with the different concentrations.

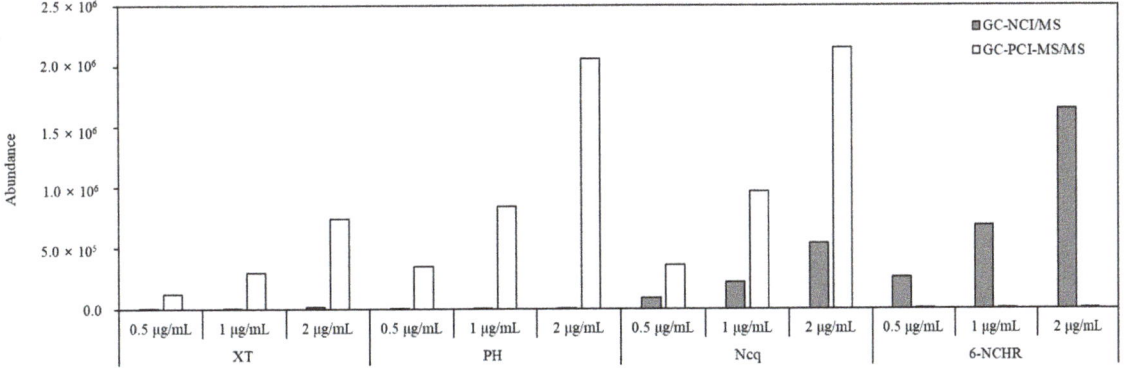

Figure 3. Comparison of the sensitivity of XT, PH, Ncq, and 6-NCHR at three different concentrations obtained by GC-NCI/MS and GC-PCI-MS/MS.

2.1.3. Chromatographic Separation

For the quantification of the target PAH derivatives from multiple components in the extracts of actual samples, the separation between the individual PAH derivatives and the neighborhood interferences obtained by GC-NCI/MS and GC-PCI-MS/MS were investigated. Even though NCI has the advantage of a high selectivity for NPAHs, 9-nitroanthracene (9-NANT) was not completely separated from nearby interferences in the actual aerosol samples when GC-NCI/MS was used. In contrast, a complete separation was achieved when GC-PCI-MS/MS was used (Figure 4b). GC-NCI/MS was superior for most NPAH congeners except for 9-NANT in terms of the sensitivity and selectivity. A typical case where the peak of 2-nitrofluorene (2-NFLUO) was only confirmed when GC-NCI/MS was used at a concentration of 0.034 µg/m^3 is shown in Figure 4a. In the case of the OPAH congeners, GC-PCI-MS/MS could offer a higher sensitivity and selectivity than GC-NCI/MS such as for the PH shown in Figure 4c through the use of the SRM mode.

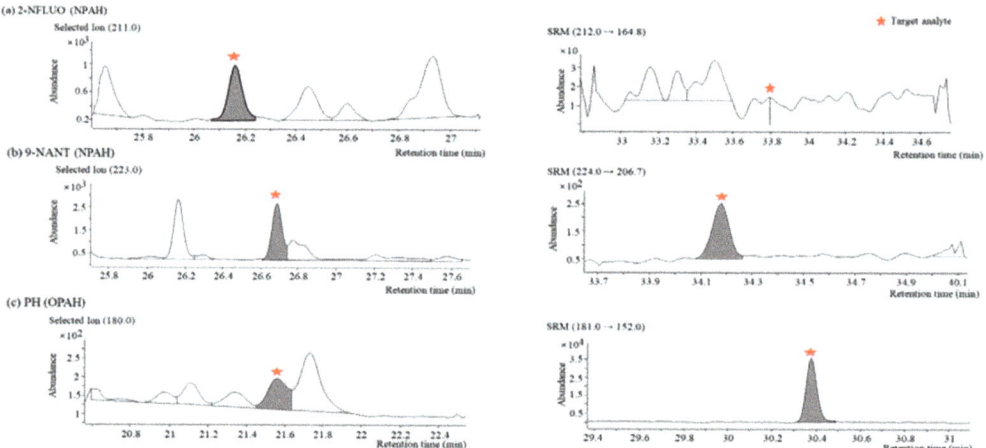

Figure 4. Comparison of separation characteristics of target PAH derivatives in actual aerosol samples obtained by GC-NCI/MS (**left**) and GC-PCI-MS/MS (**right**): (**a**) SIM (*m/z* = 211) and SRM (*m/z* 212→164.8) chromatograms of 2−NFLUO (NPAH); (**b**) SIM (*m/z* = 223) and SRM (*m/z* 224→206.7) chromatograms of 9−NANT (NPAH); (**c**) SIM (*m/z* = 180) and SRM (*m/z* 181→152) chromatograms of PH (OPAH).

2.2. Method Validation

The optimized conditions for the quantification of individual NPAH and OPAH congeners through a consideration for the sensitivity and selectivity in actual samples is shown in Table 1. For an internal calibration, fluoranthene-d$_{10}$ (Fla-d$_{10}$), which has a high efficiency for ionization and separation, was selected among the label standards used in the analysis of PAHs.

Table 1. Selected ions for the quantification of individual NPAH and OPAH congeners in NCI-SIM mode and transitions in PCI-SRM mode for MS detection.

NCI Conditions			PCI Conditions		
Compound	Qualifier Ion (*m/z*)	Quantifier Ion (*m/z*)	Compound	SRM (*m/z*)	Collision Energies (eV)
1-NNAP	157	173	9-Flu	181.0 → 152.0	30
2-NNAP	157	173	XT	197.0 → 115.1	50
Fla-d10 (IS)	213	212	PH	181.0 → 152.0	30

Table 1. Cont.

NCI Conditions			PCI Conditions		
2-NFLUO	188	211	Anq	209.0 → 151.9	30
3-NFL	231	247	1,8-NA	199.0 → 115.0	30
4-NPYR	231	247	Fla-d10	213.0 → 182.8	30
1-NPYR	231	247	2-Maq	223.0 → 152.0	30
6-NCHR	257	273	9-NANT	224.0 → 206.7	5
			BbFLU	231.0 → 202.0	30
			BZA	231.0 → 202.0	30
			BAQ	259.0 → 202.0	50
			Ncq	259.0 → 202.1	50

The range of the calibration curve was determined within a quantifiable range of individual NPAH and OPAH congeners in actual samples. The concentration of NPAHs was at least 10 times lower than that of OPAHs in the actual samples and this difference of concentration was reflected in the spiking experiments for the method validation. The calibration curves were linear in the range of each congener presented in Table 2 and their correlation coefficients were all greater than 0.99. The MDLs were measured by analyzing seven replicate samples spiked with 0.01–0.06 ng of the NPAHs and OPAHs per a control sample free of the target analytes. MDL was defined as the lowest concentration of each NPAH and OPAH congener that provided a greater than 99% confidence when the analytical method was used [42]. The recovery experiment was evaluated with two expected concentrations in the actual aerosol samples. MDLs for the NPAH and OPAH congeners ranged from 0.272 to 3.494 pg/m^3 and 0.977 to 13.345 pg/m^3, respectively.

Table 2. Calibration linear regression value, MDLs, and recoveries of individual NPAH and OPAH congeners obtained by the analytical method.

NPAHs	Calibration		MDL(pg/m^3) [a]	Recovery ± RSD% (n = 3)	
	Range (μg/mL)	R^2		0.02 ng/m^3	0.04 ng/m^3
1-NNAP	0.01–0.4	0.997	0.551	103.8 ± 11.3	104.8 ± 4.0
2-NNAP	0.01–0.4	0.997	0.527	111.7 ± 13.3	102.3 ± 9.6
2-NFLUO	0.01–0.2	0.996	0.339	91.4 ± 3.5	116.5 ± 0.7
9-NANT	0.05–1	0.995	3.494	94.6 ± 1.6	98.3 ± 1.2
3-NFL	0.01–0.1	0.995	0.272	104.3 ± 10.2	102.0 ± 2.1
4-NPYR	0.01–0.2	0.997	0.494	94.8 ± 9.1	92.7 ± 2.9
1-NPYR	0.01–0.1	0.999	0.494	92.8 ± 12.6	103.9 ± 2.3
6-NCHR	0.01–0.2	0.997	0.298	110.0 ± 8.6	105.7 ± 2.6

OPAHs	Calibration		MDL(pg/m^3)	Recovery ± RSD% (n = 3)	
	Range (μg/mL)	R^2		0.2 ng/m^3	0.9 ng/m^3
9-Flu	0.1–5	0.996	0.977	104.0 ± 3.9	92.8 ± 11.6
XT	0.2–2	0.998	5.731	107.9 ± 1.3	100.1 ± 7.3
PH	0.1–4	0.998	6.216	98.2 ± 2.3	98.6 ± 5.3
Anq	0.1–4	0.994	3.128	106.8 ± 2.1	110.0 ± 6.5
1,8-NA	0.1–5	0.999	0.873	103.3 ± 1.5	102.9 ± 2.2
2-Maq	0.1–2	0.995	2.341	104.4 ± 1.0	89.1 ± 2.0
BbFLU	0.2–4	0.994	2.789	108.6 ± 2.3	106.9 ± 1.1
BZA	0.2–4	0.996	4.666	107.7 ± 0.2	101.7 ± 2.1
BAQ	0.2–4	0.995	4.328	104.6 ± 3.4	108.8 ± 1.4
Ncq	0.2–4	0.998	13.345	102.9 ± 1.9	101.7 ± 2.4

[a] MDL = t(n − 1, 1 − α = 0.99) × SD, t(6, 0.99) = 3.14 (n = 7).

2.3. Application to Actual Aerosol Samples

For the aerosol samples collected in Noto, Seoul, and Mongolia, an analysis was only carried out when the correlation coefficients of the calibration curves for the NPAH and OPAH congeners were above 0.99. Quality control samples prepared at the middle concentration of the calibration curve were analyzed together every time after each 15-sample analysis. The QC results used to validate the analytical results obtained for every batch during the period of the sample analysis are shown in Figure 5.

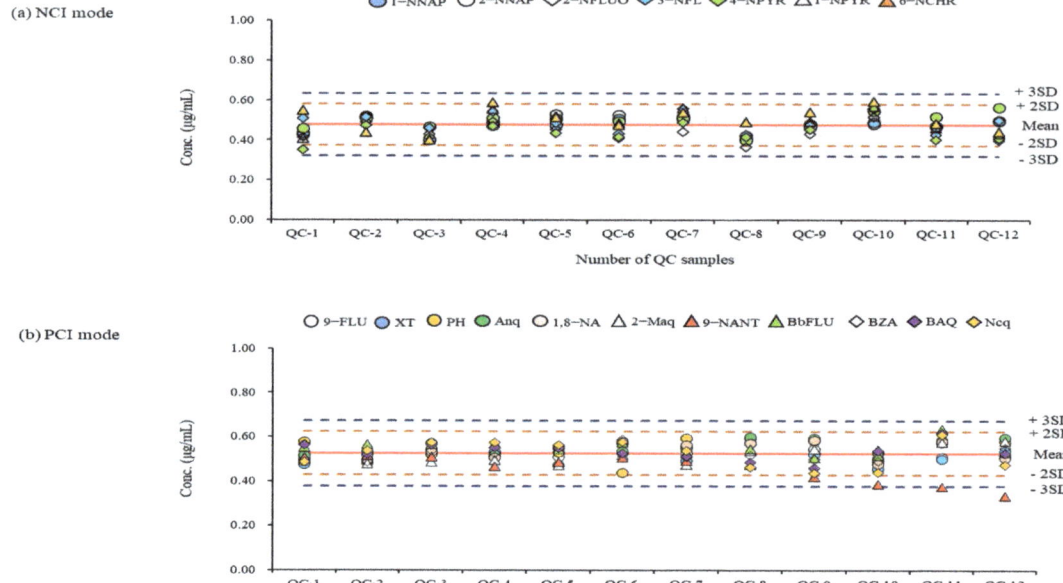

Figure 5. QC results during the period of sample analysis: (**a**) QC chart of analytes measured in NCI mode; (**b**)) QC chart of analytes measured in PCI mode. The graph indicates that upper and lower warning and rejection limits for reference standard materials (0.5 ug/mL).

A re-analysis was performed when the result value was outside the ± 2SD (standard deviation) limit through the QC result confirmation for individual NPAH and OPAH congeners obtained from their specific detection method. Most NPAH and OPAH congeners did not go out of the ±2SD limit, except for 9-NANT. When analyzing high concentration levels (such as the samples collected in Mongolia), the QC result was out of the acceptable limit. It was found that a replacement of the GC liner or the cleaning of the ion source were required for these times. From the results of measuring the NPAHs and OPAHs in actual aerosol samples (n = 84) using a validated analytical method, the total mean concentrations (NPAHs + OPAHs) in the three regions were 0.009 ± 1.009 ng/m^3 for Noto, 15.394 ± 3.134 ng/m^3 for Seoul, and 313.803 ± 176.349 ng/m^3 for Ulaanbaatar, respectively. The difference between the mean concentrations of the NPAHs and OPAHs was about 20–530-fold, depending on the region.

The ranges of the minimum and maximum concentrations of the NPAHs and OPAHs in all regions where the concentration was above the MDL are shown in Figure 6. The concentration range, mean value (ng/m^3), and detection frequency of each NPAH and OPAH congener are presented in Table S2.

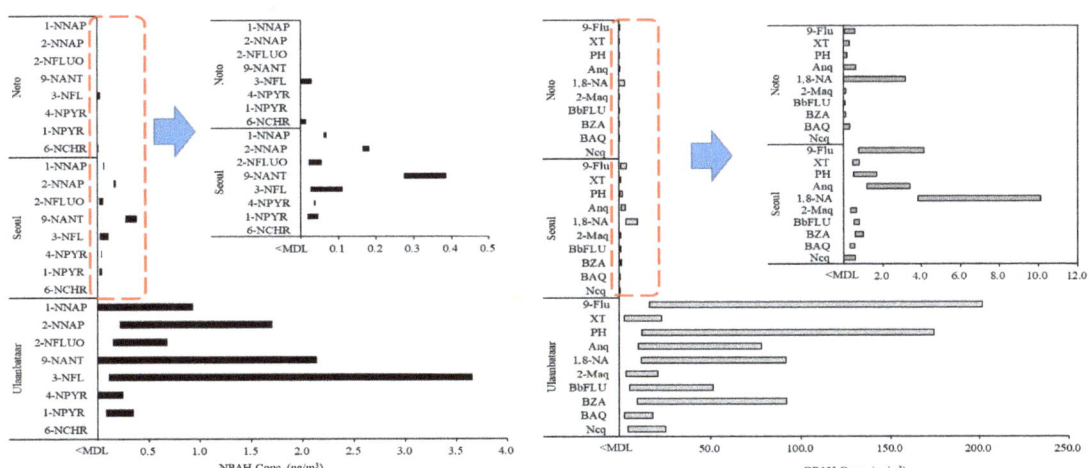

Figure 6. Concentration ranges of individual NPAH and OPAH congeners in PM$_{2.5}$ from real samples.

The measured concentrations of individual NPAH and OPAH congeners ranged from 0.002 to 201.704 ng/m^3. The ranges of the measured concentration in the aerosol samples using GC-NCI/MS and GC-PCI-MS/MS in this study were 0.016–3.659 ng/m^3 and 0.002–201.704 ng/m^3, respectively. The detection rate of the number of target analytes (18 NPAH and OPAH congeners) whose concentration was calculated to be higher than the MDL in the actual samples based on the total number of analytes per sample was found to be 27% in the samples from Noto compared with 90% from Ulaanbaatar and 94% from Seoul. When OPAHs with a high detection rate in the air were compared from the two urban areas of Ulaanbaatar and Seoul (except for Noto, where 73% was not detected), the difference in the maximum average concentration was 55 times. Among the detected 10 OPAH congeners in the samples from Seoul and Ulaanbaatar, the major congeners were 9-fluorenone (9-Flu), anthraquinone (Anq), 1,8-naphthalic anhydride (1,8-NA), and PH. Among these four congeners, the concentrations ranged from 16.703–201.704 ng/m^3 for 9-Flu and 12.549–174.388 ng/m^3 for PH. From the results for the NPAHs, the maximum concentration in the sample from Seoul was 0.387 ng/m^3 for 9-NANT. The analytical results of the samples from Ulaanbaatar showed that 3-nitrofluoranthene (3-NFL) had the highest concentration of 3.659 ng/m^3. 3-NFL, 9-NTNT, and 2-nitronaphthalene (2-NNAP) were the main congeners detected in the samples. 9-NANT was the main congener detected in the samples from both Seoul and Ulaanbaatar. It has been reported that 9-NANT is emitted from direct sources, predominantly diesel engines, and is also formed in the atmosphere as a result of secondary reactions [43,44]. Ulaanbaatar generally uses biomass fuel and coal fuel (lignite) for household cooking and heating It is an urban area with high levels of PM and PAH pollution because of the use of solid fuels [45]. In Ulaanbaatar, the concentration of PAHs is high, and this causes high concentrations of NPAHs and OPAHs because of secondary formations. Additionally, the main causes for the primary emissions of NPAHs in urban regions are diesel engine vehicles and incomplete combustion by residential heating [44]. As biomass combustion is the main source of OPAHs [46], biomass and heating fuel-use in Ulaanbaatar could contribute to high NPAH and OPAH concentrations. The contribution of individual NPAH and OPAH congeners to the regional total concentration was different because of the different characteristics of regional occurrence. The proposed analytical method for the monitoring of aerosol samples could be of practical use to reveal such a source of PAH derivatives.

3. Materials and Methods

3.1. Sampling Sites and Method

The aerosol samples were collected at three sites in winter from 15 December 2020 to 17 January 2021. Two sites were the capital cities of South Korea (Seoul) and Mongolia (Ulaanbaatar), and the other site was a background region in Japan (Noto). The sampling locations are shown in Figure S1. The sampling time was 24 h, from 10:00 a.m. to 9:00 p.m. the next day, and a high-volume air sampler (Shibata, Tokyo, Japan, HV-1000RW) equipped with an impactor for PM2.5 was used. Quartz fiber filters (QFFs, Tissuequartz 2500QAT-UP, 8 × 10 in, Pall, USA) were baked at 550 °C for 12 h before sampling to remove the organic matter. All sampled filters were wrapped in pre-baked aluminum foil and stored at −20 °C until the analysis.

3.2. Sample Extraction and Analysis

The analytes collected on the QFFs were extracted twice for 30 min each with a mixture of dichloromethane and methanol (DCM:MeOH at a ratio of 3:1 v/v) by sonication. Before the extraction, Flu-d10 was added as an internal standard and spiked at 780 ng. The extract was filtered through a 0.45 μm nylon membrane filter and concentrated to 1 mL using TurboVap (TurboVap LV, Zymark, Germany). After being transferred to a vial, nitrogen was concentrated to 0.5 mL at 35 °C or less and MS was performed. The QFF field blanks were analyzed for each field sample and processed using the same procedure.

All standard reagents for the analytes were purchased from AccuStandard Chemical (USA); the target analytes are listed in Table S3. The NPAHs and OPAHs were analyzed by gas chromatography (7890B GC) coupled with 5977A MS (Agilent Technologies, USA) and 7010 triple quadrupole MS (Agilent Technologies, USA) with interchangeable EI and CI ion sources. Chemical ionization was used for the target substances of 8 NPAHs and 10 OPAHs. The quantitative conditions for NCI and PCI were optimized using standard reagents for each analyte. NCI and PCI were quantified by SIM and SRM methods, respectively. The analytes were classified by a DB-5MS UI column and analyzed in NCI-GC/MS and PCI-GC-MS/MS modes. High-purity helium (99.999%) was used as the carrier gas, with a flow rate of 1 mL/min. The reagent gas was high-purity methane (99.999%) and the flow rate was 1 mL/min for the NCI mode and 2 mL/min for the PCI mode. All samples were injected with 2 μL in a splitless mode at 300 °C. The GC oven conditions were as follows: 60 °C → 145 °C (10 °C/min) → 220 °C (4 °C/min) → 320 °C (10 °C/min, hold for 20 min). Detailed information on the analysis conditions are summarized in Table S4.

3.3. Quality Assurance and Quality Control

Linearity, MDL, and recovery experiments of the calibration curve were performed to evaluate the QA/QC for the assay method. For the MDL, the lowest concentration was added to the filter and the experiment was repeated 7 times in the same way; the MDL was obtained from the standard deviation of the measured concentration. Thus, the procedure was carried out through three repeated experiments. In addition, in order to reduce analytical errors that may have occurred in the continuous analytical measurements and to accurately measure the results, the QC substances were periodically measured for every 15 sample batches. The analysis was performed whilst taking necessary precautions so that the accuracy of the measured value and the expected value of the QC material for which the exact value was known did not deviate from the allowable limit. QA/QC was validated for each analyte according to the optimized NCI and PCI analytical methods.

4. Conclusions

Congener-specific methods were evaluated for the quantification of NPAHs and OPAHs, which are PAH derivatives. For the NPAHs, GC-NCI/MS generated strong molecular ions and showed a better sensitivity than GC-PCI-MS/MS. However, for 9-NANT, it was difficult to obtain a complete separation from nearby interferences in the actual aerosol samples; thus, GC-PCI-MS/MS was used. In the case of all OPAHs, the background was reduced in the sample matrix by means of only the product ion being quantified from the specific parent ion for an individual OPAH congener using GC-PCI-MS/MS with the SRM mode. This enabled a highly sensitive and selective quantitative analysis for the OPAHs. For the method validation, the linearity of the calibration curve, MDL, and recovery were evaluated for each individual congener. In the actual aerosol samples collected at Noto, Seoul, and Ulaanbaatar, the analytical stability of each batch was confirmed from the results of the QC samples. A re-analysis was performed if the result was out of the allowable range. Among the three regions, the total mean concentrations of NAPHs and OPAHs were highest in Ulaanbaatar, followed, in order, by Seoul and Noto. The most abundant NPAHs in Seoul and Ulaanbaatar were 9-NANT and 3-NFL; among the OPAHs, there were high concentrations of 1,8-NA and 9-Flu. With the analysis method in this study, the concentrations of NPAHs and OPAHs in the samples of atmospheric PM2.5 could be quantified in the range of 0.016 to 3.659 ng/m^3 and 0.002 to 201.704 ng/m^3, respectively. This method allowed the quantification of NPAHs and OPAHs over a wide concentration range by region, and could be of help in the field of aerosol research.

Supplementary Materials: The following supporting information can be downloaded at: https://www.mdpi.com/article/10.3390/molecules28020775/s1. Table S1: Comparison of molecular ion intensities of NPAHs when using EI and NCI modes at a concentration of 5 μg/mL; Table S2: Regional concentration range (ng/m^3), mean value, and detection frequency of NPAH and OPAH congeners in PM$_{2.5}$; Table S3: List of target compounds; Table S4: Analytical conditions of GC/MS and GC-QqQ (MS/MS); Figure S1: Map of sampling locations.

Author Contributions: Conceptualization, J.J., J.-Y.L., and Y.-G.A.; methodology, J.J., K.-S.J., and Y.-G.A.; validation, J.J., J.-Y.L., and Y.-G.A.; sample collection, A.M., A.N., J.-Y.L., and K.-S.J.; data curation, J.J., K.-S.J., and Y.-G.A.; writing—original draft preparation, J.J., J.-Y.L., K.-S.J., and Y.-G.A.; writing—review and editing, Y.-G.A. All authors have read and agreed to the published version of the manuscript.

Funding: This research was supported by the FRIEND (Fine Particle Research Initiative in East Asia Considering National Differences) project through the National Research Foundation of Korea (NRF) funded by the Ministry of Science and ICT (2020M3G1A1114537, 2020M3G1A1114551) and KBSI (C280200) grants.

Institutional Review Board Statement: Not applicable.

Informed Consent Statement: Not applicable.

Data Availability Statement: The data presented in this study are available on request from the corresponding author.

Conflicts of Interest: The authors declare no conflict of interest.

Sample Availability: Samples of the compounds are available from the authors.

References

1. Rabha, S.; Saikia, B.K. Advanced micro-and nanoscale characterization techniques for carbonaceous aerosols. In *Handbook of Nanomaterials in Analytical Chemistry*; Elsevier: Amsterdam, The Netherlands, 2020; pp. 449–472.
2. Ravindra, K.; Sokhi, R.; Van Grieken, R. Atmospheric polycyclic aromatic hydrocarbons: Source attribution, emission factors and regulation. *Atmos. Environ.* **2008**, *42*, 2895–2921. [CrossRef]
3. Bläsing, M.; Amelung, W.; Schwark, L.; Lehndorff, E. Inland navigation: PAH inventories in soil and vegetation after EU fuel regulation 2009/30/EC. *Sci. Total Environ.* **2017**, *584*, 19–28. [CrossRef] [PubMed]
4. Keith, L.H. The source of US EPA's sixteen PAH priority pollutants. *Polycycl. Aromat. Compd.* **2015**, *35*, 147–160. [CrossRef]
5. Walgraeve, C.; Demeestere, K.; Dewulf, J.; Zimmermann, R.; Van Langenhove, H. Oxygenated polycyclic aromatic hydrocarbons in atmospheric particulate matter: Molecular characterization and occurrence. *Atmos. Environ.* **2010**, *44*, 1831–1846. [CrossRef]
6. Bandowe, B.A.M.; Meusel, H. Nitrated polycyclic aromatic hydrocarbons (nitro-PAHs) in the environment—A review. *Sci. Total Environ.* **2017**, *581*, 237–257. [CrossRef]
7. Bandowe, B.A.M.; Meusel, H.; Huang, R.-J.; Ho, K.; Cao, J.; Hoffmann, T.; Wilcke, W. PM2.5-bound oxygenated PAHs, nitro-PAHs and parent-PAHs from the atmosphere of a Chinese megacity: Seasonal variation, sources and cancer risk assessment. *Sci. Total Environ.* **2014**, *473*, 77–87. [CrossRef] [PubMed]

8. Dos Santos, R.R.; de Lourdes Cardeal, Z.; Menezes, H.C. Phase distribution of polycyclic aromatic hydrocarbons and their oxygenated and nitrated derivatives in the ambient air of a Brazilian urban area. *Chemosphere* **2020**, *250*, 126223. [CrossRef]
9. Hayakawa, K. Recent Research Progress on Nitropolycyclic Aromatic Hydrocarbons in Outdoor and Indoor Environments. *Appl. Sci.* **2022**, *12*, 11259. [CrossRef]
10. Abbas, I.; Badran, G.; Verdin, A.; Ledoux, F.; Roumié, M.; Courcot, D.; Garçon, G. Polycyclic aromatic hydrocarbon derivatives in airborne particulate matter: Sources, analysis and toxicity. *Environ. Chem. Lett.* **2018**, *16*, 439–475. [CrossRef]
11. Tomaz, S.; Jaffrezo, J.-L.; Favez, O.; Perraudin, E.; Villenave, E.; Albinet, A. Sources and atmospheric chemistry of oxy-and nitro-PAHs in the ambient air of Grenoble (France). *Atmos. Environ.* **2017**, *161*, 144–154. [CrossRef]
12. Li, W.; Wang, C.; Shen, H.; Su, S.; Shen, G.; Huang, Y.; Zhang, Y.; Chen, Y.; Chen, H.; Lin, N. Concentrations and origins of nitro-polycyclic aromatic hydrocarbons and oxy-polycyclic aromatic hydrocarbons in ambient air in urban and rural areas in northern China. *Environ. Pollut.* **2015**, *197*, 156–164. [CrossRef]
13. Kojima, Y.; Inazu, K.; Hisamatsu, Y.; Okochi, H.; Baba, T.; Nagoya, T. Influence of secondary formation on atmospheric occurrences of oxygenated polycyclic aromatic hydrocarbons in airborne particles. *Atmos. Environ.* **2010**, *44*, 2873–2880. [CrossRef]
14. Srivastava, D.; Favez, O.; Bonnaire, N.; Lucarelli, F.; Haeffelin, M.; Perraudin, E.; Gros, V.; Villenave, E.; Albinet, A. Speciation of organic fractions does matter for aerosol source apportionment. Part 2: Intensive short-term campaign in the Paris area (France). *Sci. Total Environ.* **2018**, *634*, 267–278. [CrossRef]
15. Wei, S.; Huang, B.; Liu, M.; Bi, X.; Ren, Z.; Sheng, G.; Fu, J. Characterization of PM2. 5-bound nitrated and oxygenated PAHs in two industrial sites of South China. *Atmos. Res.* **2012**, *109*, 76–83. [CrossRef]
16. Niederer, M. Determination of polycyclic aromatic hydrocarbons and substitutes (nitro-, oxy-PAHs) in urban soil and airborne particulate by GC-MS and NCI-MS/MS. *Environ. Sci. Pollut. Res.* **1998**, *5*, 209–216. [CrossRef] [PubMed]
17. Tomaz, S.; Shahpoury, P.; Jaffrezo, J.-L.; Lammel, G.; Perraudin, E.; Villenave, E.; Albinet, A. One-year study of polycyclic aromatic compounds at an urban site in Grenoble (France): Seasonal variations, gas/particle partitioning and cancer risk estimation. *Sci. Total Environ.* **2016**, *565*, 1071–1083. [CrossRef] [PubMed]
18. Albinet, A.; Leoz-Garziandia, E.; Budzinski, H.; Villenave, E. Simultaneous analysis of oxygenated and nitrated polycyclic aromatic hydrocarbons on standard reference material 1649a (urban dust) and on natural ambient air samples by gas chromatography–mass spectrometry with negative ion chemical ionisation. *J. Chromatogr. A* **2006**, *1121*, 106–113. [CrossRef]
19. Domeno, C.; Canellas, E.; Alfaro, P.; Rodriguez-Lafuente, A.; Nerin, C. Atmospheric pressure gas chromatography with quadrupole time of flight mass spectrometry for simultaneous detection and quantification of polycyclic aromatic hydrocarbons and nitro-polycyclic aromatic hydrocarbons in mosses. *J. Chromatogr. A* **2012**, *1252*, 146–154. [CrossRef]
20. Nowakowski, M.; Rykowska, I.; Wolski, R.; Andrzejewski, P. Polycyclic Aromatic Hydrocarbons (PAHs) and their Derivatives (O-PAHs, N-PAHs, OH-PAHs): Determination in Suspended Particulate Matter (SPM)—A Review. *Environ. Process.* **2022**, *9*, 2. [CrossRef]
21. Tanner, R.L.; Fajer, R. Determination of nitro-polynuclear aromatics in ambient aerosol samples. *Int. J. Environ. Sci. Technol.* **1983**, *14*, 231–241. [CrossRef]
22. Teixeira, E.C.; Garcia, K.O.; Meincke, L.; Leal, K.A. Study of nitro-polycyclic aromatic hydrocarbons in fine and coarse atmospheric particles. *Atmos. Res.* **2011**, *101*, 631–639. [CrossRef]
23. Oezel, M.Z.; Hamilton, J.F.; Lewis, A.C. New sensitive and quantitative analysis method for organic nitrogen compounds in urban aerosol samples. *Environ. Sci. Technol.* **2011**, *45*, 1497–1505. [CrossRef] [PubMed]
24. Chuang, J.C.; Mack, G.A.; Kuhlman, M.R.; Wilson, N.K. Polycyclic aromatic hydrocarbons and their derivatives in indoor and outdoor air in an eight-home study. *Atmos. Environ. Part B Urban Atmos.* **1991**, *25*, 369–380. [CrossRef]
25. Vyskocil, V.; Barek, J. Electroanalysis of nitro and amino derivatives of polycyclic aromatic hydrocarbons. *Curr. Org. Chem.* **2011**, *15*, 3059–3076. [CrossRef]
26. Galceran, M.; Moyano, E. Determination of oxygenated and nitro-substituted polycyclic aromatic hydrocarbons by HPLC and electrochemical detection. *Talanta* **1993**, *40*, 615–621. [CrossRef]
27. Reisen, F.; Arey, J. Atmospheric reactions influence seasonal PAH and nitro-PAH concentrations in the Los Angeles basin. *Environ. Sci. Technol.* **2005**, *39*, 64–73. [CrossRef]
28. Dimashki, M.; Harrad, S.; Harrison, R.M. Measurements of nitro-PAH in the atmospheres of two cities. *Atmos. Environ.* **2000**, *34*, 2459–2469. [CrossRef]
29. Bamford, H.A.; Bezabeh, D.Z.; Schantz, M.M.; Wise, S.A.; Baker, J.E. Determination and comparison of nitrated-polycyclic aromatic hydrocarbons measured in air and diesel particulate reference materials. *Chemosphere* **2003**, *50*, 575–587. [CrossRef]
30. Mueller, A.; Ulrich, N.; Hollmann, J.; Sanchez, C.E.Z.; Rolle-Kampczyk, U.E.; von Bergen, M. Characterization of a multianalyte GC-MS/MS procedure for detecting and quantifying polycyclic aromatic hydrocarbons (PAHs) and PAH derivatives from air particulate matter for an improved risk assessment. *Environ. Pollut.* **2019**, *255*, 112967. [CrossRef]
31. Drventić, I.; Šala, M.; Vidović, K.; Kroflič, A. Direct quantification of PAHs and nitro-PAHs in atmospheric PM by thermal desorption gas chromatography with electron ionization mass spectroscopic detection. *Talanta* **2023**, *251*, 123761. [CrossRef]
32. Zielinska, B.; Samy, S. Analysis of nitrated polycyclic aromatic hydrocarbons. *Anal. Bioanal. Chem.* **2006**, *386*, 883–890. [CrossRef]
33. Yadav, I.C.; Devi, N.L.; Singh, V.K.; Li, J.; Zhang, G. Concentrations, sources and health risk of nitrated-and oxygenated-polycyclic aromatic hydrocarbon in urban indoor air and dust from four cities of Nepal. *Sci. Total Environ.* **2018**, *643*, 1013–1023. [CrossRef] [PubMed]

34. Naing, N.N.; Yeo, K.B.; Lee, H.K. A combined microextraction procedure for isolation of polycyclic aromatic hydrocarbons in ambient fine air particulate matter with determination by gas chromatography-tandem mass spectrometry. *J. Chromatogr. A* **2020**, *1612*, 460646. [CrossRef] [PubMed]
35. Sun, C.; Qu, L.; Wu, L.; Wu, X.; Sun, R.; Li, Y. Advances in analysis of nitrated polycyclic aromatic hydrocarbons in various matrices. *Trends Anal. Chem.* **2020**, *127*, 115878. [CrossRef]
36. Vu-Duc, N.; Phung Thi, L.A.; Le-Minh, T.; Nguyen, L.-A.; Nguyen-Thi, H.; Pham-Thi, L.-H.; Doan-Thi, V.-A.; Le-Quang, H.; Nguyen-Xuan, H.; Thi Nguyen, T. Analysis of polycyclic aromatic hydrocarbon in airborne particulate matter samples by gas chromatography in combination with tandem mass spectrometry (GC-MS/MS). *J. Anal. Methods Chem.* **2021**, *2021*, 6641326. [CrossRef] [PubMed]
37. Pandey, S.K.; Kim, K.-H.; Brown, R.J. A review of techniques for the determination of polycyclic aromatic hydrocarbons in air. *Trends Anal. Chem.* **2011**, *30*, 1716–1739. [CrossRef]
38. Bezabeh, D.Z.; Bamford, H.A.; Schantz, M.M.; Wise, S.A. Determination of nitrated polycyclic aromatic hydrocarbons in diesel particulate-related standard reference materials by using gas chromatography/mass spectrometry with negative ion chemical ionization. *Anal. Bioanal. Chem.* **2003**, *375*, 381–388. [CrossRef]
39. Nicol, S.; Dugay, J.; Hennion, M.-C. Determination of oxygenated polycyclic aromatic compounds in airborne particulate organic matter using gas chromatography-tandem mass spectrometry. *Chromatographia* **2001**, *53*, S464–S469. [CrossRef]
40. Clench, M.; Tetler, L. CHROMATOGRAPHY: GAS | Detectors: Mass Spectrometry. *Encycl. Sep. Sci.* **2000**, 448–455.
41. Sleno, L.; Volmer, D.A. Ion activation methods for tandem mass spectrometry. *J. Mass Spectrom.* **2004**, *39*, 1091–1112. [CrossRef]
42. USEPA. *Definition and Procedure for the Determination of the Method Detection Limit, Revision 2*; EPA: Washington, DC, USA, 2016.
43. Valle-Hernández, B.; Mugica-Álvarez, V.; Salinas-Talavera, E.; Amador-Muñoz, O.; Murillo-Tovar, M.; Villalobos-Pietrini, R.; De Vizcaya-Ruíz, A. Temporal variation of nitro-polycyclic aromatic hydrocarbons in PM10 and PM2.5 collected in Northern Mexico City. *Sci. Total Environ.* **2010**, *408*, 5429–5438. [CrossRef] [PubMed]
44. Wang, T.; Zhao, J.; Liu, Y.; Peng, J.; Wu, L.; Mao, H. PM2.5-Bound Polycyclic Aromatic Hydrocarbons (PAHs), Nitrated PAHs (NPAHs) and Oxygenated PAHs (OPAHs) in Typical Traffic-Related Receptor Environments. *J. Geophys. Res. Atmos.* **2022**, *127*, e2021JD035951.
45. Sainnokhoi, T.-A.; Kováts, N.; Gelencsér, A.; Hubai, K.; Teke, G.; Pelden, B.; Tserenchimed, T.; Erdenechimeg, Z.; Galsuren, J. Characteristics of particle-bound polycyclic aromatic hydrocarbons (PAHs) in indoor PM2.5 of households in the Southwest part of Ulaanbaatar capital, Mongolia. *Environ. Monit. Assess.* **2022**, *194*, 665. [CrossRef] [PubMed]
46. Ding, J.; Zhong, J.; Yang, Y.; Li, B.; Shen, G.; Su, Y.; Wang, C.; Li, W.; Shen, H.; Wang, B. Occurrence and exposure to polycyclic aromatic hydrocarbons and their derivatives in a rural Chinese home through biomass fuelled cooking. *Environ. Pollut.* **2012**, *169*, 160–166. [CrossRef] [PubMed]

Disclaimer/Publisher's Note: The statements, opinions and data contained in all publications are solely those of the individual author(s) and contributor(s) and not of MDPI and/or the editor(s). MDPI and/or the editor(s) disclaim responsibility for any injury to people or property resulting from any ideas, methods, instructions or products referred to in the content.

Article

Markers of Chemical and Microbiological Contamination of the Air in the Sport Centers

Justyna Szulc [1,*], Małgorzata Okrasa [2], Małgorzata Ryngajłło [3], Katarzyna Pielech-Przybylska [4] and Beata Gutarowska [1]

[1] Department of Environmental Biotechnology, Lodz University of Technology, 90-530 Łódź, Poland
[2] Department of Personal Protective Equipment, Central Institute for Labour Protection—National Research Institute, 90-133 Łódź, Poland
[3] Institute of Molecular and Industrial Biotechnology, Lodz University of Technology, 90-573 Łódź, Poland
[4] Institute of Fermentation Technology and Microbiology, Lodz University of Technology, 90-530 Łódź, Poland
* Correspondence: justyna.szulc@p.lodz.pl

Abstract: This study aimed to assess the markers of chemical and microbiological contamination of the air at sport centers (e.g., the fitness center in Poland) including the determination of particulate matter, CO_2, formaldehyde (DustTrak™ DRX Aerosol Monitor; Multi-functional Air Quality Detector), volatile organic compound (VOC) concentration (headspace solid-phase microextraction coupled with gas chromatography–mass spectrometry), the number of microorganisms in the air (culture methods), and microbial biodiversity (high-throughput sequencing on the Illumina platform). Additionally the number of microorganisms and the presence of SARS-CoV-2 (PCR) on the surfaces was determined. Total particle concentration varied between 0.0445 mg m^{-3} and 0.0841 mg m^{-3} with the dominance (99.65–99.99%) of the $PM_{2.5}$ fraction. The CO_2 concentration ranged from 800 ppm to 2198 ppm, while the formaldehyde concentration was from 0.005 mg/m^3 to 0.049 mg m^{-3}. A total of 84 VOCs were identified in the air collected from the gym. Phenol, D-limonene, toluene, and 2-ethyl-1-hexanol dominated in the air at the tested facilities. The average daily number of bacteria was 7.17 × 10^2 CFU m^{-3}–1.68 × 10^3 CFU m^{-3}, while the number of fungi was 3.03 × 10^3 CFU m^{-3}–7.34 × 10^3 CFU m^{-3}. In total, 422 genera of bacteria and 408 genera of fungi representing 21 and 11 phyla, respectively, were detected in the gym. The most abundant bacteria and fungi (>1%) that belonged to the second and third groups of health hazards were: *Escherichia-Shigella*, *Corynebacterium*, *Bacillus*, *Staphylococcus*, *Cladosporium*, *Aspergillus*, and *Penicillium*. In addition, other species that may be allergenic (*Epicoccum*) or infectious (*Acinetobacter*, *Sphingomonas*, *Sporobolomyces*) were present in the air. Moreover, the SARS-CoV-2 virus was detected on surfaces in the gym. The monitoring proposal for the assessment of the air quality at a sport center includes the following markers: total particle concentration with the $PM_{2.5}$ fraction, CO_2 concentration, VOCs (phenol, toluene, and 2-ethyl-1-hexanol), and the number of bacteria and fungi.

Keywords: air contamination; VOCs; particulate matter; CO_2; bioaerosol; sport center

Citation: Szulc, J.; Okrasa, M.; Ryngajłło, M.; Pielech-Przybylska, K.; Gutarowska, B. Markers of Chemical and Microbiological Contamination of the Air in the Sport Centers. *Molecules* 2023, 28, 3560. https://doi.org/10.3390/molecules28083560

Academic Editors: Stefan Tsakovski and Tony Venelinov

Received: 27 March 2023
Revised: 11 April 2023
Accepted: 17 April 2023
Published: 18 April 2023

Copyright: © 2023 by the authors. Licensee MDPI, Basel, Switzerland. This article is an open access article distributed under the terms and conditions of the Creative Commons Attribution (CC BY) license (https://creativecommons.org/licenses/by/4.0/).

1. Introduction

In the modern world, great attention is paid to a healthy lifestyle that includes regular sporting activities that contribute to maintaining a healthy body weight, feeling good, and sustaining energy and a youthful appearance [1,2]. Physical activity can also prevent hypertension and non-communicable diseases (e.g., heart disease, stroke, diabetes, and site-specific cancers) [3].

According to the World Health Organization (WHO) guidelines, adults need at least 2.5 h of moderate-intensity physical activity weekly [3]. The Deloitte report "Sports Retail Study 2020" mentions that almost 65% of Europeans practice at least one sport discipline, devoting 8.6 h a week to physical activity [4]. Although physical activity has been documented as beneficial to human health, using sports facilities has raised concerns during the COVID-19 pandemic as contributing to the spread of SARS-CoV-2.

Various factors influence air quality in sports facilities (e.g., building construction, materials used, ventilation, air humidity and temperature, number of users, and type of physical activity) [5,6]. Many of these factors can favor the spread and multiplication of microorganisms (i.e., high air humidity from the intense sweat discharge of the users, high particulate matter concentration from the resuspension of particles sedimented on the surfaces, and regular contact between the users and sports equipment) [1].

Most of the air is inhaled through the mouth during physical activities excluding the normal nasal mechanisms for filtration. The increased airflow velocity carries airborne contaminants deeper into the respiratory tract. Thus, increased concentrations of microorganisms, their fragments and metabolites can be introduced into the respiratory tract of exercising individuals and pose a considerable health risk to them [1]. Moreover, the research shows that physical activity increases aerosol emissions due to elevated ventilation and dehydration of the airways, further elevating the bioaerosol concentrations. Furthermore, the air quality in sports facilities depends on CO_2 and other gases and volatile organic compound (VOC) concentrations [7]. With the increase in the intensity of physical activity (and thus breathing), the concentration of CO_2 in sports halls increases. According to the American Society of Heating, Refrigerating, and Air-Conditioning Engineers (ASHRAE), the maximum level of CO_2 in sports facilities is 1000 ppm. Higher CO_2 concentrations indicate poor air quality and acute health symptoms in room users (e.g., headaches and irritation of mucous membranes) and slower work efficiency [8]. Volatile organic compounds (VOCs) are gaseous and can originate from building materials and equipment (furniture, installations, electronics) [8,9].

Moreover, they are also associated with cleaning, disinfection, and using chemicals and cosmetics. Monocyclic aromatic hydrocarbons (MAH) are particularly important in the VOC group. VOCs can cause serious health effects as many of them exhibit toxic, carcinogenic, mutagenic, or neurotoxic properties. Many VOCs are odorous [8,9].

Bioaerosols are one of the main transmission routes for infectious diseases [10]. Moreover, human exposure to bioaerosols is associated with a wide range of acute and chronic health problems such as asthma, hay fever, bronchitis, chronic lung failure, diseases of the cardiovascular system, catarrh of the gastrointestinal tract, tuberculosis, legionellosis and allergic reactions as well as sinus and conjunctivitis [10,11]. Toxins of microbial origin (endotoxins and mycotoxins) play a significant role in inflammatory responses and contribute to the deterioration of lung function, causing other infections [12]. It has been found that over 80 types of fungi (mostly belonging to *Cladosporium, Alternaria, Aspergillus*, and *Fusarium* genera) can cause respiratory allergy symptoms and over 100 severe human and animal infections as well as plant diseases [13].

Because humans carry 1012 microorganisms in their epidermis and 1014 microorganisms in the digestive tract, they can pose as the primary source of microorganisms in fitness facilities [14]. Therefore, surfaces in sports facilities can also be a source of pathogenic microorganisms such as methicillin-resistant *Staphylococcus aureus* (MRSA). It was found that skin-to-skin contact is a primary route of MRSA transmission between athletes, especially in football, wrestling, rugby, and soccer players. Moreover, poor hygiene in equipment has also been implicated in the spread of contagious diseases [15]. Infections caused by MRSA are often aggressive, necrotizing, antibiotic-resistant, and sometimes fatal [16]. Sports facilities have already been the subject of microbiological research. The literature indicates different microbiological air contamination of these types of facilities, ranging from 5.80×10^1 to 1.02×10^3 CFU m^{-3} for the number of bacteria and from 2.10×10^1 to 1.44×10^2 CFU m^{-3} for the number of fungi [1,5,17,18]. Conversely, in the case of the bacterial contamination of surfaces, concentrations from 3.9×10^2 CFU cm^{-2} to 3.7×10^3 CFU cm^{-2} were observed [5,7]. Bacteria of the genera *Bacillus, Corynebacterium, Kocuria, Micrococcus*, and *Pseudomonas Staphylococcus* were characteristic of bioaerosols in sports facilities, while on surfaces, the dominance of *Staphylococcus, Bacillus, Klebsiella, Escherichia, Enterococcus, Serratia, Aerococcus*, and *Erwinia* has been described [19,20]. The environment of fitness centers, however, has not yet been comprehensively investigated in terms of the number and species of microorganisms present in these places.

Therefore, this study aimed to assess the markers of chemical and microbiological contamination of the air in sports centers. The research included the evaluation of microclimate parameters, particulate matter concentration, selected chemical contaminations, the number of microorganisms in the air and on surfaces, the diversity of microorganisms, and the presence of SARS-CoV-2 in fitness center environments. This is the first study to assess the biodiversity of sports facilities using the metagenome analysis of settled dust. The results are discussed in the context of pathogen transmission and the overall health effects of exposure to the detected contaminants. Moreover, guidelines for maintaining good air quality in sports facilities are proposed.

2. Results and Discussion

2.1. Microclimate and Particulate Matter Concentration

The mean values of microclimatic conditions and PM concentrations are presented in Table 1. The microclimate parameters were also analyzed as daily averages (Figure S1a). The measurements were also taken depending on the time of day (Figure S1b) and the sampling location (Figure S1c).

Table 1. Air quality parameters at the tested locations.

Room No.	Description	Parameter					
		Temperature, °C	Relative Humidity, %	Air Velocity m/s	Total PM Concentration, mg/m^3	CO_2 Concentration, ppm	HCN Concentration, mg/m^3
1	Reception	M: 25.63 [a] SD: 3.57	M: 58.52 [ab] SD: 10.23	M: 0.092 [bdef] SD: 0.153	M: 0.058 [a] SD: 0.009	M: 800 [ab] SD: 94	M: 0.005 [a] SD: 0.006
2	The gym	M: 25.28 [ab] SD: 0.95	M: 53.73 [b] SD: 6.99	M: 0.072 [acf] SD: 0.057	M: 0.057 [a] SD: 0.010	M: 2198 [c] SD: 111	M: 0.049 [c] SD: 0.002
3	Fitness room on the first floor	M: 24.82 [b] SD: 0.81	M: 57.62 [ab] SD: 8.16	M: 0.067 [aceg] SD: 0.053	M: 0.057 [a] SD: 0.009	M: 1773 [ad] SD: 51	M: 0.042 [ab] SD: 0.001
4	Fitness room on the second floor	M: 25.49 [ab] SD: 0.75	M: 62.31 [a] SD: 6.07	M: 0.025 [bdg] SD: 0.024	M: 0.059 [a] SD: 0.010	M: 2017 [cd] SD: 32	M: 0.047 [bc] SD: 0.001
5	Women's cloakroom	M: 26.18 [a] SD: 0.67	M: 60.82 [ab] SD: 5.84	M: 0.019 [b] SD: 0.008	M: 0.057 [a] SD: 0.010	M: 1925 [bcd] SD: 67	M: 0.046 [abc] SD: 0.001
6	Atmospheric air (external background)	M: 26.40 *[ab] SD: 2.30	M: 70.00 *[a] SD: 5.24	M: 5.125 *[a] SD: 0.978	M: 0.024 **[b] SD: 0.004	M: 593 [a] SD: 44	M: 0.067 [c] SD: 0.027

M—mean; SD—standard deviation; statistically different samples were marked with different letters within the same column (Kruskal–Wallis test followed by Dunn's post hoc tests at a significance level of 0.05); data sources for the atmospheric air: (*) https://www.ekologia.pl/pogoda/polska/lodzkie/zdunska-wola/archiwum,zakres (accessed on 26 July 2012), (**) https://powietrze.gios.gov.pl/pjp/current/station_details/archive/350# (accessed on 26 July 2012).

Airflow velocity measured during the experiments ranged from 0 m s^{-1} (no ventilation or air conditioning, no windows open, minimal number of people present at the sampling site) to 0.69 m s^{-1} (near an open window). The temperature was between 12.8 °C and 29.8 °C, and the relative humidity was between 44.3% and 78.3%. Microclimatic conditions are essential for achieving the optimal performance and comfort during exercise. The literature shows that an effective temperature below 22 °C degrades exercise performance among women, while an air temperature of 24 °C, with moderate RH, low air velocity, and weak radiation, is recommended at gyms to support exercise, comfort, and energy conservation [21]. The International Fitness Association sets different recommendations for temperature and humidity at commercial gyms [22]. A room temperature below 20 °C degrees and 50% humidity are recommended for aerobic classes, while for aerobics, cardio, weight training and Pilates areas, temperatures should be between 18 and 20 °C with a humidity between 40% and 60%.

The relation between thermal comfort and air movement at elevated activity levels was also investigated [23]. Air movement with higher temperatures produced equal or better comfort and perceived air quality below the reference condition for every temperature up to 26 °C. In our study, the air velocities and temperatures did not differ daily; the only difference was observed in the average daily humidity between Friday and the rest of the week (Figure S1a). Moreover, the air velocities did not depend on the time of day; the measurement was carried out while the temperature rose by the hour and the relative humidity first dropped (while the air conditioning was running) and then increased (Figure S1b).

The temperature conditions at different locations were similar (no statistical differences in average temperature were detected; Figure S1c). Statistically significant differences were detected for the average air velocity and average humidity (Figure S1c), corresponding to the number of windows open, the air conditioning running, and the number of people in the facility. The diversified values of these microclimatic parameters might lead to different development conditions for microorganisms between the tested locations.

The fitness center environment is very unstable, and many factors affect the parameters of temperature, relative humidity and airflow velocity. In the present study, significantly different air velocity and air humidity conditions were noted, which implies that the microclimate parameters strongly depend on the specific location. There was no correlation between the individual microclimate parameters and the number of windows open, air conditioning running, and the number of people in the facility.

Previous research has shown that the high number of people who exercise at closed sports facilities can contribute to the air quality issues inside them. The factors determining exercise intensity also affect air quality [24]. The concentration of fine particles in the indoor air fluctuates depending on the weather conditions. Furthermore, ventilation and air filtration systems at such facilities are essential for proper air exchange and purification. The present study confirmed these results as significantly higher PM concentrations were observed indoors than outdoors (Table 1).

The size distributions of airborne particles for each separate sampling variant are presented in Table S1 (Supplementary Materials). The $PM_{2.5}$ constituted almost all measured dust at the tested locations. Its share in the total quantity of the measured airborne particles was between 99.65% and 99.99%, and the range of the number of particles per size dropped with an increasing particle size. The total suspended PM concentration varied between 0.0445 mg m^{-3} and 0.0841 mg m^{-3} and differed significantly between all of the tested locations (Figure S2c). The daily averages were higher in the first three days of the experiment and significantly lower in the last two (Figure S2a). Considerably higher concentrations were observed at the beginning of each day and just before the facility was closed (Figure S2b). Based on full-factorial ANOVA, the main effects and all interactions were confirmed at a significance level of 0.05.

According to EU legislation, the annual average concentration of dust with dimensions below 2.5 μm (i.e., the fraction containing the PM_1 fraction) should not exceed 0.025 mg m^{-3} [25]. In our case, the measured $PM_{2.5}$ concentration was more than twice as high as the environmental threshold, independent of the sampling site [25]. This agrees with the literature suggesting that the air quality inside training facilities is often worse than that outdoors [26].

The presence of airborne particulate matter can affect the users' health and decrease their physical performance by around 5% [27,28]. Studies show that exposure to high PM concentrations can increase the risk of suffering from various respiratory and circulatory diseases [29,30]. Moreover, some studies have suggested that people who regularly exercise are more prone to experiencing the effects of air pollutants than those who do not participate in sports [31,32].

Carbon dioxide (CO_2) is the main gaseous air pollutant in sports facilities, connected to a natural product of human respiration [8]. The CO_2 concentration in the fitness club ranged from 800 ppm (reception desk) to 2198 ppm (gym), and for most rooms, it was statistically significantly higher than in atmospheric air. In previous studies, the CO_2 concentrations in the air of sports halls were lower and ranged from 294.8 to 1529 ppm [8]. However, the current studies have not shown any exceedance of the CO_2 concentration limits in the air developed by the WHO and the U.S. Environmental Protection Agency [33,34].

Formaldehyde concentration in the tested sports facilities ranged from 0.005 mg/m^3 to 0.049 mg/m^3, and for two out of five rooms, it was statistically significantly lower than in the control (atmospheric) air. This is probably due to the heavy traffic in the parking lot adjacent to the building and the busy street. According to EPA guidelines, the detected values of formaldehyde do not exceed the limits for this compound in the air [34].

2.2. Volatile Compounds Contamination

The volatile compounds were extracted from the air sample by SPME, followed by desorption and analysis with GC-MS. A total of 85 compounds were identified, of which 84 were present in the air collected from the gym, while only 47 were identified in the control sample (background) (Table S1). Detected compounds were divided into ten groups based on their chemical structures: hydrocarbons (30), terpenes and terpenoids (20), alcohols (10), aldehydes (eight), ketones (six), esters (six), furanes (two), phenols (one), ethers (one), and acids (one).

The sources of the compounds that were identified in the indoor air from the gym can differ; for example, the breath air exhaled by people in the gym [35] (e.g., acetone, ethanol, 1-propanol, butyl acetate, acetic acid, acetoin, and 2,3-butanedione), alcohol-based hand disinfectants and equipment cleaning agents [36], air fresheners and cosmetics [37] (e.g., ethanol, phenol, benzaldehyde, 2-propanol, α-pinene, eucalyptol, linalool, 3-carene, D-limonene, γ-terpinene, α-thujene, β-myrcene, camphene, butane, pentane, acetone, furfural, 3-methyl-1-butanol, 2-methyl-1-butanol, dihydromyrcenol, citronellal, verbenone, menthol), building materials and room finishing materials [38], and gym equipment (e.g., furfural, toluene, benzene, styrene, xylenes, ethylbenzene, heptane, decane, benzaldehyde, hexanal).

The relative amount (%) results showed that phenol was the dominant compound in both the control air sample and air sample from the gym. The presence of phenol results from the widespread use of it and its derivatives, among others, in the production of resins, detergents, medicinal products, disinfectants, and dyes, and thus are found in many common materials including antiseptics, medical preparations, plastics, cosmetics, and health care products [39]. Phenol also gets into the air through car exhaust. Phenol is not classified as a carcinogen but as a toxic substance [40]. Due to its hydrophilic and lipophilic properties, phenol easily penetrates cell membranes and dissolves in cell fractions, causing interaction with specific cellular and tissue structures [41].

Three other identified compounds, D-limonene, toluene, and 2-ethyl-1-hexanol, were characterized by over 1–2% of the relative amount in the indoor air from the gym. Among the indoor air VOCs, terpenes are a common group. Cleaning agents and cosmetics contain essential oils rich in terpenes [42].

Toluene, along with benzene, ethylbenzene, and xylenes from the BTEX group, are classified as toxic compounds, while benzene is also classified as a carcinogenic substance (Group 1). Due to their application for various purposes such as the production of plastics, synthetic fibers, floor coverings, chipboard, oils, greases, and paint, the presence of BTEX is common in indoor air. Long-term exposure to BTEX increases the risk of adverse health consequences [43].

In turn, 2-ethyl-1-hexanol is a common component of fragrances. Moreover, it is commonly used for the production of plasticizers (e.g., diethylhexyl phthalate for polyvinyl chloride resins) as well as in coating products, greases, fillers, and putties. The presence of 2-ethyl-1-hexanol may irritate the mucous membranes of the eyes and nose in humans [44,45].

2.3. Determination of Airborne Microorganism Number

The average daily number of bacteria in the facilities during the working week ranged from 7.17×10^2 CFU m^{-3} (Wednesday) to 1.68×10^3 CFU m^{-3} (Friday), while the number of fungi ranged from 3.03×10^3 CFU m^{-3} (Wednesday) to 7.34×10^3 CFU m^{-3} (Friday) (Table S3, Figure 1a). The lowest number of bacteria was recorded at 08:00 (4.48×10^2 CFU m^{-3}) and the highest at 20:00 (2.39×10^3 CFU m^{-3}). In turn, the concentration of fungi was the lowest at 16:00 (4.53×10^3 CFU m^{-3}) and the highest at 08:00 (7.42×10^3 CFU m^{-3}) (Table S3, Figure 1b). The most contaminated air was observed in Room no. 2 (the gym), where the bacteria count was 1.66×10^3 CFU m^{-3}, and in Room no. 1 (the reception), where the number of fungi was 7.30×10^3 CFU m^{-3} (daily mean).

It is noteworthy that, at the same time, in Room no. 2, the lowest number of fungi (1.90×10^3 CFU m^{-3}) among the analyzed rooms was recorded. In turn, the lowest number of bacteria was found in Room no. 4 (fitness room on the second floor) (Table S3, Figure 1c).

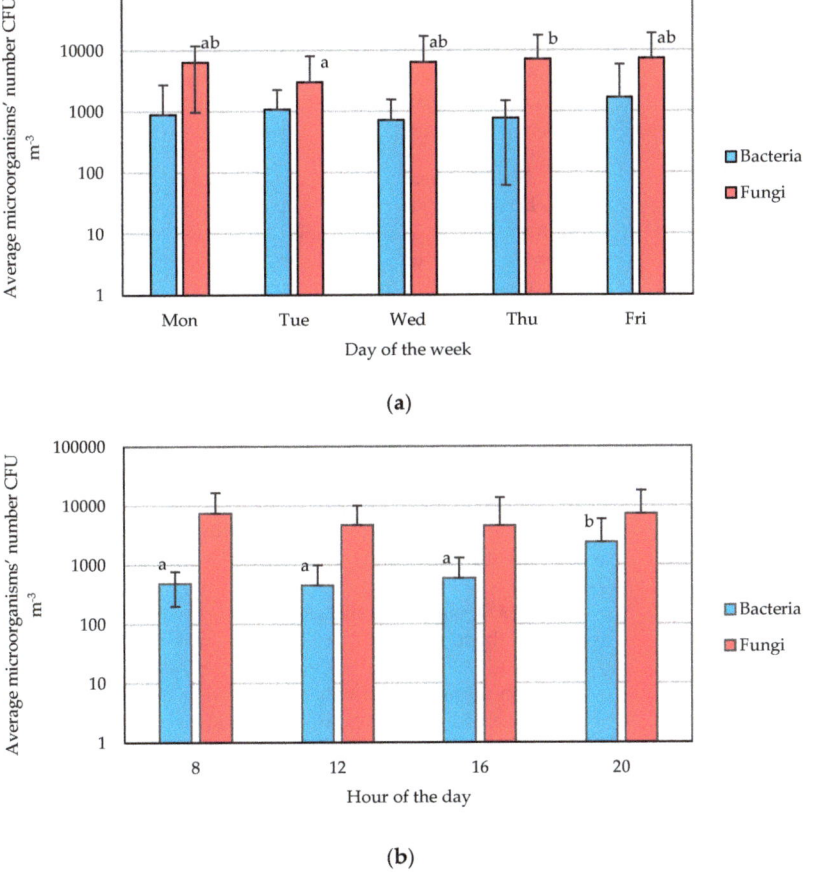

Figure 1. Cont.

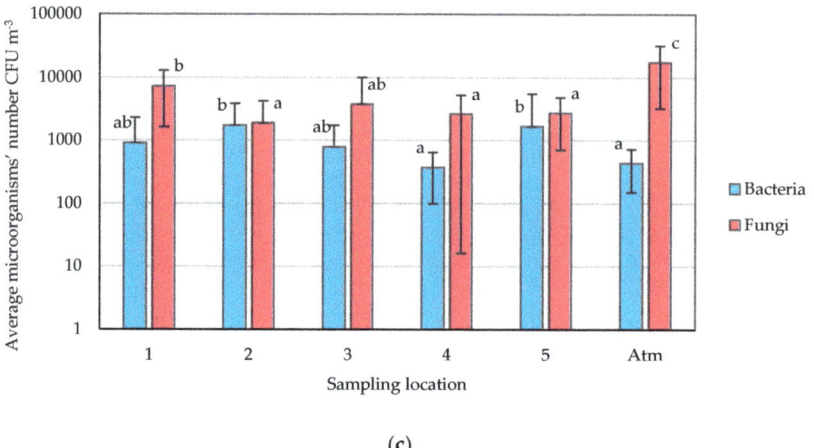

(c)

Figure 1. Averaged microorganism numbers (means with SD): (**a**) daily, (**b**) over sampling time of the day, (**c**) over sampling location; statistically different samples are marked with different letters; no letters indicate no statistical differences (Tukey's test, $\alpha = 0.05$).

No statistically significant differences were found in the fitness club's mean daily numbers of bacteria. In the case of fungi, significant differences among days were observed, with the lowest concentration on Tuesday and the highest on Thursday (Figure 1a). Considering the influence of the test hour on the number of microorganisms in the air; it can be concluded that the number of fungi is constant while the number of bacteria changes, which is most likely related to the activity of people in these facilities. Statistically higher numbers of bacteria were recorded at the end of the day—at 20:00 (Figure 1b). It was also shown that the number of fungi in the atmospheric air was statistically significantly higher than in the samples collected at the fitness club (Figure 1c).

The obtained aggregate results from the week of air quality monitoring in the fitness club were subjected to detailed statistical analysis, which showed a very weak correlation between the average number of bacteria and fungi in the air, and the airflow, temperature, relative humidity, and the number of particles in the air (Figure 2a–d). Moreover, the correlation between the number of microorganisms in the air, the number of persons present at the sampling location, and the number of open windows were also very weak (Figure 2e,f).

It is worth noting that the present research showed higher microbiological air contamination (bacteria: 7.17×10^2–1.68×10^3 CFU m^{-3}, fungi: 03×10^3 CFU m^{-3}–7.34×10^3 CFU m^{-3}) than that in previously published studies. In addition, previous research focused primarily on assessing microbial contamination in sports facilities in schools and universities. Brągoszewska et al. recorded the number of bacteria in the air in a Polish high school gym from 4.20×10^2 to 8.75×10^2 CFU m^{-3} depending on the activity of the students [17]. Additionally, other studies conducted in Europe (gyms, fitness rooms, and different facilities in academic sport centers) have shown lower microbial contamination (i.e., 5.80×10^1 to 2.00×10^4 CFU m^{-3} for the number of bacteria in the air, and 2.10×10^1 to 3.75×10^2 CFU m^{-3} for the number of fungi in sports facilities) [1,7].

Recently, Boonrattanakij et al. investigated microbial contamination in a bicycle room at a fitness center in Taiwan using the same type of air sampler and culture media as the current study [5]. The authors obtained a lower number of bacteria (4.01×10^2–7.61×10^2 CFU m^{-3}) and fungi (2.26×10^2–8.37×10^2 CFU m^{-3}).

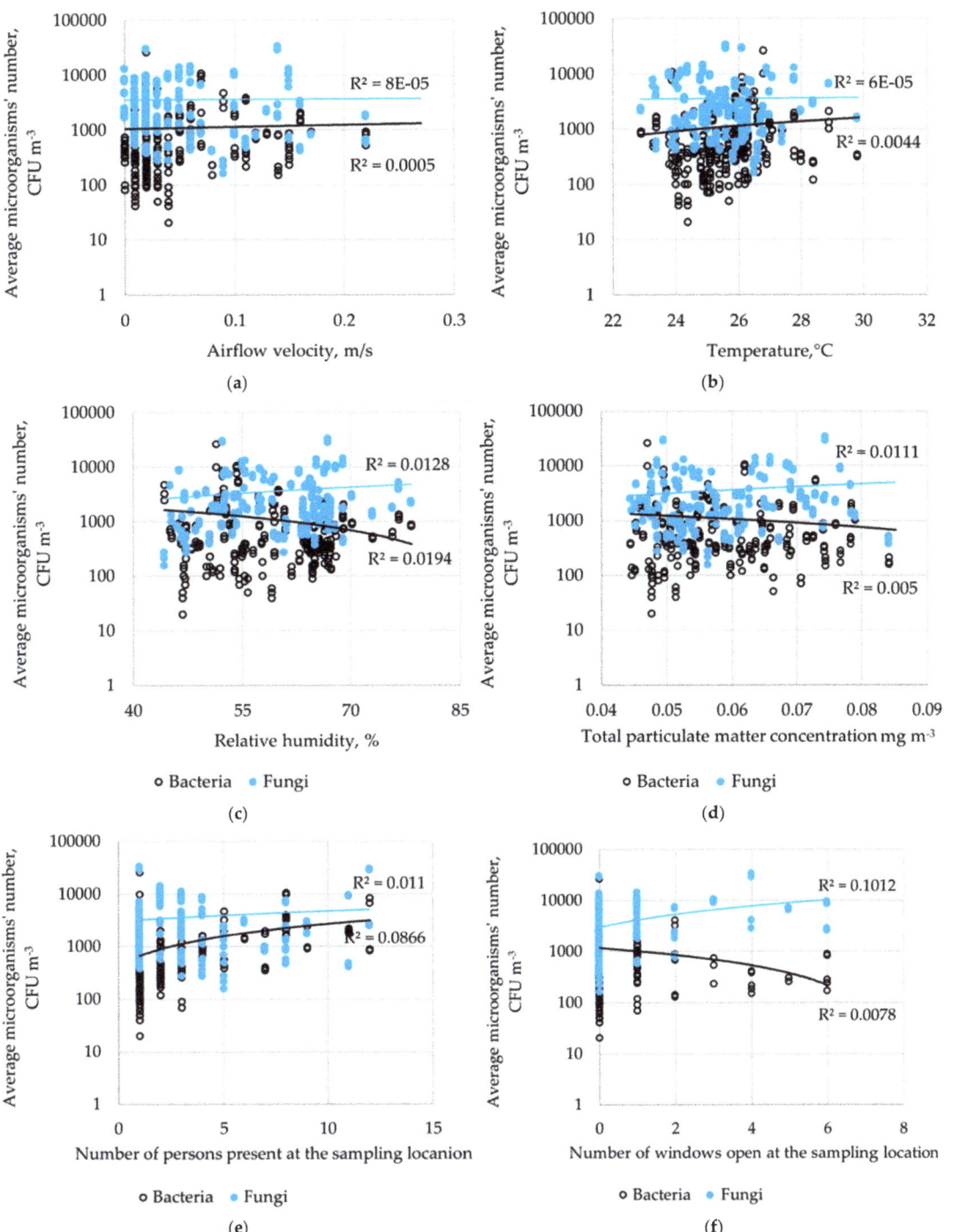

Figure 2. Correlation between microorganism numbers and the: (**a**) airflow velocity, (**b**) temperature, (**c**) relative humidity, (**d**) total particulate matter concentration, (**e**) the number of persons present at the sampling location, and (**f**) the number of windows open at the sampling location.

It should be noted that many factors can be responsible for the differences in microbial contamination shown in current and previous studies, such as building construction and materials, ventilation systems and environmental factors (season, air humidity, temperature) and others [5,6].

Unfortunately, there are no legal limits on the number of microorganisms in indoor air to which the results obtained in the present study could be referred. The WHO suggests that the total number of microorganisms should not exceed 1.0×10^3 CFU m^{-3}. At the same time, the Polish Commission for Maximum Admissible Concentrations and Intensities for Agents Harmful to Health in the Working Environment has developed limits of 5.0×10^3 CFU m^{-3} for the total number of mesophilic bacteria, and the total number of fungi for residential and public utility facilities [46].

Considering the average daily values of the number of microorganisms obtained in the present study, the number of bacteria in Rooms no. 2 and 5 was exceeded. In addition, the number of fungi in all rooms exceeded the WHO's recommendations [47]. Referring to the Polish guidelines, only the number of fungi in Room no. 1 (reception) was exceeded, where, most often, there was an open window allowing for the inflow of atmospheric air, which was highly contaminated with fungi during the research period.

Statistical analysis showed that atmospheric air could be the fungi source in the fitness club rooms tested. This hypothesis could be tested by comparing the indoor air's fungal composition to the outdoor air. In contrast, the source of bacteria in the indoor air was probably of human origin, as suggested by the highest values observed in the room where the most intense exercises were performed. This conclusion is supported by previous studies showing that bacteria can be over two times more abundant indoors than outdoors, especially in poorly ventilated and heavily occupied premises [14,48].

Although it is known that microclimate conditions, especially temperature and relative air humidity, as a rule, correlate with the number of microorganisms in the air in the rooms [49,50], this was not observed in the current research. This is probably because the environment of fitness clubs is very specific and unstable, which is mainly related to the varying number of people, who are carriers of specific microbiota, perform exercises of varying intensity, enter/leave rooms, open/close doors, open/close windows, turn on/off fans and air-conditioners, etc. These overlapping factors have unpredictable results; future research should introduce systems for continuously monitoring the microbiological air quality in sports facilities.

2.4. Determination of Surface Microbial Contamination

The highest number of bacteria was found in the shoe cabinet and on the table in the reception area, which was used by people waiting (3.8 CFU cm^{-2}). No bacteria were found on the exercise bike saddle and at the bottom of the locker used to store personal belongings (0 CFU cm^{-2}). The highest number of fungi was found on the MMA training bag (6.2 CFU cm^{-2}), while no fungi were present at the bottom of the storage locker (Figure 3). Significant differences were observed in the concentration of bacteria and fungi between the tested surfaces ($p < 0.05$).

Few studies have presented a quantitative assessment of microbial surface contamination in sports facilities. In the present study, surface microbiological contamination was lower than that in the previously published studies. Boonrattanakij et al. conducted microbiological tests on sports equipment (i.e., bicycle handle, dumbbell, and sit-up bench) [5]. The number of bacteria on the examined surfaces (bicycle handle, dumbbell, and sit-up bench) ranged from 3.9×10^2 CFU cm^{-2} to 3.7×10^3 CFU cm^{-2}.

Notably, guidance was posted in the cloakroom and gym for users to disinfect the exercise equipment and cabinets for personal belongings to prevent the spread of COVID-19. Based on the obtained results, it can be concluded that the users did not follow the recommendations in all cases and/or disinfection was ineffective.

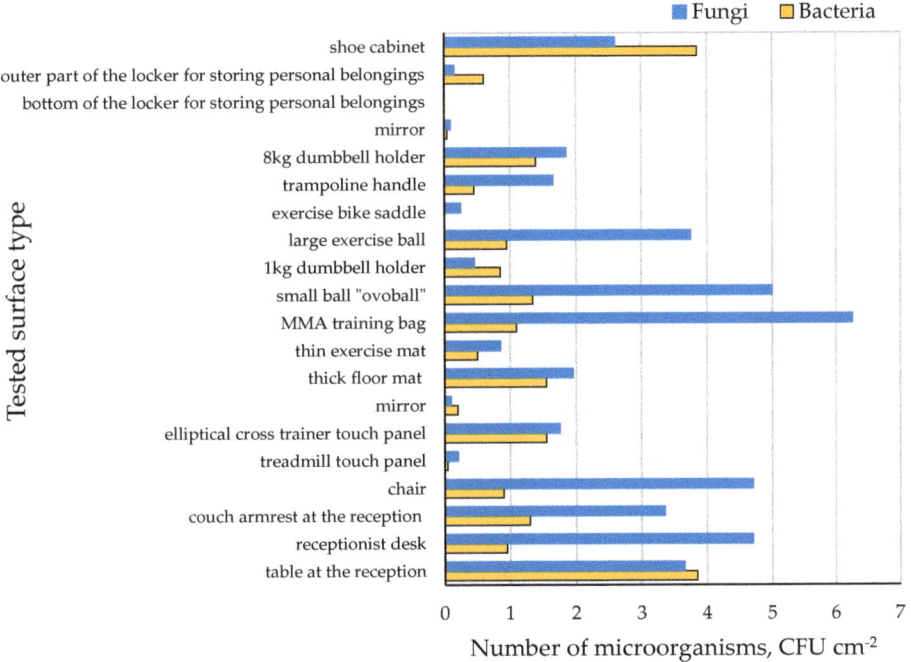

Figure 3. Microbiological contamination of surfaces.

2.5. Diversity of Microorganisms in the Fitness Center Environment

The results from the high-throughput DNA sequencing of the settled dust sample collected at the fitness club revealed a high diversity of microorganisms (Figure 4a). In total, four hundred and twenty-two (422) genera of bacteria representing 21 phyla were detected in the dust. Although the number of phyla was high, most shared a minimal number of classified reads. The most abundant phyla belonged to Cyanobacteria (46%), Proteobacteria (30%), Actinobacteriota (14%), Firmicutes (6%), and Bacteroidota (2%) (Figure 4a). The high number of reads from the Cyanobacteria phylum was a surprise. A closer analysis of the reads based on a sequence similarity search employing the NCBI Nucleotide collection database revealed that these sequences were mainly derived from pine (*Pinus* spp.) chloroplast DNA, which suggests that the dust sample was primarily contaminated with pollen.

The most abundant bacteria identified in the settled dust from the gym in question belonged to the genus *Paracoccus* (5.8%), *Sphingomonas* (3.9%), *Micrococcus* (3.8%), *Escherichia-Shigella* (2%), *Acinetobacter* (1.5%), *Enhydrobacter* (1.5%), *Corynebacterium* (1.5%), *Kocuria* (1.5%), 1174-901-12 (*Rhizobiales*; 1.2%), *Bacillus* (1.1%), and *Rubellimicrobium* (1.1%). The presence of mitochondrial DNA was most probably due to the contamination of the dust sample with pine pollen.

Following this, the presence of potentially hazardous bacterial genera sequences, according to Directive 2019/1833/EC [51], was checked among the classified reads of the dust sample. Twenty-eight hazardous genera (Groups 2 or 3) were identified; however, their share in the total number was very low (<7.5% of all classified reads). Of these, the most abundant genera were *Escherichia-Shigella* (2%), *Corynebacterium* (1.4%), *Bacillus* (1%), and *Staphylococcus* (0.8%) (Figures 4 and S3).

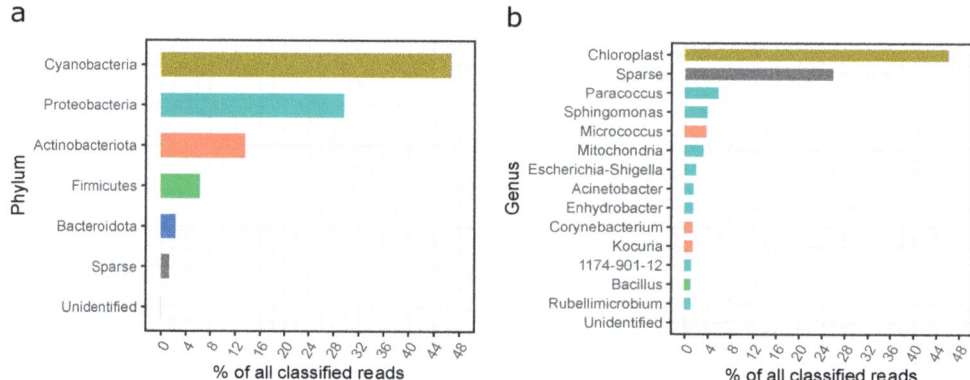

Figure 4. Phylogenetic distribution of bacteria sequences in the settled dust sample; (**a**) assigned to the phyla, (**b**) assigned to the genera; unidentified—unidentified sequences; sparse—reads assigned to low abundant phyla (less than 1% of all classified reads).

So far, the bacteria *Bacillus*, *Corynebacterium*, *Kocuria*, *Micrococcus*, and *Pseudomonas Staphylococcus* have been reported as characteristic of the environment of sports facilities, identified by classical (culture) methods [1,17,52]. Turkskani et al. isolated bacteria from two Saudi Arabian gyms and identified them based on gene sequences of their 16S rRNA [53]. The authors determined the phylogenetic affiliation of the detected bacteria to the following genera: *Bacillus*, *Brachybacterium*, *Geobacillus*, *Microbacterium*, *Micrococcus*, and *Staphylococcus*. Moreover, Haghverdian et al. demonstrated the prevalence and transmissibility of *S. aureus* on the surfaces (floor, balls, hands) in sports facilities [15]. The authors observed the viability of *S. aureus* on sequestered sports balls for 72 h, while another work demonstrated the survival of *S. aureus* strains for up to 12 days on inanimate surfaces [54]. Recently, Szulc et al., (2023) published the results of the first metagenomic analysis of a bioaerosol from a sports center (a room with a climbing wall). The authors identified bacteria mainly belonging to the genus *Cellulosimicrobium*, *Stenotrophomonas*, *Acinetobacter*, *Escherichia*, and *Lactobacillus* in these environments [7].

The present study detected bacteria of the *Paracoccus*, *Sphingomonas*, *Enhydrobacter*, *Rubellimicrobium* and 1174-901-12 genus, with a share of more than 15%, which have never previously been identified in sports facilities.

Paracoccus was isolated from various environments including soils, salines, marine sediments, wastewater, and biofilters. Most include saprophytes, but one species of *P. yeei* is known to be associated with opportunistic infections in humans [55,56]. Additionally, *Rubellimicrobium* are environmental bacteria observed in the soil, air, and slime on industrial machines [57]; therefore, their presence in a fitness club is unsurprising.

Sphingomonas has also been isolated from many environmental samples (soil, sediment, water) including samples chemically contaminated with azo dyes, phenols, dibenzofurans, insecticides, and herbicides [58]. Many *Sphingomonas* strains have been isolated from human clinical specimens and hospital environments where *Sphingomonas paucimobilis*, *S. mucosissima*, and *S. adhesiva* are most associated with human infections [59].

Genus 1174-901-12 has previously been isolated from soil, ceramic roofs, and photovoltaic panels [60], indicating that its source may be the external environment or building materials in the fitness club building.

So far, only one species of *Enhydrobacter* is known (*E. aerosaccus*), which was isolated from a eutrophic lake. These bacteria are rare and poorly described in the literature; therefore, it is challenging to conclude their source in the studied fitness club and their potential effects [61].

The ITS-based analysis revealed that, in total, four hundred and eight (408) genera of fungi representing 11 phyla were detected in the dust. The most abundant phyla belonged to *Ascomycota* (36.4%), *Basidiomycota* (28.4%), *Arthropoda* (11.4%), and *Anthophyta* (8.7%) (Figure 5a).

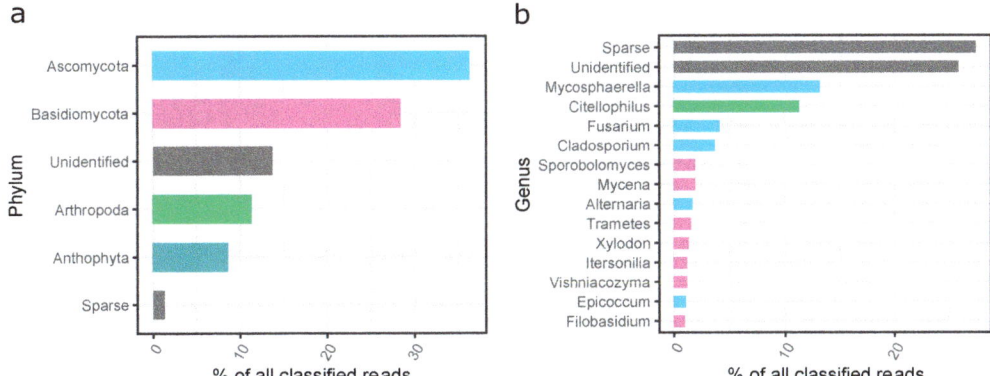

Figure 5. Phylogenetic distribution of fungi sequences in the settled dust sample; (**a**) assigned to the phyla, (**b**) assigned to the genera; unidentified—unidentified sequences; sparse—reads assigned to low abundant phyla (less than 1% of all classified reads).

The most abundant fungi identified in the settled dust from the gym in question belonged to the genus *Mycosphaerella* (13.2%), *Citellophilus* (11.3%), *Fusarium* (4.1%), *Cladosporium* (3.7%), *Sporobolomyces* (1.9%), *Mycena* (1.9%), *Alternaria* (1.7%), *Trametes* (1.6%), *Xylodon* (1.4%), *Itersonilia* (1.2%), *Vishaniacozyma* (1.2%), *Epicoccum* (1.1%), and *Filobasidium* (1.0%) (Figure 5b).

Seven genera of the hazardous category (Groups 2 or 3), according to Directive 2019/1833/EC were found, and their share in the total number was very low (less than 1.5% of all classified reads). Of these, the most abundant genera were *Cladosporium* (3.7%), *Aspergillus* (0.9%), and *Penicillium* (0.4%) (Figure S4).

Żyrek et al. indicated the presence of the yeast *Candida* sp. Małecka-Adamowicz et al. found fungi from the genus *Cladosporium*, and to a lesser extent, *Penicillium*, *Fusarium*, *Acremonium*, *Alternaria*, and *Aureobasidium* [1,52]. The occurrence of potentially allergenic molds of the genera *Aspergillus* and *Cladosporium* in Czech sports facilities was described in [18]. Viegas et al. identified 25 species of fungi occurring in ten gymnasia and found mainly molds from the following genera: *Cladosporium*, *Penicillium*, *Aspergillus*, *Mucor*, *Phoma* and *Chrysonilia* as well as yeasts from the genera: *Rhodotorula*, *Trichosporon mucoides* and *Cryptococcus uniguttulattus* [62]. Szulc et al. indicated the dominance of fungi: *Mycosphaerella*, *Botrytis*, *Chalastospora*, *Cladosporium*, *Itersonilia*, *Malassezia*, *Naganishia*, *Saccharomyces*, *Sporobolomyces*, *Trichosporon*, and *Udeniomyces* in sports facilities for climbing activities [7].

The results obtained in the present study differed from the literature data. The genera of fungi: *Acremonium*, *Aureobasidium*, *Penicillium*, *Aspergillus*, *Candida*, *Mucor*, *Phoma*, *Chrysonilia*, *Rhodotorula*, *Trichosporon* and *Cryptococcus*, which dominated in earlier studies, were found in low quantities, from 0.01% to 0.8% [52,62], which may be related to the seasonal variability of the types of fungi dominant in the atmospheric air shaping the qualitative composition of indoor fungi.

Moreover, in the present research, the following fungi: *Citellophilus*, *Mycena*, *Tramates*, *Xylodon*, *Vishniacozyma*, *Epicoccum*, and *Filobasidium* were identified for the first time in gym facilities. These fungi are common genera and likely come from the outdoor air. They are known as plant parasites but can also be allergic to humans, are often linked to decreased pulmonary function and asthma admissions, and may cause infections, particularly in immunosuppressed patients [63–70].

It should be mentioned that the use of high-throughput sequencing on the Illumina platform made it possible to identify a greater variety of microorganisms found in sports facilities than that previously described in the literature. Metagenomic analysis is increasingly used to study various environmental samples such as soil, water, technical materials (e.g., cardboard, cellulosic materials, collagen), settled dust, and many others [71–73]. The advantage of this method is the identification of microorganisms directly from the test sample, skipping the cultivation stage, which prevents the loss of species of microorganisms that cannot grow under laboratory conditions [74].

2.6. Assessment of SARS-CoV-2 Virus Presence in the Fitness Center Environment

Selected surfaces in Room no. 2 (gym) were tested for the SARS-CoV-2 virus. In the case of the treadmill touch panel, the result was positive (Table 2). The results suggest a real risk of the spread of the COVID-19 pandemic in gyms and fitness clubs.

Table 2. Detection of SARS-CoV-2 on the fitness center surfaces.

Sample No.	Description	RNA Concentration, µg/mL	SARS-CoV-2 RNA
1	Treadmill touch panel	66	Present
2	Panel and grips of elliptical cross trainer	63	Absent
3	Multi-gym grips panel	63	Absent

It is worth mentioning that between 74 and 164 cases of COVID-19 per day were recorded in Poland between 26 and 30 of July 2021. In the province where the fitness center in question was located, cases ranged between 1 and 13 per day [75]. The detection of the SARS-CoV-2 virus suggests a real risk of the spread of COVID-19 in gyms and fitness clubs. However, studies involving a larger number of tested samples are needed to confirm this hypothesis.

The risk of the transmission of COVID-19 may arise from close contact, the emission of droplets, or through fomites. Intensive physical activities in a fitness center favor these factors, mainly due to the increased physical contact, increased concentration of exhaled respiratory droplets in a confined space because of vigorous breathing, and shared communal space and equipment [76]. No RNA of SARS-CoV-2 was detected in the previously performed air and surface studies at a fitness center in the U.S. [77].

SARS-CoV-2 transmission in sports facilities has been previously proven in positive PCR tests of infected users and workers [78]. Conversely, in Norway, Helsingen et al. tested 3764 individuals divided into two groups (with and without access to training at a fitness center) [79]. They found a difference of 0.05% (one versus zero cases) in SARS-CoV-2 RNA test positivity between training and non-training individuals. The authors stated that good hygiene and physical distance in fitness centers did not increase the infection risk of SARS-CoV-2 for individuals without COVID-19-relevant comorbidities in such spaces. Therefore, it is essential to make the users and employees of these facilities aware of the principles of sanitary safety and the proper disinfection of hands and sports equipment.

2.7. Directions for Minimizing Microbiological and Chemical Threats in the Sports Facilities

The benefits of physical activities should be strengthened by reducing the exposure to physicochemical and microbiological contamination and consequently by minimizing the risk of possible adverse health effects for users in sports facilities [80].

In the tested fitness club, we found a high concentration of dust, microorganisms, and the SARS-CoV-2 virus. It is worth mentioning that the performed studies had some limitations resulting from (a) the uniqueness of the sample (only one fitness club was tested); (b) the season in which the samples were taken due to the influence of external bioaerosols on the amount and composition of internal bioaerosols; (c) the holiday/vacation season, which meant that the number of users was lower than the rest of the year; (d) the small number of samples taken for analysis for the detection of SARS-CoV-2 and metagenomic analysis. However, studies suggest that air purification systems with proven effectiveness are needed for continuous operation during opening hours in sports facilities.

Various chemical and physical methods are currently known and tested for air disinfection. Air disinfection includes filtration, ozonation, exposure to ultraviolet radiation, photocatalysis, and cold plasma [81]. Recently, it has been proposed to use strong electric fields in which the destruction or electroporation of microorganisms occurs [82]. Among these disinfection techniques, chemical fogging, ozonation, and UV radiation of the air are the main solutions available on the market [81]. These methods are currently used in clinical and pharmaceutical objects; however, they seem suitable for sports facilities.

One of the ways to prevent the spread of viruses and pathogenic microorganisms in sports facilities is to use floors that feature antibacterial properties and other materials (e.g., clothing and towels) with biostatic properties.

It is worth noting that one of the practices related to preventing the spread of the COVID-19 pandemic was the introduction of spray bottles filled with a disinfectant solution in sports centers for wiping the exercise equipment after use. It should be noted that these practices have their weaknesses.

Rarely are surface disinfectants at sports facilities in their original packaging, allowing the center to control their composition and the concentration of active substances. Therefore, it is crucial to use EPA-approved disinfectants, consider the type of disinfected surfaces (metal, plastic, leather, etc.), and prepare working solutions of disinfectants following the manufacturer's guidelines, labeling them properly, and providing detailed instructions for use for the end users. This is important because the effectiveness of disinfection will depend on the contact time of the preparation used with the surface. Common misconduct is spraying and immediately wiping off sports equipment. Such disinfection will not be effective and will even become dangerous for the user and the environment. Therefore, the staff at sports clubs must be properly prepared (trained) to use appropriate safety procedures and personal protective equipment (if necessary) during disinfection. An alternative to sprayed disinfectants can be disinfectant-impregnated wipes, consisting of towels saturated with diluted disinfectant and other compounds (i.e., surfactants, preservatives, enzymes, and perfumes) [83]. Staff and the users of exercise facilities should wash their hands with water and plain soap before entering and leaving and before and after any contact with other people and equipment in sports facilities and avoid sharing towels (preferably use disposable paper towels) and other personal items. Wounds, cuts, scrapes, etc., should be covered with a clean, dry dressing to prevent contamination. The World Health Organization (WHO) recommends alcohol-based formulations to disinfect hands; such formulations have been shown to inactivate SARS-CoV-2 efficiently.

Moreover, hydrogen peroxide or povidone-iodine and other biocides possess antiviral properties and can be used to disinfect biological surfaces [84]. The sharing of exercise equipment should be avoided if possible. If this is not possible, the use of a towel is recommended, or, for example, gloves that provide a barrier between the skin and such equipment. After the entire working day, the facility staff should wash and disinfect all common exercise equipment used on a given day. Moreover, objects inside a sports facility that require special attention include countertops, light switches, faucet handles, and doorknobs. Staff should be excluded from the use of damaged equipment (e.g., torn upholstery) that cannot be properly disinfected due to damage.

Future research should aim at introducing Internet of Things (IoT) technology systems of constant air quality monitoring in sports facilities (e.g., using multiple sensors including microfluidic chips as well as developing warning systems against exceeding the concentration of suspended dust, or the recommended number of microorganisms in the air).

3. Materials and Methods
3.1. Tested Fitness Center and Sampling Strategy

The research was conducted in a fitness club in Zduńska Wola (central Poland). The tested fitness center is located in a service and commercial building built in the1990s and operates from Monday to Friday from 8:00 to 22:00 and on the weekends from 9:00 to 15:00. The characteristics of the rooms under study are presented in Table 3. Samples of the air were collected from five fitness center rooms equipped with an occasional air conditioning system.

Table 3. Characteristics of the sampling sites at points in the tested fitness center.

Room No.	Description	Area, m²	Surface Sampling Sites	Number of Users	Number of Ceiling Fans/Opening Windows	Airconditioning	Number of Samples
1	Reception	18	Table for the waiting, receptionist's desk, armrest on the couch, chair	2–12	1/1	No	N = 60 n = 4
2	The gym	220	Treadmill touch panel, the touch panel of the cross trainer, mirror, thick mat on the floor	1–9	3/6	Yes (2 units)	N = 60 n = 4
3	Fitness room on the first floor	100	Thin exercise mat, MMA bag, small "ovoball", 1 kg dumbbell holder	1–11	2/5	Yes (1 unit)	N = 60 n = 4
4	Fitness room on the second floor	85	Large exercise ball, exercise bike saddle, trampoline handle, 8 kg dumbbell holder	1–12	2/6	Yes (1 unit)	N = 60 n = 4
5	Women's cloakroom	26	Mirror, the bottom of the locker, the external part of the locker, shoe cabinet	1–5	1/0	No	N = 60 n = 4
6	Atmospheric air (external background)	-	Parking lot located 10 m from the entrance to the fitness club building	-	-	-	N = 60 n = 0

"-"—not applicable; N—number of air samples collected during the whole working week, collected in triplicate; n—total number of surface samples.

Moreover, control samples (atmospheric air) in front of the building were collected simultaneously. Samples were collected during the entire working week (Monday–Friday) at 8:00, 12:00, 16:00 and 20:00 under normal operating conditions. At the same time, the microclimate and particulate matter concentrations were analyzed. The microbial contamination was also assessed for 20 surfaces in the fitness center (Table 3). The chemical contamination of the air was checked in the gym (Room no. 2) in comparison to the control atmospheric air (Room no. 6).

Additionally, three samples were taken from the surface of Room no. 2 (gym) to verify the presence of the SARS-CoV-2 virus. A pooled sample of settled dust was also collected to determine the biodiversity of the microorganisms.

3.2. Microclimate, Particulate Matter Concentration Carbon Dioxide, and Formaldehyde Analysis

A VelociCalc® Multi-Function Velocity Meter 9545 (TSI, Dallas, TX, USA) thermo-anemometer was used to establish the temperature, relative humidity, and airflow rate at the selected workstations. The measurements were taken over 2 min at 1 s intervals; averages were logged for each sampling variant (day/hour/location).

The concentration of particulate matter (PM_1; $PM_{2.5}$; PM_4; PM_{10}; PM_{total}) was measured using a DustTrak™ DRX Aerosol Monitor 8533 portable laser photometer (TSI, USA). The detection range for particles with diameters ranging from 0.1 to 15 μm was between 0.001 and 150 mg m^{-3}. The measurements were carried out in triplicate for each location at 1.5 m from the ground level. The sampling rate was set to 3 L min^{-1} and the sampling interval to 5 s. The total sampling time was 3 min. The carbon dioxide and formaldehyde concentrations were measured using a M200 Multi-functional Air Quality Detector (Temtop, China).

3.3. Volatile Compounds Analysis

Detailed analysis of the volatile compounds was carried out using headspace solid-phase microextraction coupled to gas chromatography-mass spectrometry (HS-SPME-GC). Tedlar bags (5 L) were used for the collection of air samples from Rooms no. 2 and 6 (gym and external background). For extraction of the volatile compounds from the air samples, the solid-phase microextraction technique was used with the fiber covered with 50/30 μm divinylbenzene/carboxen/polydimethylsiloxane (DVB/CAR/PDMS) phase (length 1 cm). The SPME fiber was inserted via the sampling port, followed by exposure for 60 min at 20 °C. After the adsorption of volatiles, the fiber was retracted into the needle and transferred to the inlet of the GC apparatus for the desorption of analytes. Desorption was carried out for 5 min at 250 °C. Before each extraction, the fiber was heated for 10 min in the inlet of the GC apparatus at 260 °C for cleaning. A GC-MS system was used for the volatile compound analysis (GC Agilent 7890A and MS Agilent MSD 5975C, Agilent Technologies, Santa Clara, CA, USA). The compounds were separated on a capillary column DB-1ms 60 m × 0.25 mm × 0.25 μm (Agilent Technologies, Santa Clara, CA, USA). All injections were performed in a splitless mode. As a carrier gas, helium was used with a flow rate of 1.1 mL/min. The GC oven temperature was programmed to increase from 30 °C (10 min) to 70 °C at a rate of 2 °C/min and kept for 2 min, then to 235 °C at a rate of 10 °C/min, and finally kept for 3.5 min. The MS ion source, transfer line, and quadrupole analyzer temperatures were 230, 250, and 150 °C, respectively. The electron impact energy was set at 70 eV. The mass spectrometer was operated in full scan mode (SCAN). The qualification of volatiles was performed by a comparison of the obtained spectra with the reference mass spectra from the NIST/EPA/NIH mass spectra library (2012; Version 2.0 g) or with mass spectra obtained from the GC standards and confirmed with the use of the deconvolution procedure. Then, retention indices (RI) were calculated according to the formula proposed by van den Dool and Kratz [85] relative to a homologous series of n-alkanes from C5 to C20. Retention indices were compared with the literature data [86]. Data processing was conducted with Mass Hunter Workstation Software (Agilent, Santa Clara, CA, USA). The relative amounts of volatile compounds were calculated by the individual peak area relative to the total peak areas.

3.4. Determination of Airborne Microorganism Number

Air samples were collected in triplicate (20–100 L) per sampling site at the height of about 1.5 m from ground level with an airflow rate of 100 L min^{-1} using a MAS-100 Eco Air Sampler (Merck, Darmstadt, Germany), according to EN 13098 [87]. The microbiological contamination of the air was determined using: TSA (tryptic soy agar, Merck, Germany) with (0.2%) nystatin to determine the number of bacteria and MEA (MALT EXTRACT agar, Merck, Germany) medium with (0.1%) chloramphenicol to determine the fungi number. The samples were incubated at either 25 ± 2 °C for 5–7 days (fungi) or 30 ± 2 °C for 48 h (bacteria). After incubation, the colonies were counted and corrected based on Feller's statistical correction table. The results were calculated as the arithmetic mean of three independent repetitions and expressed in CFU m^{-3}.

3.5. Determination of Surface Microbial Contamination

Samples from 20 different surfaces throughout the facility (two independent repetitions) were collected on the first day of testing (Monday) between 8:00 and 10:00 using Hygicult® TPC (Orion Diagnostica Oy, Espoo, Finland) with the Total Plate Count medium. The collected samples were incubated at 30 ± 2 °C for 3–5 days. Next, the colonies were counted, and the results (arithmetic mean of two independent repetitions) were expressed in CFU cm^{-2}.

3.6. Detection of SARS-CoV-2

Swabs were taken from approx. 100 cm^2 from three surfaces (treadmill touch panel, panel and grips of elliptical cross trainer, and multi-gym grips) located in the gym (Room no. 2) using R9F buffer (A&A Biotechnology, Gdańsk, Poland). The surfaces were selected based on the highest frequency of use, the presence of direct contact with the user's hands, and their vicinity to the breathing zone. RNA isolation was performed with the CoV RNA Kit (A&A Biotechnology, Poland). The presence of the SARS-CoV-2 virus RNA in the tested samples was confirmed by Real-Time PCR with Taq-Man probes. The presence of SARS-CoV-2 was tested using the MediPAN-2G+ FAST COVID test (Medicofarma, Warsaw, Poland) kit by A&A Biotechnology (Poland) according to the manufacturer's instructions. The test detects fragments of two SARS-CoV-2 genes (i.e., ORF1ab (nsp2) and gene S). A synthetic fragment of a plant virus genome was used as a control.

3.7. Determination of Biodiversity

Dust deposited on the surface of the gym equipment (10–12 devices located 0.5–2 m from the ground) was collected with steely, dry swabs and refrigerated overnight (4 °C). Then, the samples were combined into one and used for DNA extraction. According to the manufacturer's instructions, genomic DNA was extracted using the Soil DNA Purification Kit (EURX, Poland). The presence of genomic DNA in the tested samples was confirmed with fluorimetry (Qubit). The extracted DNA concentration was 1 µg mL^{-1}. Universal primers amplifying the 16S rRNA bacterial gene's fragment and fungal ITS regions were used in the reaction [88–90]. Q5 Hot Start High-Fidelity 2X Master Mix (NEB, Ipswich, MA, USA) was used for PCR according to the manufacturer's instructions. The libraries were prepared and sequenced by Genomed (Warsaw, Poland) using the paired-end technology on the Illumina MiSeq (2×300 nt) platform with the use of a v3 Kit (Illumina, San Diego, CA, USA). Automatic initial analysis was performed on the MiSeq sequencer using MiSeq Reporter (MSR) v2.6. The obtained results were next subjected to bioinformatic analysis. Adapter sequences were removed from the reads, which were next subjected to quality control with the Cutadapt program using quality (<20) and the minimal length of (30 nt) threshold [91]. Library reads 16S were further processed using the DADA2 package to separate sequences of biological origin from those generated during the sequencing process. This package was also used for selecting unique sequences of biological origin, the so-called amplicon sequence variant (ASV). Bioinformatics analysis of the reads for species-level classification was performed using the QIIME 2 program based on the Silva 138 database using a hybrid approach [92]. First, ASV sequences were compared with the database to find identical reference sequences using the VSEARCH algorithm [93]. Next, the atypical sequences left over from the previous step were classified based on machine learning, which was performed using SKLearn. ITS library reads classification at the species level was performed using QIIME based on the UNITE v8 reference database [94]. After filtering, as described above, the reads were clustered based on the reference database using the UCLUST algorithm. Chimeric sequences were removed using the USEARCH (usearch61) algorithm. Finally, the taxonomy was assigned to the reference database using the BLAST algorithm. Sequencing data files in the FASTQ format were deposited in the NCBI Sequence Read Archive (SRA) under BioProject accession number PRJNA818521 (BioSampleAcc. SAMN26866224 and Run Acc. SRR18428312 and SRR18428311).

3.8. Statistical Analysis

Statistical analysis was carried out with Statistica 13.1 (Statsoft, Tulsa, OK, USA). Descriptive statistics were calculated for all variables of interest. For the microclimate parameters and the number of microorganisms in the air, one-way analysis of variance (ANOVA) was performed for data grouped depending on the sampling day, hour, and location. ANOVA assumptions were checked with the Shapiro–Wilk and Levene tests. When a statistical difference was detected, the means were compared using Tukey's post hoc test or Dunn's post hoc tests. Full-factorial ANOVA was performed for the particulate matter concentration, followed by Tukey's post hoc test. In the case of surface microbial contamination, the Fisher–Snedecor test was carried out for the number of microorganisms averaged over the tested surfaces. The variances in the number of bacteria and fungi on the examined surfaces were heterogeneous. Thus, a *t*-test was performed for unequal variances. All tests were performed at a significance level of 0.05.

Linear regression was performed to check a correlation between the number of bacteria and fungi in the air and other measured parameters. To describe the strength of the correlation, the Evans (1996) guide for the absolute value of correlation coefficient r: 0.00–0.19 "very weak"; 0.20–0.39 "weak"; 0.40–0.59 "moderate"; 0.60–0.79 "strong"; 0.80–1.0 "very strong" [95] was used.

4. Conclusions

High particulate matter, especially the $PM_{2.5}$ concentration, was observed in fitness centers that exceeded the environmental threshold and supports statements that air quality inside sports facilities can be worse than that outdoors. Moreover, chemical markers such as CO_2 concentration and VOCs (phenol, toluene, and 2-ethyl-1-hexanol) may be useful for air quality monitoring in sports facilities.

Additionally, the concentration of airborne microorganisms was high compared to the previous research and literature recommendations. It is also noteworthy that genera of bacteria (*Escherichia-Shigella*, *Corynebacterium*, *Bacillus*, *Staphylococcus*) and fungi (*Cladosporium*, *Aspergillus*, *Penicillium*), potentially belonging to the second and third groups of health hazards following Directive 2019/1833/EC, were detected, albeit at relatively low concentrations. In addition, other species that may be allergenic (*Epicoccum*) or infectious (*Acinetobacter*, *Sphingomonas*, *Sporobolomyces*) and the SARS-CoV-2 virus were detected.

Due to the possibility of high contamination with chemicals, CO_2, bacteria and fungi, and the spread of the SARS-CoV-2 virus in sports facilities, air purification systems with proven effectiveness are also needed (e.g., UV flow lamps, photocatalytic ionizers) for continuous operation during opening hours. Future research should aim at introducing systems of constant air quality monitoring in sports facilities (e.g., using multiple sensors including microfluidic chips) and developing warning systems against exceeding the concentration of suspended dust or the recommended number of microorganisms in the air.

Supplementary Materials: The following supporting information can be downloaded at: https://www.mdpi.com/article/10.3390/molecules28083560/s1, Table S1: Number–size distribution of airborne particles detected at the tested locations. Table S2. Volatile compounds identified in the gym; Table S3: The average number of microorganisms in the tested fitness club; Figure S1. Microclimatic conditions (means with SD): (a) daily averages, (b) averaged over sampling time of the day, (c) averaged over sampling location; statistically different samples were marked with different letters; no letters indicate no statistical differences (Tukey's test, α = 0.05); Figure S2. Averaged values of the total PM concentration (means with SD): (a) daily, (b) over sampling time of the day, (c) over sampling location; statistically different samples were marked with different letters; (Tukey's test, α = 0.05); Figure S3: Phylogenetic distribution of hazardous bacteria sequences assigned to genera in the settled dust sample; Figure S4: Phylogenetic distribution of hazardous fungi sequences assigned to genera in the settled dust sample.

Author Contributions: Conceptualization, J.S. and M.O.; Methodology, J.S. and M.O.; Software, M.R. and K.P.-P.; Validation, J.S. and M.O.; Formal analysis, J.S.; Investigation, J.S., M.O., M.R. and K.P-P; Resources, J.S.; Data curation, J.S. and M.O.; Writing—original draft preparation, J.S., M.O., M.R., K.P.-P. and B.G.; Writing—review and editing, J.S. and M.O.; Visualization, M.O. and M.R.; Supervision, J.S.; Project administration, J.S.; Funding acquisition, J.S. All authors have read and agreed to the published version of the manuscript and have contributed substantially to the work reported.

Funding: The research was carried out within the "Assessment of biological hazards in sports facilities" project under the "FU2N—Young Scientists' Skills Improvement Fund" program supporting scientific excellence at Lodz University of Technology.

Institutional Review Board Statement: Not applicable.

Informed Consent Statement: Not applicable.

Data Availability Statement: The data that support the findings of this study are available on request from the corresponding author.

Conflicts of Interest: The authors declare no conflict of interest.

Sample Availability: Samples are available from the authors.

References

1. Małecka-Adamowicz, M.; Kubera, Ł.; Jankowiak, E.; Dembowska, E. Microbial diversity of bioaerosol inside sports facilities and antibiotic resistance of isolated *Staphylococcus* spp. *Aerobiologia* **2019**, *35*, 731–742. [CrossRef]
2. Ramos, C.A.; Wolterbeek, H.T.; Almeida, S.M. Exposure to indoor air pollutants during physical activity in fitness centers. *Build. Environ.* **2014**, *82*, 349–360. [CrossRef]
3. WHO Physical Activity Fact Sheet. Available online: https://www.who.int/publications/i/item/WHO-HEP-HPR-RUN-2021.2 (accessed on 22 March 2022).
4. Deloitte Sports Retail Study 2020. Findings from a Central European Consumer Survey. Available online: https://www2.deloitte.com/pl/pl/pages/consumer-business/articles/sports-retail-study-2020.html (accessed on 29 March 2022).
5. Boonrattanakij, N.; Yomchinda, S.; Lin, F.-J.; Bellotindos, L.M.; Lu, M.-C. Investigation and disinfection of bacteria and fungi in sports fitness center. *Environ. Sci. Pollut. Res.* **2021**, *28*, 52576–52586. [CrossRef]
6. Saini, J.; Dutta, M.; Marques, G. A comprehensive review on indoor air quality monitoring systems for enhanced public health. *Sustain. Environ. Res.* **2020**, *30*, 6. [CrossRef]
7. Szulc, J.; Cichowicz, R.; Gutarowski, M.; Okrasa, M.; Gutarowska, B. Assessment of Dust, Chemical, Microbiological Pollutions and Microclimatic Parameters of Indoor Air in Sports Facilities. *Int. J. Environ. Res. Public Health* **2023**, *20*, 1551. [CrossRef]
8. Bralewska, K.; Rogula-Kozłowska, W.; Bralewski, A. Indoor air quality in sports center: Assessment of gaseous pollutants. *Build. Environ.* **2022**, *208*, 108589. [CrossRef]
9. Finewax, Z.; Pagonis, D.; Claflin, M.S.; Handschy, A.V.; Brown, W.L.; Jenks, O.; Nault, B.A.; Day, D.A.; Lerner, B.M.; Jimenez, J.L.; et al. Quantification and source characterization of volatile organic compounds from exercising and application of chlorine-based cleaning products in a university athletic center. *Indoor Air* **2021**, *31*, 1323–1339. [CrossRef]
10. Mbareche, H.; Morawska, L.; Duchaine, C. On the interpretation of bioaerosol exposure measurements and impacts on health. *J. Air Waste Manag. Assoc.* **2019**, *69*, 789–804. [CrossRef]
11. Kim, K.-H.; Kabir, E.; Jahan, S.A. Airborne bioaerosols and their impact on human health. *J. Environ. Sci.* **2018**, *67*, 23–35. [CrossRef] [PubMed]
12. Maus, R.; Goppelsröder, A.; Umhauer, H. Survival of bacterial and mold spores in air filter media. *Atmos. Environ.* **2001**, *35*, 105–113. [CrossRef]
13. Kalogerakis, N.; Paschali, D.; Lekaditis, V.; Pantidou, A.; Eleftheriadis, K.; Lazaridis, M. Indoor air quality—Bioaerosol measurements in domestic and office premises. *J. Aerosol Sci.* **2005**, *36*, 751–761. [CrossRef]
14. Prussin, A.J.; Marr, L.C. Sources of airborne microorganisms in the built environment. *Microbiome* **2015**, *3*, 78. [CrossRef] [PubMed]
15. Haghverdian, B.A.; Patel, N.; Wang, L.; Cotter, J.A. The sports ball as a fomite for transmission of Staphylococcus aureus. *J. Environ. Health* **2018**, *80*, 8–13.
16. CDC Centers for Disease Control and Prevention Methicillin-Resistant Staphylococcus Aureus (MRSA). Available online: https://www.cdc.gov/mrsa/index.html (accessed on 29 March 2022).
17. Bragoszewska, E.; Biedroń, I.; Mainka, A. Microbiological air quality in a highschool gym located in an urban area of Southern Poland-preliminary research. *Atmosphere* **2020**, *11*, 797. [CrossRef]
18. Kic, P.; Ruzek, L.; Popelářová, E. Concentration of air-borne microorganisms in sport facilities. *Agron. Res.* **2018**, *16*, 1720–1727. [CrossRef]
19. Fadare, O.S.; Durojaye, O.B. Antibiotic Susceptibility Profile of Bacteria Isolated from Fitness Machines in Selected Fitness Centers at Akure and Elizade University in Ondo State Nigeria. *Microbiol. Res. J. Int.* **2019**, *26*, 1–9. [CrossRef]

20. Mukherjee, N.; Dowd, S.; Wise, A.; Kedia, S.; Vohra, V.; Banerjee, P. Diversity of Bacterial Communities of Fitness Center Surfaces in a U.S. Metropolitan Area. *Int. J. Environ. Res. Public Health* **2014**, *11*, 12544–12561. [CrossRef]
21. Huang, C.; Que, J.; Liu, Q.; Zhang, Y. On the gym air temperature supporting exercise and comfort. *Build. Environ.* **2021**, *206*, 108313. [CrossRef]
22. International Fitness Association Gym Temperature and Noise Standards. Available online: https://www.ifafitness.com/health/temperature.htm (accessed on 29 March 2022).
23. Zhai, Y.; Elsworth, C.; Arens, E.; Zhang, H.; Zhang, Y.; Zhao, L. Using air movement for comfort during moderate exercise. *Build. Environ.* **2015**, *94*, 344–352. [CrossRef]
24. Bralewska; Rogula-Kozłowska; Bralewski Size-Segregated Particulate Matter in a Selected Sports Facility in Poland. *Sustainability* **2019**, *11*, 6911. [CrossRef]
25. Directive 2008/50/EC of the European Parliament and of the Council of 21 May 2008 on ambient air quality and cleaner air for Europe. *Off. J. Eur. Union* **2008**, *L152*, 1–44.
26. Andrade, A.; Dominski, F.H.; Coimbra, D.R. Scientific production on indoor air quality of environments used for physical exercise and sports practice: Bibliometric analysis. *J. Environ. Manag.* **2017**, *196*, 188–200. [CrossRef] [PubMed]
27. Rundell, K.W.; Caviston, R. Ultrafine and Fine Particulate Matter Inhalation Decreases Exercise Performance in Healthy Subjects. *J. Strength Cond. Res.* **2008**, *22*, 2–5. [CrossRef] [PubMed]
28. Cutrufello, P.T.; Smoliga, J.M.; Rundell, K.W. Small Things Make a Big Difference. *Sport. Med.* **2012**, *42*, 1041–1058. [CrossRef]
29. Pope, C.A.; Burnett, R.T.; Thurston, G.D.; Thun, M.J.; Calle, E.E.; Krewski, D.; Godleski, J.J. Cardiovascular Mortality and Long-Term Exposure to Particulate Air Pollution. *Circulation* **2004**, *109*, 71–77. [CrossRef] [PubMed]
30. Daigle, C.C.; Chalupa, D.C.; Gibb, F.R.; Morrow, P.E.; Oberdörster, G.; Utell, M.J.; Frampton, M.W. Ultrafine Particle Deposition in Humans During Rest and Exercise. *Inhal. Toxicol.* **2003**, *15*, 539–552. [CrossRef] [PubMed]
31. Castro, A.; Calvo, A.I.; Alves, C.; Alonso-Blanco, E.; Coz, E.; Marques, L.; Nunes, T.; Fernández-Guisuraga, J.M.; Fraile, R. Indoor aerosol size distributions in a gymnasium. *Sci. Total Environ.* **2015**, *524–525*, 178–186. [CrossRef]
32. Phalen, R.F. *Inhalation Studies: Foundations and Techniques*, 2nd ed.; Informa Healthcare: New York, NY, USA, 2009.
33. World Health Organization. *WHO Guidelines for Air Quality: Selected Pollutants*; WHO: Copenhagen, Denmark, 2010; ISBN 9789289002134.
34. Environmental Protection Agency National Ambient Air Quality Standards (NAAQS). Available online: https://www.epa.gov/sites/default/files/2015-02/documents/criteria.pdf (accessed on 29 March 2022).
35. Issitt, T.; Wiggins, L.; Veysey, M.; Sweeney, S.T.; Brackenbury, W.J.; Redeker, K. Volatile compounds in human breath: Critical review and meta-analysis. *J. Breath Res.* **2022**, *16*, 024001. [CrossRef] [PubMed]
36. Jahn, L.G.; Tang, M.; Blomdahl, D.; Bhattacharyya, N.; Abue, P.; Novoselac, A.; Ruiz, L.H.; Misztal, P.K. Volatile organic compound (VOC) emissions from the usage of benzalkonium chloride and other disinfectants based on quaternary ammonium compounds. *Environ. Sci. Atmos.* **2023**, *3*, 363–373. [CrossRef]
37. Singal, M.; Vitale, D.; Smith, L. Fragranced Products and VOCs. *Environ. Health Perspect.* **2011**, *119*, 17–38. [CrossRef]
38. Sherzad, M.; Jung, C. Evaluating the emission of VOCs and HCHO from furniture based on the surface finish methods and retention periods. *Front. Built Environ.* **2022**, *8*. [CrossRef]
39. Phenol and Phenolic Compound. Available online: https://cpcb.nic.in/uploads/News_Letter_Phenols_Phenolic_Compounds_2017.pdf (accessed on 29 March 2022).
40. European Chemicals Agency Phenol. Available online: https://echa.europa.eu/pl/substance-information/-/substanceinfo/100.003.303 (accessed on 29 March 2022).
41. Michałowicz, J.; Duda, W. Phenols—Sources and toxicity. *Polish J. Environ. Stud.* **2007**, *16*, 347–362.
42. Pytel, K.; Marcinkowska, R.; Zabiegała, B. Investigation on air quality of specific indoor environments—Spa salons located in Gdynia, Poland. *Environ. Sci. Pollut. Res.* **2021**, *28*, 59214–59232. [CrossRef] [PubMed]
43. Gonçalves, A.D.; Martins, T.G.; Cassella, R.J. Passive sampling of toluene (and benzene) in indoor air using a semipermeable membrane device. *Ecotoxicol. Environ. Saf.* **2021**, *208*, 111707. [CrossRef]
44. Wakayama, T.; Ito, Y.; Sakai, K.; Miyake, M.; Shibata, E.; Ohno, H.; Kamijima, M. Comprehensive review of 2-ethyl-1-hexanol as an indoor air pollutant. *J. Occup. Health* **2019**, *61*, 19–35. [CrossRef]
45. European Commission—Employment Social Affairs & Inclusion Recommendation from the Scientific Committee on Occupational Exposure Limits for 2-Ethylhexanol. Available online: https://www.google.com/url?sa=t&rct=j&q=&esrc=s&source=web&cd=&cad=rja&uact=8&ved=2ahUKEwiX7PGYnLP-AhXis4sKHREOAH8QFnoECA0QAQ&url=https%3A%2F%2Fec.europa.eu%2Fsocial%2FBlobServlet%3FdocId%3D6660%26langId%3Den&usg=AOvVaw0sUX_qCunAQmBjIZhWuHFe (accessed on 29 March 2022).
46. Skowroń, J.; Górny, R.L. Harmful biological agents. In *The Interdepartmental Commission for Maximum Admissible Concentrations and Intensities for Agents Harmful to Health in the Working Environment: Limit Values 2020*; Pośniak, M., Skowroń, J., Eds.; CIOP-PIB: Warsaw, Poland, 2020.
47. World Health Organization. *Air Quality Guidelines for Particulate Matter, Ozone, Nitrogen Dioxide and Sulfur Dioxide: Global Update 2005: Summary of Risk Assessment*; WHO: Geneva, Switzerland, 2006.
48. Adams, R.I.; Bhangar, S.; Pasut, W.; Arens, E.A.; Taylor, J.W.; Lindow, S.E.; Nazaroff, W.W.; Bruns, T.D. Chamber bioaerosol study: Outdoor air and human occupants as sources of indoor airborne microbes. *PLoS ONE* **2015**, *10*, e0128022. [CrossRef] [PubMed]

49. Frankel, M.; Bekö, G.; Timm, M.; Gustavsen, S.; Hansen, E.W.; Madsen, A.M. Seasonal Variations of Indoor Microbial Exposures and Their Relation to Temperature, Relative Humidity, and Air Exchange Rate. *Appl. Environ. Microbiol.* **2012**, *78*, 8289–8297. [CrossRef]
50. Skóra, J.; Gutarowska, B.; Pielech-Przybylska, K.; Stępień, Ł.; Pietrzak, K.; Piotrowska, M.; Piotrowski, P. Assessment of microbiological contamination in the work environments of museums, archives and libraries. *Aerobiologia* **2015**, *31*, 389–401. [CrossRef] [PubMed]
51. European Commission. *Commission Directive (EU) 2019/1833 of 24 October 2019 Amending Annexes I, III, V and VI to Directive 2000/54/EC of the European Parliament and of the Council as Regards Purely Technical Adjustments*; European Commission: Brussels, Belgium, 2019; pp. 54–79.
52. Żyrek, D. Ocena skażenia mikrobiologicznego powierzchni sprzętu do ćwiczeń w siłowniach. *Forum Zakażeń* **2019**, *10*, 219–225. [CrossRef]
53. Turkstani, M.A.; Sultan, R.M.S.; Al-Hindi, R.R.; Ahmed, M.M.M. Molecular identification of microbial contaminations in the fitness center in Makkah region. *Biosci. J.* **2021**, *37*, e37020. [CrossRef]
54. Boa, T.T.; Rahube, T.O.; Fremaux, B.; Levett, P.N.; Yost, C.K. Prevalence of methicillin-resistant staphylococci species isolated from computer keyboards located in secondary and postsecondary schools. *J. Environ. Health* **2013**, *75*, 50–58. [PubMed]
55. Lasek, R.; Szuplewska, M.; Mitura, M.; Decewicz, P.; Chmielowska, C.; Pawłot, A.; Sentkowska, D.; Czarnecki, J.; Bartosik, D. Genome Structure of the Opportunistic Pathogen *Paracoccus yeei* (Alphaproteobacteria) and Identification of Putative Virulence Factors. *Front. Microbiol.* **2018**, *9*, 2553. [CrossRef]
56. Daneshvar, M.I.; Hollis, D.G.; Weyant, R.S.; Steigerwalt, A.G.; Whitney, A.M.; Douglas, M.P.; Macgregor, J.P.; Jordan, J.G.; Mayer, L.W.; Rassouli, S.M.; et al. *Paracoccus yeeii* sp. nov. (Formerly CDC Group EO-2), a Novel Bacterial Species Associated with Human Infection. *J. Clin. Microbiol.* **2003**, *41*, 1289–1294. [CrossRef] [PubMed]
57. Cao, Y.-R.; Jiang, Y.; Wang, Q.; Tang, S.-K.; He, W.-X.; Xue, Q.-H.; Xu, L.-H.; Jiang, C.-L. *Rubellimicrobium roseum* sp. nov., a Gram-negative bacterium isolated from the forest soil sample. *Antonie Van Leeuwenhoek* **2010**, *98*, 389–394. [CrossRef]
58. Leys, N.M.E.J.; Ryngaert, A.; Bastiaens, L.; Verstraete, W.; Top, E.M.; Springael, D. Occurrence and Phylogenetic Diversity of Sphingomonas Strains in Soils Contaminated with Polycyclic Aromatic Hydrocarbons. *Appl. Environ. Microbiol.* **2004**, *70*, 1944–1955. [CrossRef]
59. El Beaino, M.; Fares, J.; Malek, A.; Hachem, R. Sphingomonas paucimobilis-related bone and soft-tissue infections: A systematic review. *Int. J. Infect. Dis.* **2018**, *77*, 68–73. [CrossRef]
60. Moura, J.B.; Delforno, T.P.; do Prado, P.F.; Duarte, I.C. Extremophilic taxa predominate in a microbial community of photovoltaic panels in a tropical region. *FEMS Microbiol. Lett.* **2021**, *368*, fnab105. [CrossRef] [PubMed]
61. Premalatha, N.; Gopal, N.O.; Jose, P.A.; Anandham, R.; Kwon, S.W. Optimization of cellulase production by *Enhydrobacter* sp. ACCA2 and its application in biomass saccharification. *Front. Microbiol.* **2015**, *6*, 1046. [CrossRef]
62. Viegas, C.; Alves, C.; Carolino, E.; Rosado, L.; Silva Santos, C. Prevalence of Fungi in Indoor Air with Reference to Gymnasiums with Swimming Pools. *Indoor Built Environ.* **2010**, *19*, 555–561. [CrossRef]
63. Zhu, L.; Li, T.; Xu, X.; Shi, X.; Wang, B. Succession of Fungal Communities at Different Developmental Stages of Cabernet Sauvignon Grapes from an Organic Vineyard in Xinjiang. *Front. Microbiol.* **2021**, *12*, 718261. [CrossRef] [PubMed]
64. Tsuji, M.; Tanabe, Y.; Vincent, W.F.; Uchida, M. *Vishniacozyma ellesmerensis* sp. nov., a psychrophilic yeast isolated from a retreating glacier in the Canadian High Arctic. *Int. J. Syst. Evol. Microbiol.* **2019**, *69*, 696–700. [CrossRef]
65. Riebesehl, J.; Yurchenko, E.; Nakasone, K.K.; Langer, E. Phylogenetic and morphological studies in Xylodon (Hymenochaetales, Basidiomycota) with the addition of four new species. *MycoKeys* **2019**, *47*, 97–137. [CrossRef]
66. Markson, A.A.; Akwaji, P.I.; Umana, E.J. Mushroom Biodiversity of Cross River National Park (Oban Hills Division), Nigeria. *World Sci. News* **2017**, *65*, 59–80.
67. Aronsen, A.; Læssøe, T. *Fungi of Northern Europe, Volume 5: The Genus Mycena s.l.*; Svampetryk: Hornbæk, Denmark, 2016.
68. de Hoog, G.S.; Guarro, J.; Gené, J.; Ahmed, S.; Al-Hatmi, A.M.S.; Figueras, M.J.; Vitale, R.G. *Atlas of Clinical Fungi*, 2nd ed.; Centraalbureau voor Schimmelcultures: Utrecht, The Netherlands, 2001.
69. Damji, R.; Mukherji, A.; Mussani, F. Sporobolomyces salmonicolor: A case report of a rare cutaneous fungal infection. *SAGE Open Med. Case Rep.* **2019**, *7*, 2050313X1984415. [CrossRef] [PubMed]
70. Ilinsky, Y.; Lapshina, V.; Verzhutsky, D.; Fedorova, Y.; Medvedev, S. Genetic Evidence of an Isolation Barrier between Flea Subspecies of *Citellophilus tesquorum* (Wagner, 1898) (Siphonaptera: Ceratophyllidae). *Insects* **2022**, *13*, 126. [CrossRef]
71. Rybitwa, D.; Wawrzyk, A.; Rahnama, M. Application of a Medical Diode Laser (810 nm) for Disinfecting Small Microbiologically Contaminated Spots on Degraded Collagenous Materials for Improved Biosafety in Objects of Exceptional Historical Value From the Auschwitz-Birkenau State Museum and Prot. *Front. Microbiol.* **2020**, *11*, 596852. [CrossRef] [PubMed]
72. Rybitwa, D.; Wawrzyk, A.; Wilczyński, S.; Łobacz, M. Irradiation with medical diode laser as a new method of spot-elimination of microorganisms to preserve historical cellulosic objects and human health. *Int. Biodeterior. Biodegrad.* **2020**, *154*, 105055. [CrossRef]
73. Gutarowska, B.; Szulc, J.; Nowak, A.; Otlewska, A.; Okrasa, M. Dust at various workplaces-microbiological and toxicological threats. *Int. J. Environ. Res. Public Health* **2018**, *15*, 877. [CrossRef] [PubMed]
74. Edet, U.; Antai, S.; Brooks, A.; Asitok, A.; Enya, O.; Japhet, F. An Overview of Cultural, Molecular and Metagenomic Techniques in Description of Microbial Diversity. *J. Adv. Microbiol.* **2017**, *7*, 1–19. [CrossRef]

75. Polish Ministry of Health Coronavirus Infections Report (SARS-CoV-2). Available online: https://www.gov.pl/web/koronawirus/wykaz-zarazen-koronawirusem-sars-cov-2 (accessed on 26 April 2022).
76. Chu, D.K.W.; Gu, H.; Chang, L.D.J.; Cheuk, S.S.Y.; Gurung, S.; Krishnan, P.; Ng, D.Y.M.; Liu, G.Y.Z.; Wan, C.K.C.; Tsang, D.N.C.; et al. SARS-CoV-2 Superspread in Fitness Center, Hong Kong, China, March 2021. *Emerg. Infect. Dis.* **2021**, *27*, 2230–2232. [CrossRef] [PubMed]
77. Li, H.; Shankar, S.N.; Witanachchi, C.T.; Lednicky, J.A.; Loeb, J.C.; Alam, M.M.; Fan, Z.H.; Mohamed, K.; Eiguren-Fernandez, A.; Wu, C.-Y. Environmental Surveillance and Transmission Risk Assessments for SARS-CoV-2 in a Fitness Center. *Aerosol Air Qual. Res.* **2021**, *21*, 210106. [CrossRef] [PubMed]
78. Lendacki, F.R.; Teran, R.A.; Gretsch, S.; Fricchione, M.J.; Kerins, J.L. COVID-19 Outbreak Among Attendees of an Exercise Facility—Chicago, Illinois, August–September 2020. *MMWR. Morb. Mortal. Wkly. Rep.* **2021**, *70*, 321–325. [CrossRef] [PubMed]
79. Helsingen, L.M.; Løberg, M.; Refsum, E.; Gjøstein, D.K.; Wieszczy, P.; Olsvik, Ø.; Juul, F.E.; Barua, I.; Jodal, H.C.; Herfindal, M.; et al. COVID-19 transmission in fitness centers in Norway—A randomized trial. *BMC Public Health* **2021**, *21*, 2103. [CrossRef]
80. Salonen, H.; Salthammer, T.; Morawska, L. Human exposure to air contaminants in sports environments. *Indoor Air* **2020**, *30*, 1109–1129. [CrossRef] [PubMed]
81. Masotti, F.; Cattaneo, S.; Stuknytė, M.; De Noni, I. Airborne contamination in the food industry: An update on monitoring and disinfection techniques of air. *Trends Food Sci. Technol.* **2019**, *90*, 147–156. [CrossRef]
82. Vasilyak, L.M. Physical Methods of Disinfection (A Review). *Plasma Phys. Rep.* **2021**, *47*, 318–327. [CrossRef]
83. Song, X.; Vossebein, L.; Zille, A. Efficacy of disinfectant-impregnated wipes used for surface disinfection in hospitals: A review. *Antimicrob. Resist. Infect. Control* **2019**, *8*, 139. [CrossRef]
84. Viana Martins, C.P.; Xavier, C.S.F.; Cobrado, L. Disinfection methods against SARS-CoV-2: A systematic review. *J. Hosp. Infect.* **2022**, *119*, 84–117. [CrossRef] [PubMed]
85. van Den Dool, H.; Kratz, P.D. A generalization of the retention index system including linear temperature programmed gas—Liquid partition chromatography. *J. Chromatogr. A* **1963**, *11*, 463–471. [CrossRef]
86. NIST Chemistry WebBook. Available online: https://webbook.nist.gov/chemistry/ (accessed on 25 March 2023).
87. CEN EN 13098:2019; Workplace Exposure—Measurement of Airborne Microorganisms and Microbial Compounds—General Requirements. CEN: Brussels, Belgium, 2019.
88. Schmidt, P.-A.; Bálint, M.; Greshake, B.; Bandow, C.; Römbke, J.; Schmitt, I. Illumina metabarcoding of a soil fungal community. *Soil Biol. Biochem.* **2013**, *65*, 128–132. [CrossRef]
89. Vilgalys, R.; Gonzalez, D. Organization of ribosomal DNA in the basidiomycete Thanatephorus praticola. *Curr. Genet.* **1990**, *18*, 277–280. [CrossRef]
90. Klindworth, A.; Pruesse, E.; Schweer, T.; Peplies, J.; Quast, C.; Horn, M.; Glöckner, F.O. Evaluation of general 16S ribosomal RNA gene PCR primers for classical and next-generation sequencing-based diversity studies. *Nucleic Acids Res.* **2013**, *41*, e1. [CrossRef]
91. Martin, M. Cutadapt removes adapter sequences from high-throughput sequencing reads. *EMBnet J.* **2011**, *17*, 10–12. [CrossRef]
92. Bolyen, E.; Rideout, J.R.; Dillon, M.R.; Bokulich, N.A.; Abnet, C.C.; Al-Ghalith, G.A.; Alexander, H.; Alm, E.J.; Arumugam, M.; Asnicar, F.; et al. Reproducible, interactive, scalable and extensible microbiome data science using QIIME 2. *Nat. Biotechnol.* **2019**, *37*, 852–857. [CrossRef] [PubMed]
93. Rognes, T.; Flouri, T.; Nichols, B.; Quince, C.; Mahé, F. VSEARCH: A versatile open source tool for metagenomics. *PeerJ* **2016**, *4*, e2584. [CrossRef]
94. Kõljalg, U.; Nilsson, H.R.; Schigel, D.; Tedersoo, L.; Larsson, K.-H.; May, T.W.; Taylor, A.F.S.; Jeppesen, T.S.; Frøslev, T.G.; Lindahl, B.D.; et al. The Taxon Hypothesis Paradigm—On the Unambiguous Detection and Communication of Taxa. *Microorganisms* **2020**, *8*, 1910. [CrossRef] [PubMed]
95. Evans, J.D. *Straightforward Statistics for the Behavioral Sciences*; Duxbury Press: London, UK, 1995; ISBN 0534231004.

Disclaimer/Publisher's Note: The statements, opinions and data contained in all publications are solely those of the individual author(s) and contributor(s) and not of MDPI and/or the editor(s). MDPI and/or the editor(s) disclaim responsibility for any injury to people or property resulting from any ideas, methods, instructions or products referred to in the content.

Article

ICP-MS Measurement of Trace and Rare Earth Elements in Beach Placer-Deposit Soils of Odisha, East Coast of India, to Estimate Natural Enhancement of Elements in the Environment

Nimelan Veerasamy [1,2], Sarata Kumar Sahoo [1,*], Rajamanickam Murugan [1], Sharayu Kasar [1], Kazumasa Inoue [2], Masahiro Fukushi [2] and Thennaarassan Natarajan [1,2]

1 National Institute of Radiological Sciences, National Institutes for Quantum Sciences and Technology (QST), 4-9-1 Anagawa, Inage-ku, Chiba 263-8555, Japan; nimelanveerasamy@gmail.com (N.V.); murugan.rajamanickam@qst.go.jp (R.M.); kasar.sharayu@qst.go.jp (S.K.); nthennarassan@gmail.com (T.N.)
2 Department of Radiological Sciences, Tokyo Metropolitan University, 7-2-10 Higashiogu, Arakawa-ku, Tokyo 116-8551, Japan; kzminoue@tmu.ac.jp (K.I.); fukushi@tmu.ac.jp (M.F.)
* Correspondence: sahoo.sarata@qst.go.jp

Abstract: Inductively coupled plasma mass spectrometry (ICP-MS) has been used to measure the concentration of trace and rare earth elements (REEs) in soils. Geochemical certified reference materials such as JLk-1, JB-1, and JB-3 were used for the validation of the analytical method. The measured values were in good agreement with the certified values for all the elements and were within 10% analytical error. Beach placer deposits of soils mainly from Odisha, on the east coast of India, have been selected to study selected trace and rare earth elements (REEs), to estimate enrichment factor (EF) and geoaccumulation index (I_{geo}) in the natural environment. Enrichment factor (EF) and geoaccumulation index (I_{geo}) results showed that Cr, Mn, Fe, Co, Zn, Y, Zr, Cd and U were significantly enriched, and Th was extremely enriched. The total content of REEs (ΣREEs) ranged from 101.3 to 12,911.3 $\mu g\ g^{-1}$, with an average 2431.1 $\mu g\ g^{-1}$ which was higher than the average crustal value of ΣREEs. A high concentration of Th and light REEs were strongly correlated, which confirmed soil enrichment with monazite minerals. High ratios of light REEs (LREEs)/heavy REEs (HREEs) with a strong negative Eu anomaly revealed a felsic origin. The comparison of the chondrite normalized REE patterns of soil with hinterland rocks such as granite, charnockite, khondalite and migmatite suggested that enhancement of trace and REEs are of natural origin.

Keywords: soils; trace elements; rare earth elements; geoaccumulation index; enrichment factor; ICP-MS

1. Introduction

Environmental pollution has pervaded many parts of the world due to anthropogenic activities such as urbanization, exploration, mining of natural resources, industrialization, etc., which has resulted in contamination of trace elements (TEs) and REEs into the environment directly or indirectly [1–3]. Natural contents of REEs in soil are highly influenced by their parent materials, weathering and pedogenesis processes [4]. In soil, the enrichment of REEs is mainly controlled by the abundance of REE-bearing minerals such as apatite, allanite, bastnaesite, monazite, xenotime and zircon [5]. There are a few reports showing a gradual increase in REEs in soil by anthropogenic activities [6,7]. The REEs background data could be used as baselines to identify contamination level as well as quantitative risk assessment in soils. Therefore, monitoring of TEs and REEs is essential for the establishment of baselines from the viewpoint of environmental pollution or contamination. Geochemical analyses of natural materials (soils, sand, etc.) are necessary to determine the level of contamination, and to elucidate whether it is from geogenic or anthropogenic sources [8]. Environmental contaminations have been evaluated using two pollution indices such as the enrichment factor (EF) and geoaccumulation index (I_{geo}), to identify the degree of contamination in soil and sediments and their origin [9].

Beach placer deposits are formed by sediments produced through weathering and erosion of rocks (i.e., igneous, sedimentary and metamorphic rocks) that are transported by rivers and streams to coastal areas. During these processes heavy minerals (specific gravity, $\rho > 2.89$ g/cm^3) such as monazite, ilmenite, zircon, rutile, garnet, and sillimanite are accumulated along the beaches [10]. Monazite [(Ce, La, Nd, Th) PO$_4$] is an important heavy mineral containing a high concentration of Th and rare earth elements (REEs), especially light REEs (LREEs) [11,12].

Recently, increasing attention has been paid towards not only environmental radioactivity studies but also to the origin of beach placer deposits in the southwest coast of Sri Lanka [13], Sithonia Peninsula, Greece [14], Calabria, Italy [15], Langkawi, Malaysia [16], Chittagong, Bangladesh [17], and Mandena, Madagascar [18]. Several Indian coastal areas, well-known as high background radiation areas (HBRAs), with beach placer deposits have been investigated; these areas are in Karnataka [19]; Andhra Pradesh [20]; Kerala [21]; Tamil Nadu [22–24] and Odisha [25–28].

The Odisha state is an important littoral state on the eastern coast of India, and the coastal stretch between the Rushikulya river and Gopalpur town is known as the Chhatrapur–Gopalpur beach placer deposit. The total weight percentage of heavy minerals in this beach placer deposit ranges from 2.9 to 20.4%. It includes heavy minerals such as garnet, hornblende, ilmenite, magnetite, monazite, pyroxene, rutile, sillimanite, sphene, tourmaline and zircon [29,30]. Due to the high accumulation of monazites, ilmenites and rutiles minerals in the beach sand, this region has been explored by Indian Rare Earth Limited (IREL) and an extensive exploration process is in progress [10]. Eventually, this will lead to the possibility of anthropogenic contamination in the environment. Therefore, environmental monitoring studies with respect to pollution and contamination are necessary.

In Odisha's coastal soils, there is a lack of TEs and REEs data of bulk sand and soil in the Chhatrapur–Gopalpur beach placer deposits. The REEs background data could be used as baselines to identify contamination level as well as to conduct a quantitative risk assessment in soils. Therefore, analyses of TEs and REEs in soils have been carried out using inductively coupled plasma mass spectrometry (ICP-MS) to evaluate two pollution indices, the enrichment factor (EF) and geoaccumulation index (I_{geo}), to identify the degree of environmental contamination.

(1) To validate analysis of TEs and REEs with certified reference materials using ICP-MS;
(2) Determination of TEs and REEs in Chhatrapur–Gopalpur beach placer-deposit soils;
(3) Estimation of EF and I_{geo} of TEs to evaluate natural enrichment and anthropogenic contamination in soils;
(4) To understand the origin/source of TEs and REEs in beach placer-deposit soils.

2. Results and Discussion
2.1. Analytical Validation of TEs and REEs

In this study, geochemical certified reference materials (CRMs) such as Japan lake sediment (JLk-1) and Japan basalts (JB-1 and JB-3), supplied by the Geological Survey of Japan, were used to the validate analytical method for TEs and REEs using ICP-MS. The concentrations (μg g^{-1}) of TEs such as Cr, Mn, Fe, Co, Ni, Cu, Zn, Rb, Sr, Zr, Cd, Cs, Ba, Pb, Th and U and REEs (Y, La, Ce, Pr, Nd, Sm, Eu, Gd, Tb, Dy, Ho, Er, Tm, Yb, Lu) are given in Table 1. The TEs and REEs results were compared with the certified values of CRMs [31,32]. The recovery of the mean measured values of JLk-1, JB-1 and JB-3 for TEs and REEs ranged from 90 to 110%.

Table 1. Analytical results of TEs and REEs (µg g^{-1}) for JLk-1, JB-1 and JB-3.

Elements	JLk-1				JB-1				JB-3			
	Mean (µg g^{-1})	SD	CV (µg g^{-1})	Recovery (%)	Mean (µg g^{-1})	SD	CV (µg g^{-1})	Recovery (%)	Mean (µg g^{-1})	SD	CV (µg g^{-1})	Recovery (%)
Cr	71.4	0.3	69	103	433.3	1.3	425	102	55.3	0.2	58.1	95
Mn	2358	7	2092	103	1217	4	1200	101	1453	3	1400	104
Fe	47,538	94	46,738	102	63,605	362	62,900	101	82,355	70	82,700	100
Co	18.5	0.1	18	103	37.9	0.2	38.2	99	34.9	0.2	34.3	102
Ni	38.0	0.2	35	109	133.4	1.2	133	100	36.0	0.4	36.2	99
Cu	69.1	0.3	62.9	110	51.3	0.3	55.1	93	183.9	0.7	194	95
Zn	166.2	0.3	152	109	83.7	4.6	85.2	98	108.0	2.2	100	108
Rb	160.6	0.6	147	109	40.6	0.8	41.3	98	15.6	0.2	15.1	103
Sr	68.8	1.6	67.5	102	431.9	0.1	444	97	424.7	5.3	403	105
Y	43.1	0.2	40	108	23.5	0.7	24.3	97	25.9	0.2	26.9	96
Zr	125.1	0.7	137	91	135.8	1.0	141	96	94.1	1.0	97.8	96
Cd	0.61	0.06	0.57	107	0.12	0.01	0.11	109	0.082	0.008	0.081	101
Cs	10.7	0.2	10.9	98	1.3	0.1	1.2	108	0.92	0.09	0.94	98
Ba	595.7	4.7	574	104	518.2	0.5	493	105	254.5	3.0	245	104
Pb	46.7	0.4	43.7	107	10.0	0.2	10	100	5.6	0.2	5.6	100
La	44.0	0.2	40.6	108	37.7	0.2	38.6	98	8.3	0.1	8.8	94
Ce	89.8	0.6	87.9	102	65.5	0.6	67.8	97	21.0	0.3	21.5	98
Pr	9.3	0.2	8.5	109	6.9	0.3	7	99	3.1	0.3	3.1	100
Nd	35.0	0.3	35.7	98	25.9	0.4	26.8	97	15.1	0.2	15.6	97
Sm	8.1	0.2	7.9	103	5.0	0.2	5.1	98	4.3	0.1	4.3	100
Eu	1.3	0.1	1.3	100	1.6	0.1	1.5	107	1.3	0.1	1.3	100
Gd	6.3	0.3	6	105	4.5	0.4	4.9	90	4.7	0.3	4.7	100
Tb	1.2	0.1	1.2	100	0.80	0.02	0.82	98	0.80	0.03	0.73	110
Dy	6.2	0.2	6.6	94	4.0	0.2	4.1	98	4.6	0.1	4.5	102
Ho	1.1	0.1	1.1	100	0.80	0.04	0.79	101	0.80	0.07	0.80	100

Table 1. Cont.

Elements	JLk-1				JB-1				JB-3			
	Mean (µg g^{-1})	SD	CV (µg g^{-1})	Recovery (%)	Mean (µg g^{-1})	SD	CV (µg g^{-1})	Recovery (%)	Mean (µg g^{-1})	SD	CV (µg g^{-1})	Recovery (%)
Er	3.8	0.2	3.6	106	2.3	0.1	2.3	100	2.6	0.2	2.5	104
Tm	0.57	0.01	0.53	108	0.38	0.02	0.35	109	0.46	0.02	0.42	110
Yb	3.9	0.2	4	98	2.1	0.1	2.1	100	2.5	0.2	2.6	96
Lu	0.55	0.02	0.57	96	0.31	0.02	0.31	100	0.41	0.01	0.39	105
Th	19.4	0.4	19.5	99	9.3	0.3	9.3	100	1.3	0.1	1.3	100
U	3.7	0.1	3.8	97	1.7	0.1	1.7	100	0.50	0.01	0.48	104

Errors of analysis are represented as standard deviation (SD) which refers to the precision [33]. The accuracy as a relative bias (RB%) of the measurement of TEs and REEs was ≤10%. This states that the reproducibility as a measure of precision of the analytical method is in good agreement with the certified values for TEs and REEs, i.e., within analytical error of 10%. The same method was applied to all soils.

2.2. TEs in Beach Placer-Deposit Soils

The mean concentration of TEs ($\mu g\ g^{-1}$) of each sample location from the study area are summarized in Table 2. The results showed that the mean concentration of elements in the soils are in the following order: Fe > Mn > Th > Ba > Zr > Y > Cr > Zn > Pb > U > Rb > Co > Sr > Ni > Cu > Cs > Cd.

The Fe (iron) concentration in samples varied from 19,000 to 150,000 $\mu g\ g^{-1}$ with an average of 57,508 $\mu g\ g^{-1}$, i.e., higher than World Health Organization (WHO) global limit (50,000 $\mu g\ g^{-1}$) [34]. The Fe concentration was high in the samples collected from Aryapalli, Boxipalli, Kanamana, Gopalpur and Matikhalo. The Mn concentration varied from 460 to 3700 $\mu g\ g^{-1}$ with an average value of 1300 $\mu g\ g^{-1}$ and was less than the WHO critical value (2000 $\mu g\ g^{-1}$) [34]. However, Aryapalli samples showed Mn concentration more than 2000 $\mu g\ g^{-1}$.

Concentration of Th ranged from 35.0 to 900 $\mu g\ g^{-1}$ with a mean value of 390 $\mu g\ g^{-1}$. The high concentration of Th in the soils is attributed to the presence of monazite minerals. U concentration varied from 1.4 to 53.2 $\mu g\ g^{-1}$ with a mean value of 14.6 $\mu g\ g^{-1}$. Pb concentration ranged from 16.2 to 65.0 $\mu g\ g^{-1}$ with a mean value of 40.0 $\mu g\ g^{-1}$. The highest Pb concentration was observed at Aryapalli, however all samples were below global limit 85 $\mu g\ g^{-1}$. The presence of Pb in the human body causes damage to bones and organs such as the liver, kidneys, brain, lungs, and central nervous system. Ba concentration varied from 3.4 to 385 $\mu g\ g^{-1}$ with a mean value of 142 $\mu g\ g^{-1}$. The highest concentration of Ba was observed at Jagnyasala.

Zn concentration varied from 27.0 to 250 $\mu g\ g^{-1}$ with a mean value of 103 $\mu g\ g^{-1}$. The highest concentration of Zn was observed at Aryapalli, Kanamana, Matikhalo and Venkatraipur. Zr concentration varied from 2.2 to 370 $\mu g\ g^{-1}$ with a mean value of 102 $\mu g\ g^{-1}$, which was less than the average upper continental crust (UCC) value of 190 $\mu g\ g^{-1}$. Cr concentration varied from 35.6 to 180 $\mu g\ g^{-1}$ with a mean value of 83 $\mu g\ g^{-1}$. The mean concentration was less than the global limit of 150 $\mu g\ g^{-1}$. The highest concentration of Cr was observed at Aryapalli. A high concentration of Cr causes skin related diseases. Co concentration varied from 10.4 to 75.0 $\mu g\ g^{-1}$ with a mean value of 27.4 $\mu g\ g^{-1}$. Ni concentration varied from 1.1 to 24.5 $\mu g\ g^{-1}$ with a mean value of 12.0 $\mu g\ g^{-1}$. Other trace elements were in very low concentrations—below the recommended global limits.

2.3. Enrichment Factor (EF) of TEs in Soil

The EF results of trace elements in soils are given in Table 3. The results showed that Th was extremely enriched in Aryapalli, highly enriched in Boxipalli, significantly enriched in Kanamana, Badaputti, Matikhalo, Gopalpur, Kalipalli, Chhatrapur and Venkatraipur, and moderately enriched in Basanaputi. U was extremely enriched in Aryapalli, highly enriched in Boxipalli, significantly enriched in Kanamana, Matikhalo, Gopalpur, Kalipalli, Chhatrapur, and Venkatraipur and moderately enriched in Badaputti. The extreme enrichment of Th and U in the soils could be explained mainly by the presence of monazite minerals and felsic-source rocks in the study area. There were no anthropogenic activities related to the enrichment of Th and U.

Table 2. Mean concentration (µg g^{-1}) of TEs in soils.

Elements	Aryapalli	Boxipalli	Kanamana	Gopalpur	Chhatrapur	Matikhalo	Kalipalli	Venkatraipur	Badaputti	Basanaputti	Jagnyasala	Kalyaballi
Cr	180.0 ± 20	48.6 ± 4.5	68.4 ± 6.5	125.0 ± 12	103.0 ± 11	130.0 ± 15	35.6 ± 3.4	87.1 ± 7.9	61.8 ± 6.3	51.4 ± 5.1	43.1 ± 3.8	55.3 ± 5.3
Mn	3700 ± 43	1600 ± 18	1500 ± 16	1600 ± 18	780 ± 9	1300 ± 14	940 ± 10	1060 ± 11	830 ± 9	700 ± 8	690 ± 7	460 ± 5
Fe	1.5 × 10^5 ± 1533	58,000 ± 777	85,000 ± 870	77,700 ± 787	43,000 ± 428	70,000 ± 661	34,000 ± 348	46,000 ± 458	36,000 ± 359	28,000 ± 261	37,000 ± 362	19,000 ± 189
Co	75.0 ± 7.3	22.5 ± 2.9	44.9 ± 4.6	39.2 ± 4.2	17.4 ± 1.7	37.6 ± 3.2	15.3 ± 1.4	17.8 ± 1.6	17.4 ± 1.6	14.3 ± 1.3	16.8 ± 1.7	10.4 ± 1.0
Ni	17.3 ± 1.6	1.1 ± 0.1	19.5 ± 1.9	14.7 ± 1.5	23.2 ± 2.3	16.6 ± 1.5	1.2 ± 0.1	1.1 ± 0.1	3.1 ± 0.3	4.4 ± 0.3	24.5 ± 2.2	15.7 ± 1.3
Cu	15.7 ± 1.4	2.1 ± 0.2	17.5 ± 1.6	18.3 ± 1.7	20.5 ± 1.9	16.6 ± 1.5	1.3 ± 0.1	1.8 ± 0.2	2.0 ± 0.2	3.4 ± 0.3	18.5 ± 1.4	11.9 ± 0.9
Zn	250.0 ± 25	93 ± 9.0	135.0 ± 12	135.0 ± 13	91.0 ± 9.0	125.0 ± 12	27.0 ± 3.0	165.0 ± 16	55.0 ± 4.0	39 ± 4.0	80 ± 8.0	43.0 ± 4.0
Rb	21.9 ± 0.8	36.6 ± 1.2	31.5 ± 1.1	54.6 ± 1.6	60.9 ± 1.7	31.3 ± 1.1	37.6 ± 1.3	20.4 ± 0.7	52.3 ± 1.5	47.5 ± 1.4	61.9 ± 1.8	26.6 ± 0.9
Sr	23.9 ± 1.5	36.6 ± 2.1	22.5 ± 1.4	44.6 ± 2.7	28.0 ± 1.4	20.5 ± 1.1	23.5 ± 2.0	20.0 ± 1.8	34.7 ± 1.6	36.3 ± 1.8	22.6 ± 1.3	14.5 ± 0.9
Y	180 ± 15	101.0 ± 9.0	44.8 ± 4.1	61.6 ± 6.4	25.7 ± 2.3	39.4 ± 3.2	31.2 ± 2.9	32.1 ± 2.6	28.4 ± 2.4	26.1 ± 2.3	40.8 ± 3.5	11.1 ± 0.9
Zr	370.0 ± 28	4.3 ± 0.1	160.0 ± 14	143.0 ± 13	125.0 ± 11	150.0 ± 15	2.4 ± 0.2	2.2 ± 0.2	57.8 ± 4.5	39.9 ± 3.3	110.0 ± 10	55.2 ± 4.4
Cd	0.11 ± 0.02	0.05 ± 0.01	0.12 ± 0.02	0.14 ± 0.02	0.05 ± 0.01	0.11 ± 0.02	0.05 ± 0.01	0.05 ± 0.01	0.05 ± 0.01	0.05 ± 0.01	0.05 ± 0.01	0.05 ± 0.01
Cs	0.8 ± 0.1	6.0 ± 0.5	1.1 ± 0.2	0.8 ± 0.1	3.6 ± 0.3	1.0 ± 0.1	11.1 ± 0.6	5.5 ± 0.4	1.1 ± 0.2	0.6 ± 0.1	1.1 ± 0.2	0.8 ± 0.1
Ba	13.5 ± 1.1	185.0 ± 15	50.0 ± 3.0	107.0 ± 6.0	285.0 ± 7.0	52.3 ± 4.1	181.0 ± 17.0	130.0 ± 7.0	117.0 ± 6.0	92.0 ± 7.0	385.0 ± 19	110.0 ± 7.0
Pb	65.0 ± 8.0	62.0 ± 6.0	49.0 ± 5.0	54.0 ± 5.0	47.4 ± 4.3	43.0 ± 4.2	23.5 ± 1.9	28.5 ± 1.9	23.2 ± 2.1	16.2 ± 1.3	46.2 ± 3.6	19.8 ± 1.7
Th	930 ± 87	830 ± 75	500 ± 49	560 ± 62	200 ± 23	560 ± 63	300 ± 31	370 ± 41	200 ± 25	100 ± 15	37 ± 7.0	35 ± 6.0
U	53.2 ± 4.8	36.3 ± 3.1	15.3 ± 1.8	15.4 ± 1.9	6.7 ± 0.4	11.8 ± 1.7	12.7 ± 1.3	12.4 ± 1.1	5.6 ± 0.3	2.4 ± 0.2	1.7 ± 0.1	1.4 ± 0.1

Table 3. Enrichment factor of TEs in soils.

Elements	Kanamana	Basanaputi	Badaputti	Matikhalo	Gopalpur	Aryapalli	Kalyaballi	Jagnyasala	Chhatrapur	Kalipalli	Boxipalli	Venkatraipur
Cr	1.8	1.2	1.8	2.9	2.8	7.7	1.1	1.1	4.2	0.6	1.1	1.5
Mn	2.4	1.1	1.5	1.8	2.2	10.0	0.6	1.1	2.0	1.0	2.2	1.2
Fe	2.5	0.8	1.2	1.8	2.0	8.7	0.5	1.0	2.0	0.7	1.5	0.9
Co	3.0	0.9	1.3	2.1	2.2	9.3	0.5	1.0	1.8	0.7	1.3	0.8
Ni	0.9	0.2	0.2	0.7	0.6	1.6	0.6	1.0	1.8	−0.4	−1.2	−0.3
Cu	1.0	0.2	−0.4	0.9	1.0	1.8	0.6	1.0	2.0	−2.3	−5.5	−2.0
Zn	1.9	0.5	0.9	1.5	1.6	6.3	0.5	1.0	2.0	0.3	1.1	1.6
Rb	0.6	0.8	1.0	0.5	0.9	0.9	0.4	1.0	1.6	0.5	0.6	0.2
Sr	1.1	1.7	1.9	0.9	2.0	2.2	0.6	1.0	2.0	0.8	1.6	0.7
Y	1.2	0.6	0.9	0.9	1.4	7.6	0.2	1.0	1.1	0.6	2.4	0.6
Zr	1.7	0.4	0.7	1.2	1.3	6.5	0.4	1.0	1.9	0.0	0.0	0.0
Cd	1.7	0.1	0.8	1.2	3.5	5.1	0.4	1.0	1.6	0.0	0.0	0.0
Cs	1.0	0.5	1.1	0.8	0.7	1.9	0.5	1.1	4.6	7.4	5.1	3.7
Ba	0.1	0.2	0.4	0.1	0.3	0.0	0.2	1.0	1.2	0.4	0.5	0.3
Pb	1.2	0.4	0.6	0.9	1.1	1.0	0.4	1.1	1.8	0.4	1.3	0.5
Th	15.5	2.7	6.9	13.5	14.0	52.4	0.8	1.0	10.7	6.8	21.7	7.5
U	10.0	1.4	4.0	6.1	8.4	51.3	0.7	1.1	6.7	5.5	20.4	5.4

Cr, Mn, Fe, Co, Zn, Y, Zr, and Cd were significantly enriched in Aryapalli samples. Cs has been significantly enriched in Chhatrapur and Boxipalli. Mn, Fe and Co were significantly enriched in Kanamana samples. Cr, Mn and Co were significantly enriched in Gopalpur samples. Cr, Mn, Fe, Cu, Sr, and Cs were significantly enriched in Chattrapur samples. Cr and Co were significantly enriched in Matikhalo samples. Mn and Y were significantly enriched in Boxipalli samples. The significant enrichment of Cr may be due to the mafic-source rock present in the study area. Mn and Fe enrichment may be due to the presence of ilmenite mineral present in the study area.

2.4. Geoaccumulation Index (I_{geo}) of TEs in Soils

The results of I_{geo} values for the elements in soils are presented in Table 4. Th was extremely enriched in Aryapalli, Boxipalli, Kanamana and Matikhalo, and highly enriched in Gopalpur, Kalipalli and Venkatraipur. Enrichment of Th in Chhatrapur was moderate to high, whereas it was moderately enriched in Badaputti and Basanaputti and slightly enriched in Jagnyasala and Kalyaballi. U is moderately to heavily enriched in Aryapalli and Boxipalli, slightly enriched in Kanamana, Matikhalo, Kalipalli, and Venkatraipur. Pb and Y were slightly enriched in Boxipalli. Mn, Co and Zn were slightly enriched in Aryapalli. The slight enrichment of Pb is due to the mining activities near the Aryapalli and Boxipalli study areas.

2.5. Geochemistry of REEs in Soils

The mean concentrations of light and heavy REEs (LREE and HREE) from all samples are given in Table 5 along with descriptive statistics. The mean $\sum$LREEs (2308.8 µg g^{-1}) concentration was about 17 times higher than the UCC value (132.5 µg g^{-1}). On the other hand, the mean $\sum$HREEs (71.2 µg g^{-1}) concentration was five times higher than the UCC value (13.9 µg g^{-1}). The total concentrations of $\sum$REEs ranged from 101.3–12911.3 µg g^{-1} with a mean value of 2431.1 µg g^{-1}. The mean $\sum$REEs concentration was 16 times higher than the UCC value (146.4 µg g^{-1}) [35].

The enrichment of REEs (µg g^{-1}) was in the following order: Ce (1121.5) > La (540.7) > Nd (458.5) > Pr (119.1) > Sm (66.9) > Gd (39.9) > Dy (13.4) > Er (5.4) > Yb (4.3) > Tb (4.2) > Ho (2.0) > Eu (1.9) > Lu (0.9) > Tm (0.9). The REEs concentrations exhibited the same order as for the Oddo-Harkins rule with two exceptions (i.e., depletion of Eu and slight enrichment of Lu). This type of small exception in the order of REE concentrations has been observed in Cuban soils [36]. The REEs concentration in the study area has been arranged in decreasing order as follows: Aryapalli > Boxipalli > Kanamana > Gopalpur > Matikhalo > Chhatrapur > Venkatraipur > Kalipalli > Badaputti > Basanaputti > Kalyaballi > Jagnyasala.

Pearson's correlation coefficients (significant at the 99% level) were used to understand the relationship between Th, U and REEs. The coefficients are presented in Table 6. The results indicate that there is a stronger correlation in LREEs than HREEs. Th showed a stronger positive correlation with LREEs (R^2 = 0.64 to 0.90) compared to HREEs (R^2 = 0.46 to 0.83). This positive correlation between Th and LREEs corroborates that Th is a high-field-strength element and strongly supports the presence of monazite minerals. REEs showed similarities in behaviour including low solubility and immobility during weathering and sedimentation [37]. U also showed a strong positive correlation with all REEs (R^2 = 0.62 to 0.99).

Table 4. Geoaccumulation index values of TEs in soils.

Elements	Kanamana	Basanaputti	Badaputti	Matikhalo	Gopalpur	Aryapalli	Kalyaballi	Jagnyasala	Chhatrapur	Kalipalli	Boxipalli	Venkatraipur
Cr	−1.0	−1.5	−1.4	−0.1	−0.4	0.3	−1.2	−1.9	−0.3	−2.0	−1.5	−0.7
Mn	0.3	−0.9	−0.6	0.2	0.1	1.8	−1.3	−0.5	−0.5	−0.3	0.4	−0.1
Fe	−0.5	−2.3	−1.9	−0.8	−1.1	0.4	−2.5	−1.9	−1.4	−1.8	−1.0	−1.4
Co	0.8	−1.0	−0.6	0.5	0.2	1.6	−1.2	−0.5	−0.5	−0.8	−0.2	−0.5
Ni	−1.9	−4.1	−3.1	−2.1	−2.4	−2.0	−2.2	−1.9	−1.7	0.0	0.0	0.0
Cu	−1.3	−3.8	−2.9	−1.4	−1.4	−1.4	−1.8	−1.4	−1.1	0.0	0.0	0.0
Zn	0.4	−1.6	−0.9	0.3	0.1	1.4	−1.1	−0.3	−0.1	−1.9	−0.1	0.7
Rb	−2.1	−1.4	−1.2	−2.0	−1.1	−3.6	−2.2	−0.7	−1.4	−1.7	−1.8	−2.6
Sr	−4.6	−3.8	−3.6	−4.5	−3.3	−4.4	−5.1	−4.2	−4.3	−4.4	−3.7	−4.6
Y	0.5	−0.5	−0.3	0.3	0.4	2.5	−1.5	0.6	−0.2	0.0	1.7	0.0
Zr	−0.9	−3.1	−2.3	−1.0	−1.5	0.5	−2.5	−1.3	−1.1	−6.9	−6.1	−7.0
Cd	−1.2	0.0	0.0	−2.1	−0.9	0.4	−2.4	−1.6	−1.8	0.0	0.0	0.0
Cs	−2.9	−3.7	−3.1	−2.9	−2.7	−5.1	−2.5	−2.6	−2.1	−0.1	0.6	−0.4
Ba	−4.3	−3.4	−3.2	−4.2	−3.1	0.0	−3.2	−1.0	−2.0	−2.4	−2.3	−2.8
Pb	0.9	−0.7	−0.2	0.7	0.6	0.0	−0.3	0.9	0.9	−0.1	1.3	0.2
Th	5.0	2.2	2.6	5.1	4.3	5.8	1.3	1.5	4.0	4.4	5.7	4.6
U	1.9	−1.1	−0.7	1.4	0.7	3.6	−1.4	−1.2	0.9	1.6	3.2	1.6

Table 5. Descriptive statistics of REEs ($\mu g\ g^{-1}$) in soils ($n = 36$).

Element	Mean	Min	Median	Max	Skewness	Kurtosis	CV
La	540.76	17.65	334.97	2770.36	1.86	3.67	1.16
Ce	1121.56	40.62	687.72	5797.34	1.90	3.81	1.16
Pr	119.09	3.91	68.66	638.91	1.94	4.09	1.20
Nd	458.58	16.10	269.25	2557.04	2.06	4.78	1.22
Sm	66.88	2.74	41.37	389.59	2.25	6.12	1.22
Eu	1.96	0.31	1.49	7.96	2.06	4.78	0.88
Gd	39.97	1.96	24.16	228.39	2.25	6.05	1.19
Tb	4.23	0.34	2.46	26.18	2.67	8.97	1.21
Dy	13.44	1.40	7.70	89.13	3.32	13.93	1.19
Ho	2.05	0.23	1.23	11.03	2.62	8.22	1.07
Er	5.37	0.64	3.67	32.50	3.23	13.28	1.08
Tm	0.93	0.09	0.51	4.99	2.37	5.71	1.19
Yb	4.30	0.65	3.12	26.32	3.61	16.44	1.05
Lu	0.95	0.10	0.46	5.66	2.50	6.65	1.26
ΣREE	2431.19	101.31	1469.72	12,911.35	1.98	4.32	1.17
ΣLREE	2308.82	85.40	1403.06	12,160.34	1.94	4.04	1.18
ΣHREE	71.24	6.48	44.25	421.50	2.65	8.88	1.14
Eu/Eu*	0.21	0.06	0.11	0.78	1.59	1.39	0.95
Ce/Ce*	1.46	0.98	1.04	4.51	2.51	4.66	0.75
$(La/Sm)_N$	5.31	2.87	4.92	10.06	2.29	4.58	0.29
$(La/Yb)_N$	84.35	5.40	82.44	211.66	0.28	−1.15	0.70
$(Gd/Yb)_N$	7.23	1.27	7.25	18.59	0.39	−0.70	0.64

CV, coefficient variant; Min, minimum; Max, maximum. Eu/Eu* and Ce/Ce* are the calculated europium and cerium anomalies, respectively. Subscript N indicates chondrite normalized values.

Table 6. Pearson correlation coefficient of Th, U and REEs in soils ($n = 36$).

	La	Ce	Pr	Nd	Sm	Eu	Gd	Tb	Dy	Ho	Er	Tm	Yb	Lu	Th	U
La	1.00															
Ce	1.00	1.00														
Pr	1.00	1.00	1.00													
Nd	1.00	1.00	1.00	1.00												
Sm	1.00	1.00	1.00	1.00	1.00											
Eu	0.79	0.80	0.79	0.77	0.77	1.00										
Gd	0.99	0.99	1.00	0.99	1.00	0.81	1.00									
Tb	0.98	0.98	0.99	0.99	0.99	0.79	1.00	1.00								
Dy	0.95	0.95	0.95	0.96	0.97	0.75	0.97	0.99	1.00							
Ho	0.88	0.89	0.89	0.87	0.88	0.95	0.91	0.91	0.89	1.00						
Er	0.91	0.91	0.91	0.91	0.93	0.86	0.94	0.96	0.97	0.97	1.00					
Tm	0.67	0.68	0.68	0.64	0.65	0.96	0.70	0.68	0.64	0.92	0.80	1.00				
Yb	0.85	0.85	0.86	0.86	0.88	0.77	0.89	0.92	0.96	0.91	0.98	0.70	1.00			
Lu	0.61	0.62	0.62	0.58	0.58	0.95	0.64	0.61	0.57	0.87	0.74	0.99	0.63	1.00		
Th	0.90	0.90	0.90	0.89	0.87	0.64	0.86	0.83	0.75	0.68	0.69	0.50	0.60	0.46	1.00	
U	0.95	0.95	0.95	0.96	0.97	0.78	0.97	0.99	0.99	0.91	0.96	0.69	0.93	0.62	0.75	1.00

In this study, Leedey chondrite values [38] were used for REEs normalization of soils. The chondrite normalized REE patterns of soils are shown in Figure 1. The soils showed enrichment of LREEs and a flat HREEs pattern with negative Eu anomaly. Although the absolute concentrations of REEs in the soils were different, the distribution of chondrite normalized REE patterns of individual samples was remarkably similar. The chondrite normalized REE patterns uniformly showed a high concentration of LREEs and a relatively high concentration of Gd, Tb and Dy in all samples.

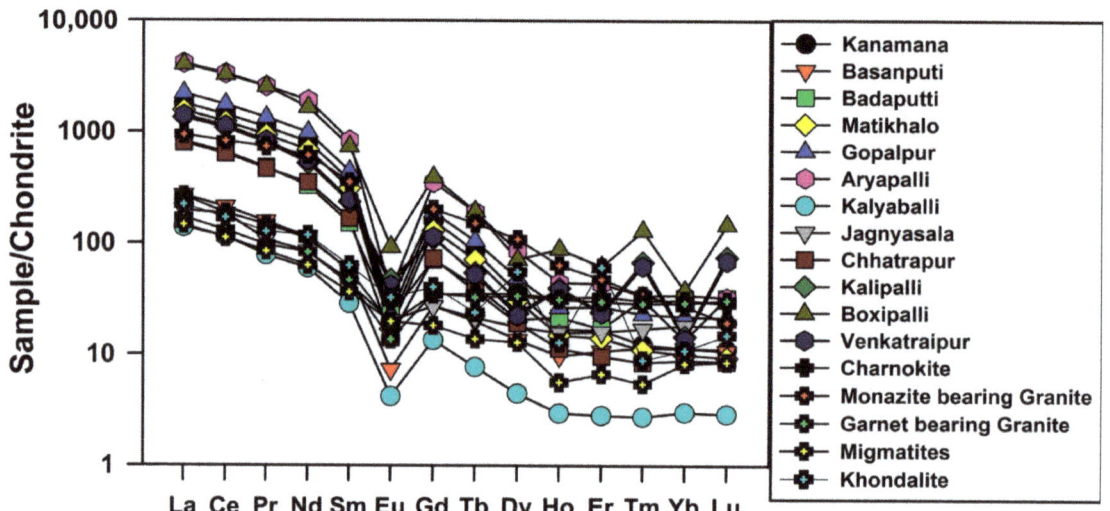

Figure 1. Chondrite normalized REE patterns of soils and hinterland rocks.

The europium (Eu_A) anomaly of the samples was estimated as follows:

$$Eu_A = \frac{Eu_N}{\sqrt{Sm_N \times Gd_N}} \quad (1)$$

here, Sm_N and Gd_N are the concentrations of samarium and gadolinium of the bulk soils normalized with respect to the chondrite value.

An Eu anomaly value equal to 1 indicates no anomaly. If the value is >1, there is a positive anomaly and if <1, there is a negative anomaly. All the samples had prominent, negative Eu anomalies (Figure 1). The Eu anomaly values of the soils ranged from 0.06 to 0.78. Similar observations in coastal sediments have been reported in the literature [39,40]. The negative Eu anomaly is a peculiar characteristic of felsic rocks, e.g., granite [41]. The soils had higher LREE/HREE ratios with a strong negative Eu anomaly, which suggested that the soils might have been derived from a felsic source.

The LREEs enrichment and positive correlation of Th in soils confirmed the presence of monazite mineral, and the relatively high concentration of Gd, Tb and Dy might be due to the presence of hornblende, pyroxene and garnet. To confirm the source rocks of the Chhatrapur–Gopalpur beach placer deposits, the REE patterns of various rock types present in the hinterland regions compared with soils are shown in Figure 1. The hinterland rocks comprised charnockite, khondalite, migmatite, monazite-bearing granite and garnet-bearing granite. The REE data on hinterland rocks were mainly granite [42], migmatite and charnockite [43] and khondalite [44]. The chondrite normalized REE patterns of charnockite, khondalite, granulite and granite were plotted to compare them with soil patterns. The obtained chondrite normalized REE patterns of soils were almost same as the chondrite normalized REE patterns of granite, migmatite, khondalite and charnockite.

Hence, granite, charnockite, and migmatite might be the major source rocks for monazite and other heavy minerals present in the soils.

2.6. Possible Source for TEs and REEs Enrichment in Soils

The TEs and REEs in soils were normalized with UCC and plotted in Figure 2. The UCC-normalized multielement diagram showed the enrichment of Mn, Fe, Co, Zn, Y, Pb, Th, U and REEs. Among these, Th and REEs are more enriched. Whereas the other elements were depleted compared to UCC values. The elements' enrichment values observed from the calculated EF, I_{geo}, and UCC normalized patterns were almost similar in the soils.

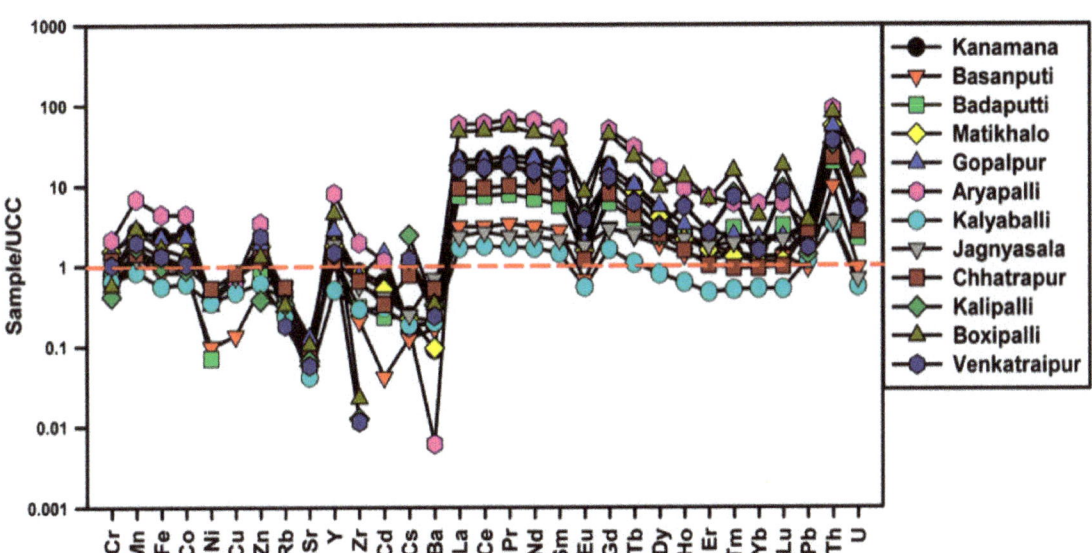

Figure 2. Plot showing UCC-normalized TEs and REEs patterns.

The EF results show the enrichment of Mn and Fe, which could be due to presence of a solid solution form of ilmenite $(Fe, Mn, Ti)O_3$. These minerals are manganiferous end members of the solid solution series [45]. The EF and I_{geo} results showed high enrichment of Th as well as high concentration of REEs, which could be assigned to the presence of monazite minerals in the soils. Therefore, it indicated that the enrichment of high Th, U and REEs are from natural origin and without involvement of any anthropogenic activities.

3. Materials and Methods

3.1. Study Area

The Chhatrapur–Gopalpur beach placer deposits are in the Ganjam district of Odisha, India. These areas extend 20 km length from Chhatrapur City in the north to Gopalpur Town in the south (19° 15′–19° 35′ N Lat; 84° 50′–85° 00′ E Long) with an average width of more than 2 km. A map showing the locations of sampling stations is given in Figure 3. The Bay of Bengal is on the south-eastern side of the study area, and the Eastern Ghats Mobile Belt (EGMB) is on the north and north-western sides. The main drainage system of this area is the Rushikulya River, which originates from the highlands of the EGMB and flows to the sea near Chhatrapur City. Many streams originate in the nearby coastal hills which are ephemeral in nature and could be major suppliers of sediments [46].

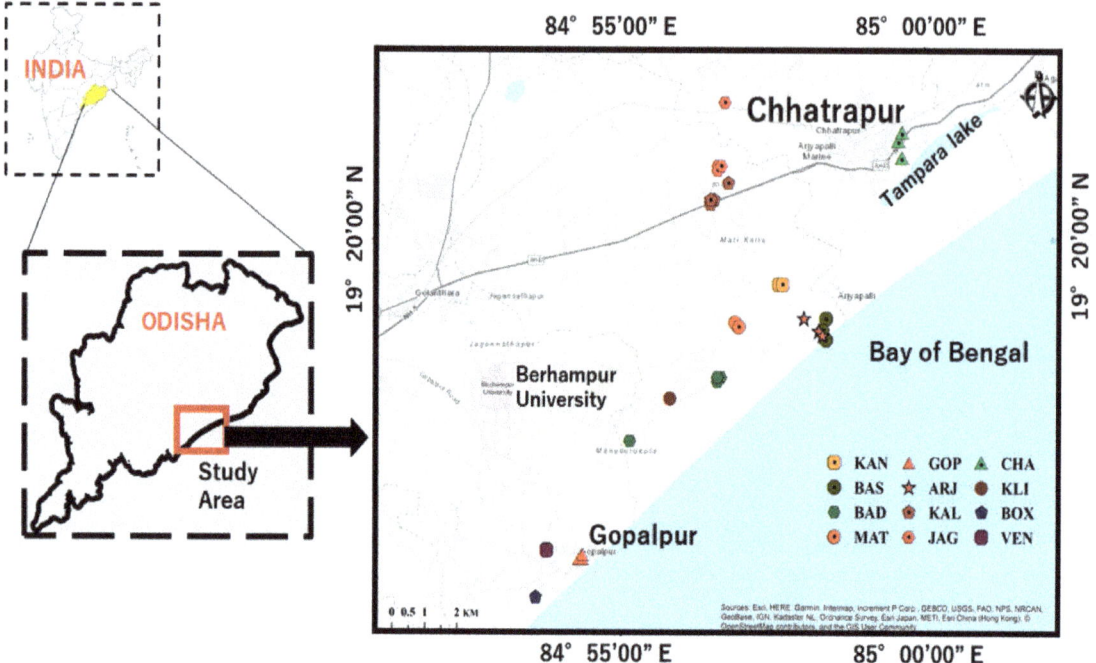

Figure 3. Map showing geographical locations of soils and ambient dose rates. (KAN—Kanamana; BAS—Basanaputti; BAD—Badaputti; MAT—Matikhalo; GOP—Gopalpur; ARJ—Aryapalli; KAL—Kalyaballi; JAG—Jagynasala; CHA—Chhatrapur; KLI—Kalipalli; BOX—Boxipalli; VEN—Venkatraipur). This map was prepared using Arc GIS 10.1. and enhanced using Coral draw software.

The Chhatrapur–Gopalpur beach placer deposits overlay high-grade granulite and intrusive rocks of the EMGB. The major litho-units of the EMGB are khondalite, charnockite and migmatite. The heavy minerals in the beach placers are ilmenite (39.01 mt), garnet (29.40 mt), sillimanite (17.91 mt), rutile (1.81 mt), zircon (1.33 mt) and monazite (1.13 mt) [11]. This study area has paleo dunes, sand bars, planted beach ridges, and red soils with heavy minerals [29].

3.2. Sampling and Sample Preparation

Soil samples were collected from a surface layer (0–10 cm depth) using a Daiki soil sampler. At each sampling point, five samples were taken from an area of about 1 m^2, and these samples were mixed to form a composite sample. Before collection, stones, grass, litter, roots, and shoots were removed from the surface layer. The sampling site selection was based on ambient dose rate, measured using a CsI (Tl) scintillation survey meter (PDR-101, Hitachi-Aloka Medical, Ltd., Tokyo, Japan). Three composite samples were obtained from each sampling location. Approximately 2 kg of each of the 36 composite samples were collected from corresponding 12 sampling locations of the study area. These were brought to the laboratory and air-dried at room temperature. After manually removing remaining roots, shoots, and stones, they were sieved using a 2 mm mesh sieve. The sieved samples were oven-dried at 110 °C for 24 h. Then, all samples were pulverized using a ball mill to less than 150 µm in size prior to chemical decomposition.

3.3. Measurement of Trace Elelements and REEs

About 250 mg of homogenized soil samples were ashed in a muffle furnace (KDF-S70, Kyoto, Japan) to decompose organic matter. In the furnace, temperature was increased sequentially as follows: 100 °C for 2 h, 200 °C for 3 h and 600 °C for 5 h. After that it was allowed to cool down for a further 7 h. The furnace-dried samples were chemically digested using a microwave (Milestone MLS 1200 Mega, Sorisole, B.G., Italy) in a closed PTFE pressure vessel with a mixture of concentrated HNO_3, HF, $HClO_4$ and HCl (Tama Pure Chemical Industries, Kawasaki, Japan). The microwave digestion was carried out in two steps. In step one, a mixture of concentrated HNO_3 (3 mL), HF (2 mL) and $HClO_4$ (0.5 mL) was added, and the digestion method was operated at a temperature of 80 °C and 600 W power for 2 h, including cooling time. In step two, a mixture of HNO_3 (3 mL) and HF (1 mL) was added and the method was similar to step one. The microwave-digested solution was followed by open digestion using aqua regia (HCl (3 mL): HNO_3 (1 mL)) at 200 °C for 2 h in a clean fume hood. After complete evaporation of aqua regia, the residue was dissolved in 10 mL of 6 M HCl and dried completely. Finally, the sample solution was prepared in 20 mL of 3% HNO_3. An experimental blank solution was also processed in the same way.

An internal standard Rh was spiked into each diluted sample to correct the signal attenuation due to the presence of various constituents in the samples (matrix effect) as well as for possible changes during ICP-MS measurement. The concentrations of TEs (Al, Cr, Mn, Fe, Co, Ni, Cu, Zn, Rb, Sr, Cd, Ba, Pb, Th and U) and REEs (Y, La, Ce, Pr, Nd, Sm, Eu, Gd, Tb, Dy, Ho, Er, Tm, Yb and Lu) in the decomposed samples were determined using an ICP-MS system (Agilent Technologies 8800 Triple Quad, Tokyo, Japan). The ICP-MS instrument was equipped with a MicroMist nebulizer and a Peltier-cooled (2 °C) Scott-type spray chamber for sample introduction. There was also an octopole-based collision/reaction cell, located between two quadrupole analyzers. The instrument was operated in a gas mode with He (flowing at 5 mL/min) to remove polyatomic ion interferences in case of multielement analysis. The analytical procedure has been described elsewhere [33]. The ICP-MS detection limit was calculated as three times the standard deviation of the calibration blank measurements ($n = 5$). The detection limits varied from (0.03 to 0.2) $\times 10^{-6}$ µg g^{-1} for all elements.

3.4. Pollution Indices

The pollution indices are an objective tool to assess the enrichment of elements in soils. The individual indices were used to obtain information on the level of soil pollution using each element's analysed data. The complex indices were used to determine the total pollution of an area. The simultaneous use of several indicators allows us to assess the pollution of soil with elements more accurately [47]. The pollution indices, namely enrichment factor (EF), and geoaccumulation index (I_{geo}), were used in the present study to evaluate the level of contamination in the soils.

In the present study, the EF was used to evaluate the influences of natural enrichment and anthropogenic contamination in the soils with respect to the reference sample in the study area. The EF was calculated using Equation (2).

$$EF_{(El)} = \frac{\frac{Conc(El)_{sample}}{Conc(X)_{sample}}}{\frac{Conc(El)_{Ref\ Sample}}{Conc(X)_{Ref\ Sample}}} \qquad (2)$$

here, "*El*" is the element under consideration, "*Conc*" is concentration (µg g^{-1}), and "*X*" stands for the reference element [48]. The subscripts "*sample*" and "*Ref. sample*" indicate their respective concentrations.

The normalized EF has been applied to differentiate element sources as anthropogenic or natural [49]. The TEs, Th, U, and Al average values of Jagnyasala samples (Table 2) are used as a reference sample for this calculation. In general, the EF was classified as unpolluted (EF < 2); moderate (2 < EF < 5); significant (5 < EF < 20); very high (20 < EF < 40), and extremely high (EF > 40). Soil samples' contamination level can be categorized based on the enrichment factor.

The I_{geo} was calculated using Equation (2), proposed by [50]. The I_{geo} classification was used to determine the level of contamination.

$$I_{geo} = Log_2[C_i/1.5B_i] \qquad (3)$$

here, C_i is the element concentration in soil, B_i is the geochemical background value of an element (average value of UCC) and 1.5 is the coefficient of variation attributed to natural rock.

The geochemical background values of Cr, Mn, Fe, Co, Ni, Cu, Zn, Rb, Sr, Y, Zr, Cd, Cs, Ba, La, Ce, Pr, Nd, Sm, Eu, Gd, Tb, Dy, Ho, Er, Tm, Yb, Lu, Pb, Th, and U are 92, 774, 79344, 17.3, 47, 28, 67, 84, 320, 21, 193, 0.09, 4.9, 624, 31, 63, 7.1, 27, 4.7, 1, 4, 0.7, 3.9, 0.83, 2.3, 0.3, 2, 0.31, 17, 10.5, and 2.7 $\mu g\ g^{-1}$, respectively [35]. There are seven classifications in this category. These are: uncontaminated ($I_{geo} \leq 0$; Class 0), uncontaminated to moderately contaminated (I_{geo} 0–1; Class 1), moderately contaminated (I_{geo} 1–2; Class 2), moderately to strongly contaminated (I_{geo} 2–3; Class 3), strongly contaminated (I_{geo} 3–4; Class 4), strongly to extremely contaminated (I_{geo} 4–5; Class 5), and extremely contaminated ($I_{geo} \geq 5$; Class 6). In this study, the contamination is considered as enrichment.

4. Conclusions

In this study, the concentration of TEs and REEs in Odisha beach placer-deposit soils were determined. EF values showed extreme enrichment of Th, U and significant enrichment of Cr, Mn, Fe, Co, Zn, Y, Zr, Cd and Cu. The extreme enrichment of Th was followed by U, Mn, Co, and Zn, Pb and Y, a slight enrichment was observed in the I_{geo} results. The enrichment of Mn, Fe, Co, Zn, Y, Pb, U, Th, and REEs was observed in the multielement diagram normalized with UCC values. The high concentrations of Fe and Mn were due to the presence of ilmenite heavy mineral, U was due to the presence of zircon, and the enrichment of LREEs and Th was due to the presence of monazite in the soils. Investigation of the REEs geochemistry revealed that the sources of monazite and other heavy minerals might have been derived from charnockite, migmatite, khondalite and granite rocks of the EGMB. The enrichment of elements in the soils is natural in origin. Consequently, the present data in this study will be used as a baseline for future monitoring of TEs and REEs levels in Chhatrapur–Gopalpur beach placer-deposits soils, where it is expected that substantial economic exploration into heavy minerals will occur in the coming decades.

Author Contributions: Conceptualization, N.V., S.K.S. and R.M.; Methodology, S.K.S., S.K. and N.V.; Investigation, R.M., N.V. and S.K.S.; Sample Collection, N.V., S.K.S. and K.I.; Data Curation, N.V., S.K., K.I. and S.K.S.; Writing—original draft preparation, N.V., R.M., S.K.S., S.K., K.I., M.F. and T.N.; Writing—review and editing, N.V., R.M., S.K.S., S.K., K.I., and T.N. All authors have read and agreed to the published version of the manuscript.

Funding: This work was supported partially by JSPS Core-to-Core Program (Grant number: JPJSCCB20210008).

Institutional Review Board Statement: Not applicable.

Informed Consent Statement: Not applicable.

Data Availability Statement: The data presented in this study are available on request from the corresponding author.

Acknowledgments: The authors thank Amulya Tripathy and Ashok Mohanty, Berhampur, Odisha, India for kind support during sample collection.

Conflicts of Interest: The authors declare no conflict of interest.

Sample Availability: Samples are available from the authors.

References

1. Govil, P.K.; Sorlie, J.E.; Murthy, N.N.; Sujatha, D.; Reddy, G.L.N.; Rudolph-Lund, K.; Krishna, A.K.; Mohan, K.R. Soil contamination of trace elements in the Katedan industrial development area, Hyderabad, India. *Environ. Monit. Assess.* **2008**, *140*, 313–323. [CrossRef]
2. Malik, R.N.; Jadoon, W.A.; Husain, S.Z. Metal contamination of surface soils of industrial city Sialkot, Pakistan: A multivariate and GIS approach. *Environ. Geochem. Health* **2010**, *32*, 179–191. [CrossRef] [PubMed]
3. Wang, L.; Liang, T. Geochemical fractions of rare earth elements in soil around a mine tailing in Baotou, China. *Sci. Rep.* **2015**, *5*, 12483. [CrossRef]
4. Hu, Z.; Haneklaus, S.; Sparovek, G.; Schnug, E. Rare earth elements in soils. *Commun. Soil Sci. Plant Anal.* **2006**, *37*, 1381–1420. [CrossRef]
5. Ramos, S.J.; Dinali, G.S.; Oliveira, C.; Martins, G.C.; Moreira, C.G.; Siqueira, J.O.; Guilherme, L.R.G. Rare earth elements in the soil environment. *Curr. Pollut. Rep.* **2016**, *2*, 28–50. [CrossRef]
6. Sá Paye, H.; Mello, J.W.V.; Mascarenhas, G.R.L.; Gasparon, M. Distribution and fractionation of the rare earth elements in Brazilian soils. *J. Geochem. Explor.* **2016**, *161*, 27–41. [CrossRef]
7. Sojka, M.; Choinski, A.; Ptak, M.; Siepak, M. Causes of variation of trace and rare earth elements concentration in lakes bottom sediments in the Bory Tucholskie National Park, Poland. *Sci. Rep.* **2021**, *11*, 244. [CrossRef] [PubMed]
8. Marques, R.; Prudêncio, M.I.; Rocha, F.; Cabral Pinto, M.M.S.; Silva, M.M.V.G.; Ferreira da Silva, E. REE and other trace and major elements in the topsoil layer of Santiago Island, Cape Verde. *J. Afr. Earth Sci.* **2012**, *64*, 20–33. [CrossRef]
9. Kowalska, J.B.; Mazurek, R.; Gąsiorek, M.; Zaleski, T. Pollution indices as useful tools for the comprehensive evaluation of the degree of soil contamination—A review. *Environ. Geochem. Health* **2018**, *40*, 2395–2420. [CrossRef]
10. Behera, P. Heavy minerals in beach sands of Gopalpur and Paradeep along Orissa coastline, east cost of India. *Indian J. Mar. Sci.* **2003**, *32*, 172–174.
11. Rao, N.S.; Misra, S. Sources of monazite sand in southern Orissa beach placer, eastern India. *J. Geol. Soc. India* **2009**, *74*, 357–362.
12. Anitha, J.K.; Joseph, S.; Rejith, R.G.; Sundararajan, M. Monazite chemistry and its distribution along the coast of Neendakara–Kayamkulam belt, Kerala, India. *SN Appl. Sci.* **2020**, *2*, 812. [CrossRef]
13. Withanage, A.P.; Mahawatte, P. Radioactivity of beach sand in the southwestern coast of Sri Lanka. *Radiat. Prot. Dosim.* **2013**, *153*, 384–389. [CrossRef]
14. Papadopoulos, A.; Christofides, G.; Koroneos, A.; Stoulos, S. Natural radioactivity distribution and gamma radiation exposure of beach sands from Sithonia Peninsula. *Cent. Eur. J. Geosci.* **2014**, *6*, 229–242. [CrossRef]
15. Caridi, F.; Marguccio, S.; Belvedere, A.; Belmusto, G.; Marcianò, G.; Sabatino, G.; Mottese, A. Natural radioactivity and elemental composition of beach sands in the Calabria region, south of Italy. *Environ. Earth Sci.* **2016**, *75*, 629. [CrossRef]
16. Shuaibu, H.K.; Khandaker, M.U.; Alrefae, T.; Bradley, D.A. Assessment of natural radioactivity and gamma-ray dose in monazite rich black Sand Beach of Penang Island, Malaysia. *Mar. Pollut. Bull.* **2017**, *119*, 423–428. [CrossRef]
17. Yasmin, S.; Barua, B.S.; Khandaker, M.U.; Kamal, M.; Rashid, M.A.; Sani, S.A.; Ahmed, H.; Nikouravan, B.; Bradley, D.A. The presence of radioactive materials in soil, sand and sediment samples of Potenga sea beach area, Chittagong, Bangladesh: Geological characteristics and environmental implication. *Results Phys.* **2018**, *8*, 1268–1274. [CrossRef]
18. Van Hao, D.; Dinh, C.N.; Jodłowski, P.; Kovacs, T. High-level natural radionuclides from the Mandena deposit, South Madagascar. *J. Radioanal. Nucl. Chem.* **2019**, *319*, 1331–1338. [CrossRef]
19. Radhakrishna, A.P.; Somasekarappa, H.M.; Narayana, Y.; Siddappa, K. A new natural background radiation area on the southwest coast of India. *Health Phys.* **1993**, *65*, 390–395. [CrossRef] [PubMed]
20. Paul, A.C.; Pillai, P.M.B.; Haridasan, P.P.; Radhakrishnan, S.; Krishnamony, S. Population exposure to airborne thorium at the high natural radiation areas in India. *J. Environ. Radioact.* **1998**, *40*, 251–259. [CrossRef]
21. Nair, M.K.; Nambi, K.S.V.; Amma, N.S.; Gangadharan, P.; Jayalekshmi, P.; Jayadevan, S.; Cherian, V.; Reghuram, K.N. Population study in the high natural background radiation area in Kerala, India. *Radiat. Res.* **1999**, *152*, 145–148. [CrossRef]
22. Kannan, V.; Rajan, M.P.; Iyngar, M.A.R.; Ramesh, R. Distribution of natural and anthropogenic radionuclides in soil and beach sand samples of Kalpakkam (India) using hyper pure germanium (HPGe) gamma ray spectrometry. *Appl. Radiat. Isot.* **2002**, *57*, 109–119. [CrossRef]
23. Singh, H.N.; Shanker, D.; Neelakandan, V.N.; Singh, V.P. Distribution patterns of natural radioactivity and delineation of anomalous radioactive zones using in situ radiation observations in southern Tamil Nadu, India. *J. Hazard. Mater.* **2007**, *141*, 264–272. [CrossRef] [PubMed]
24. Perumalsamy, C.; Bhadra, S.; Balakrishnan, S. Decoding evolutionary history of provenance from beach placer monazites: A case study from Kanyakumari coast, southwest India. *Chem. Geol.* **2016**, *427*, 83–97. [CrossRef]
25. Veerasamy, N.; Sahoo, S.K.; Inoue, K.; Arae, H.; Fukushi, M. Geochemical behaviour of uranium and thorium in sand and sandy soil samples from a natural high background radiation area of the Odisha coast, India. *Environ. Sci. Pollut. Res.* **2020**, *27*, 31339–31349. [CrossRef] [PubMed]
26. Mohanty, A.K.; Sengupta, D.; Das, S.K.; Saha, S.K.; Van, K.V. Natural radioactivity and radiation exposure in the high background area at Chhatrapur beach placer deposit of Orissa, India. *J. Environ. Radioact.* **2004**, *75*, 15–33. [CrossRef]

27. Sahoo, S.K.; Kierepko, R.; Sorimachi, A.; Omori, Y.; Ishikawa, T.; Tokonami, S.; Prasad, G.; Gusain, G.S.; Ramola, R.C. Natural radioactivity level and elemental composition of soil samples from a high background radiation area on eastern coast of India (Odisha). *Radiat. Prot. Dosim.* **2016**, *171*, 172–178. [CrossRef] [PubMed]
28. Inoue, K.; Sahoo, S.K.; Veerasamy, N.; Kasahara, S.; Fukushi, M. Distribution patterns of gamma radiation dose rate in the high background radiation area of Odisha, India. *J. Radioanal. Nucl. Chem.* **2020**, *324*, 1423–1434. [CrossRef]
29. Ghosal, S.; Singh, A.; Agrahari, S.; Sengupta, D. Delineation of heavy mineral bearing placers by electrical resistivity and radiometric techniques along coastal Odisha, India. *Pure Appl. Geophys.* **2020**, *177*, 4913–4923. [CrossRef]
30. Mohanty, A.K.; Das, S.K.; Van, K.V.; Sengupta, D.; Saha, S.K. Radiogenic heavy minerals in Chhatrapur beach placer deposit of Orissa, southeastern coast of India. *J. Radioanal. Nucl. Chem.* **2003**, *258*, 383–389. [CrossRef]
31. Imai, N.; Terashima, S.; Itoh, S.; Ando, A. 1994 compilation of analytical data for minor and trace elements in seventeen GSJ geochemical reference sample "igneous rock series". *Geostand. Newslett.* **1995**, *19*, 135–213. [CrossRef]
32. Imai, N.; Terashima, S.; Itoh, S.; Ando, A. 1996 compilation of analytical data on nine GSJ geochemical reference samples "sedimentary rock series". *Geostand. Newslett.* **1996**, *20*, 165–216. [CrossRef]
33. Kasar, S.; Murugan, R.; Arae, H.; Aono, T.; Sahoo, S.K. A microwave digestion technique for the analysis of rare earth elements, thorium and uranium in geochemical certified reference materials and soils by inductively coupled plasma mass spectrometry. *Molecules* **2020**, *25*, 5178. [CrossRef] [PubMed]
34. World Health Organization (WHO). *Permissible Limits of Trace Elements in Soil and Plants*; WHO: Geneva, Switzerland, 1996.
35. Rudnick, R.L.; Gao, S. The composition of the continental crust. In *Treatise on Geochemistry*; Rudnick, R.L., Ed.; Elsevier Science: New York, NY, USA, 2003; Volume 3, pp. 1–64.
36. Alfaro, R.M.; Nascimento, C.W.A.; Biondi, C.M.; Silva, Y.J.A.B.; Silva, Y.J.A.B.; Aguiar, A.M.; Montero, A.; Ugarte, O.M.; Estevez, J. Rare-earth-element geochemistry in soils developed in different geological settings of Cuba. *Catena* **2018**, *162*, 317–324. [CrossRef]
37. Nyakairu, G.W.A.; Koeberl, C. Mineralogical and chemical composition and distribution of rare earth elements in clay-rich sediments from central Uganda. *Geochem. J.* **2001**, *35*, 13–28. [CrossRef]
38. Masuda, A.; Nakamura, N.; Tanaka, T. Fine structures of mutually normalized rare-earth patterns of chondrites. *Geochim. Cosmochim. Acta* **1973**, *37*, 239–248. [CrossRef]
39. Armstrong-Altrin, J.S.; Machain-Castillo, M.L.; Rosales-Hoz, L.; Carranza-Edwards, A.; Sanchez-Cabeza, J.A.; Ruíz-Fernández, A.C. Provenance and depositional history of continental slope sediments in the Southwestern Gulf of Mexico unravelled by geochemical analysis. *Cont. Shelf Res.* **2015**, *95*, 15–26. [CrossRef]
40. Papadopoulos, A. Geochemistry and REE content of beach sands along the Atticocycladic coastal zone, Greece. *Geosci. J.* **2018**, *22*, 955–973. [CrossRef]
41. Rollinson, H.R. *Using Geochemical Data: Evaluation, Presentation, Interpretation*; Longman Scientific & Technical: Harlow, UK, 1993; pp. 133–140.
42. Narayana, B.L.; Rao, P.R.; Reddy, G.L.N.; Rao, V.D. Geochemistry and origin of megacrystic granitoid rocks from eastern ghats granulite belt. *Gondwana Res.* **1999**, *2*, 105–115. [CrossRef]
43. Bhadra, S.; Das, S.; Bhattacharya, A. Shear zone-hosted migmatites (Eastern India): The role of dynamic melting in the generation of REE-depleted felsic melts, and implications for disequilibrium melting. *J. Petrol.* **2007**, *48*, 435–457. [CrossRef]
44. Bhattacharya, S.; Chaudhary, A.K.; Basei, M. Original nature and source of khondalites in the Eastern Ghats Province, India. In *Geological Society*; Special Publications: London, UK, 2012; Volume 365, pp. 147–159.
45. Peter, T.S.; Chandrasekar, N.; Wilson, J.J.S.; Selvakumar, S.; Krishnakumar, S.; Magesh, N.S. A baseline record of trace elements concentration along the beach placer mining areas of Kanyakumari coast, South India. *Mar. Pollut. Bull.* **2017**, *119*, 416–422. [CrossRef] [PubMed]
46. Mohanty, A.K.; Das, S.K.; Vijayan, V.; Sengupta, D.; Saha, S.K. Geochemical studies of monazite sands of Chhatrapur beach placer deposit of Orissa, India by PIXE and EDXRF method. *Nucl. Instrum. Methods Phys. Res. B* **2003**, *211*, 145–154. [CrossRef]
47. Hołtra, A.; Zamorska-Wojdyła, D. The pollution indices of trace elements in soils and plants close to the copper and zinc smelting works in Poland's Lower Silesia. *Environ. Sci. Pollut. Res.* **2020**, *27*, 16086–16099. [CrossRef]
48. Reimann, C.; Caritat, P. Intrinsic flows of element enrichment factors (EFs) in environmental geochemistry. *Environ. Sci. Technol.* **2000**, *34*, 5084–5091. [CrossRef]
49. Zahra, A.; Hashmi, M.Z.; Malik, R.N.; Ahmed, Z. Enrichment and geo-accumulation of trace elements and risk assessment of sediments of the Kurang Nallah-feeding tributary of the Rawal Lake reservoir, Pakistan. *Sci. Total Environ.* **2014**, *470*, 925–933. [CrossRef] [PubMed]
50. Muller, G. Index of geo-accumulation in sediments of the Rhine River. *Geojournal* **1969**, *2*, 108–118.

Article

Evaluation and Assessment of Trivalent and Hexavalent Chromium on *Avena sativa* and Soil Enzymes

Edyta Boros-Lajszner, Jadwiga Wyszkowska * and Jan Kucharski

Department of Soil Science and Microbiology, University of Warmia and Mazury in Olsztyn, Plac Łódzki 3, 10-727 Olsztyn, Poland; edyta.boros@uwm.edu.pl (E.B.-L.); jan.kucharski@uwm.edu.pl (J.K.)
* Correspondence: jadwiga.wyszkowska@uwm.edu.pl

Abstract: Chromium (Cr) can exist in several oxidation states, but the two most stable forms—Cr(III) and Cr(VI)—have completely different biochemical characteristics. The aim of the present study was to evaluate how soil contamination with Cr(III) and Cr(VI) in the presence of Na$_2$EDTA affects *Avena sativa* L. biomass; assess the remediation capacity of *Avena sativa* L. based on its tolerance index, translocation factor, and chromium accumulation; and investigate how these chromium species affect the soil enzyme activity and physicochemical properties of soil. This study consisted of a pot experiment divided into two groups: non-amended and amended with Na$_2$EDTA. The Cr(III)- and Cr(VI)-contaminated soil samples were prepared in doses of 0, 5, 10, 20, and 40 mg Cr kg^{-1} d.m. soil. The negative effect of chromium manifested as a decreased biomass of *Avena sativa* L. (aboveground parts and roots). Cr(VI) proved to be more toxic than Cr(III). The tolerance indices (TI) showed that *Avena sativa* L. tolerates Cr(III) contamination better than Cr(VI) contamination. The translocation values for Cr(III) were much lower than for Cr(VI). *Avena sativa* L. proved to be of little use for the phytoextraction of chromium from soil. Dehydrogenases were the enzymes which were the most sensitive to soil contamination with Cr(III) and Cr(VI). Conversely, the catalase level was observed to be the least sensitive. Na$_2$EDTA exacerbated the negative effects of Cr(III) and Cr(VI) on the growth and development of *Avena sativa* L. and soil enzyme activity.

Keywords: Cr(III); Cr(VI); plant tolerance to chromium; chromium translocation in the plant; soil biochemical properties

Citation: Boros-Lajszner, E.; Wyszkowska, J.; Kucharski, J. Evaluation and Assessment of Trivalent and Hexavalent Chromium on *Avena sativa* and Soil Enzymes. *Molecules* **2023**, *28*, 4693. https://doi.org/10.3390/molecules28124693

Academic Editors: Stefan Tsakovski and Tony Venelinov

Received: 5 May 2023
Revised: 7 June 2023
Accepted: 8 June 2023
Published: 10 June 2023

Copyright: © 2023 by the authors. Licensee MDPI, Basel, Switzerland. This article is an open access article distributed under the terms and conditions of the Creative Commons Attribution (CC BY) license (https://creativecommons.org/licenses/by/4.0/).

1. Introduction

Chromium is one of the transition metals, and is found in group VI B of the periodic table. It occurs in several oxidation states, the most common and stable of which are Cr (III) and Cr (VI), which differ in their chemical characteristics [1–4]. Chromium (III) exists in the following species: Cr^{3+}, $Cr(OH)_2^+$, $Cr(OH)_3$, $Cr(OH)_4^-$, and $Cr(OH)_5^{2-}$, which may occur in soil or water. Chromium (III) readily combines with oxygen to form hydroxides, sulfates, and chelate organic bonds [5]. Cr(III) can also be oxidized into Cr(VI) in high-redox soils [3,6]. These properties of chromium (III) translate to a low mobility and make it significantly less bioavailable and toxic than chromium (VI) [7,8]. The primary chromium (VI) species are CrO_4^{2-}, $HCrO_4^-$, and $Cr_2O_7^{2-}$ anions, namely K_2CrO_4 and $K_2Cr_2O_7$ [1,9]. Chromium (VI) is a potent oxidant, and can be reduced to Cr(III) in the presence of organics. The more acidic the environment, the more quickly the reduction occurs [2]. Chromium (VI) is pathogenic to humans [10–12], animals [13], plants [14], and microorganisms [15]. Chromium was chosen for this study due to it being one of the most toxic metal pollutants [16,17].

In plants, the toxic effects of Cr manifest as delayed seed germination, root damage and reduced root growth, reduced biomass, reduced plant height, impaired photosynthesis, membrane damage, leaf chlorosis, necrosis, low grain yield, and, ultimately, plant death [18]. Chromium is a fairly active metal and readily reacts with environmental oxygen. Trivalent and hexavalent chromium are the most stable forms of Cr in nature. In addition, Cr(VI) exhibits higher toxicity than Cr(III) due to its higher solubility and mobility in the aqueous system [19]. Both valence states of Cr, i.e., Cr(III) and Cr(VI), are taken up by plants [10]. Cr(VI) is actively taken up into plant cells by sulfate carriers [20]. On the other hand, Cr(III) enters passively through plant cell wall cation exchange sites [21]. In addition, carboxylic acids which are present in root secretions facilitate the solubilization of Cr and, thus, its uptake by plants [22].

Chromium is released naturally in the environment through rock and soil erosion, as well as by volcanic eruptions [23,24]. Its anthropogenic sources include steelmaking, papermaking, textile manufacturing, fertilizer production, pesticide production, galvanization, tanning, pigment manufacturing, nuclear weapon production, and the electronic industry [7,24–28]. Global chromium production increased from 23.7 to 41 million tonnes during the period from 2010 to 2021. Leading chromium producers include South Africa, India, Kazakhstan, and China [29]. The total chromium emissions in the European Union amounted to 296 tonnes in 2019, of which Poland accounted for 36 tonnes [30]. Chromium—released into the atmosphere as fly ash from CHP plants and other industrial facilities—can settle on plants and soils around the emission source or be transported by wind over long distances (depending on the size of the particles), causing plant and soil pollution [24].

Phytoextraction is a technique used to effectively remove chromium from contaminated soils by harnessing hyperaccumulator plants, which can collect and accumulate heavy metals in their aboveground parts at levels 100 times higher than other plants [31,32]. Phytoextraction can be bolstered by amending the soil with chelating agents, which can desorb metals and increase their uptake through the roots of plants [33]. EDTA (ethylenediaminetetraacetic acid) is the most effective, most popular, and a relatively stable chelator [34–36]. An important application of EDTA is in fixing the ions of various metals, for example bismuth, chromium (III), zinc, zirconium, aluminum, cadmium, cobalt, magnesium, copper, nickel, lead, thorium, vanadium, and iron (III), by forming stable and soluble chelate complexes [37–40]. The chelation capacity of EDTA is strong enough to even form complexes with alkaline earth metals [41]. The most commonly used chemical compound in phytoextraction is the disodium salt of ethylenediaminetetraacetic acid—Na_2EDTA [42–44]. This substance, also known as Complexone III, can form chelate complexes with metal ions when dissolved in water [45]. Na_2EDTA has been the subject of pot experiments on induced phytoextraction [35,36,43,45]. Depending on the dosage, type of metal, species of plant, and characteristics of the soil, the effectiveness of Na_2EDTA for phytoextraction can vary considerably: from having no significant effect on metal uptake to an over 100-fold increase in phytoextraction capacity [35,36,43,45]. Na_2EDTA has non-specific chelating properties for heavy metals such as Cr, Pb, Cu, and Zn [43,45–47]. *Avena sativa* L. was selected in this study for its potential usefulness in the reclamation of heavy-metal contaminated soils [48]. Due to it having a high calorific value, its grain has also been used for energy purposes, mainly for heating, especially in Scandinavian countries, with Sweden being the primary user [49]. Oats are also often used for human and animal consumption, at least in Scandinavian countries. The results of this study are, therefore, also of relevance for uptake in humans/animals. Oat has also found many less conventional uses—it has been used as a component of cat litter and biodegradable plastics [50]. Therefore, determining the impact of growing plants on soils containing metal complexes with Na_2EDTA is a key area of research. This raises the question of what effect Na_2EDTA has on a crop such as *Avena sativa* L. and on biomass production, as well as on the biochemical and physicochemical properties of the soil in the presence and absence of Cr(III) and Cr(VI).

The aim of the present study was to evaluate how soil contamination with Cr(III) and Cr(VI) in the presence of Na2EDTA affects *Avena sativa* L. biomass, assess the remediation capacity of *Avena sativa* L. based on its tolerance index, translocation factor, and chromium accumulation, and investigate how these chromium species affect the soil enzyme activity and physicochemical properties of soil.

2. Results

2.1. Effect of Chromium on Avena sativa L. Growth and Development

Chromium phytotoxicity (expressed as the reduction in biomass yield) varied depending on the soil contamination with Cr, the oxidization state of Cr, and the Na$_2$EDTA amendment (Figure 1).

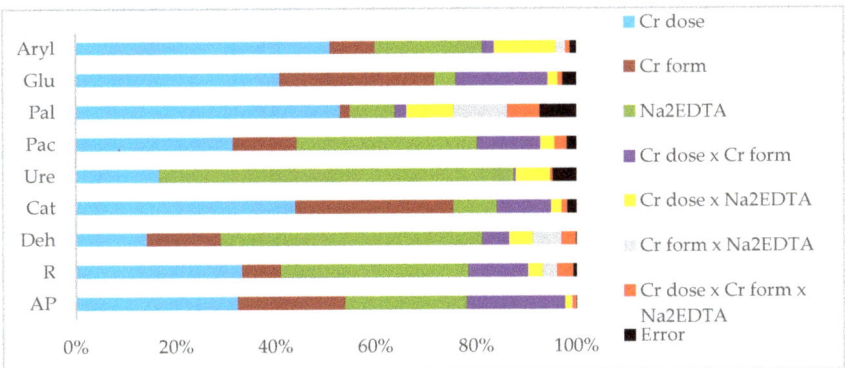

Figure 1. Percentage share of observed variability factors η^2. Explanations: AP—aboveground parts; R—roots; Deh—dehydrogenases; Cat—catalase; Ure—urease; Pac—acid phosphatase; Pal—alkaline phosphatase; Glu—β-glucosidase; Aryl—arylsulfatase.

Cr(III) and Cr(VI) stunted aboveground and root biomass growth in *Avena sativa* L. (Figure 2a,b). The aboveground biomass progressively diminished against the control as the levels of chromium (III) and chromium (VI) in the soil increased. The reduction was more pronounced in the Cr(VI)-contaminated soil than in the Cr(III)-contaminated soil samples. In sites with 40 mg Cr(VI) and Cr(III) kg^{-1} DM of soil, reductions in the biomass of the aboveground parts of *Avena sativa* L. were observed by 78% and 13%, respectively, compared to the uncontaminated sites. On the other hand, the reduction in biomass was higher for roots than for aboveground parts (Figure 1b). The greatest reduction in yield was recorded for Cr(VI) contamination. The root biomass in these objects decreased significantly by 75% compared to the control, while, for chromium (III), it decreased by 12%. Na$_2$EDTA, introduced into the soil, caused a reduction in the yield of *Avena sativa* L. (Figure 1a,b). In the series with chromium (VI), a dose of 40 mg Cr(VI) kg^{-1} caused the greatest reductions in the biomass of aboveground parts and roots, by 87% and 81%, respectively, compared to the uncontaminated sites.

The tolerance indices (TI) showed that *Avena sativa* L. was more tolerant to Cr(III) contamination than to Cr(VI) contamination. This was particularly noticeable for the highest chromium dose (40 mg kg^{-1}). In the no-Na$_2$EDTA group, the indices were: 0.871 (aerial parts) and 0.876 (roots) for Cr(III), and 0.224 and 0.254, respectively, for Cr(VI) (Figure 3). In the Na$_2$EDTA-amended group, the values were: 0.917 (aerial parts) and 0.574 (roots) for Cr(III), and 0.127 and 0.192, respectively, for Cr(VI).

Avena sativa L. (aboveground parts and roots) specimens exposed to Cr(VI) absorbed higher amounts of chromium than than those exposed to Cr(III) (Table 1). In the no-Na$_2$EDTA group, the aerial parts of *Avena sativa* L. which were grown on Cr(VI)-contaminated soil contained 6.21 mg kg^{-1} chromium, compared to the 1.66 mg kg^{-1} for Cr(III). The chromium levels in the roots were 45.40 and 41.30 mg kg^{-1}, respectively. In the Na$_2$EDTA-amended group, the Cr(VI)-contaminated specimens contained 16.30 (aboveground parts) and 86.80 (roots) chromium, compared to the 2.19 and 47.90 mg kg^{-1}, respectively, found in the Cr(III) runs. The Cr levels in the soil followed a similar pattern, with higher concentrations found in the Cr(VI)-contaminated soils than in the Cr(III) ones—61.60 and 43.30 mg kg^{-1}. Na$_2$EDTA induced higher levels of chromium in the soil.

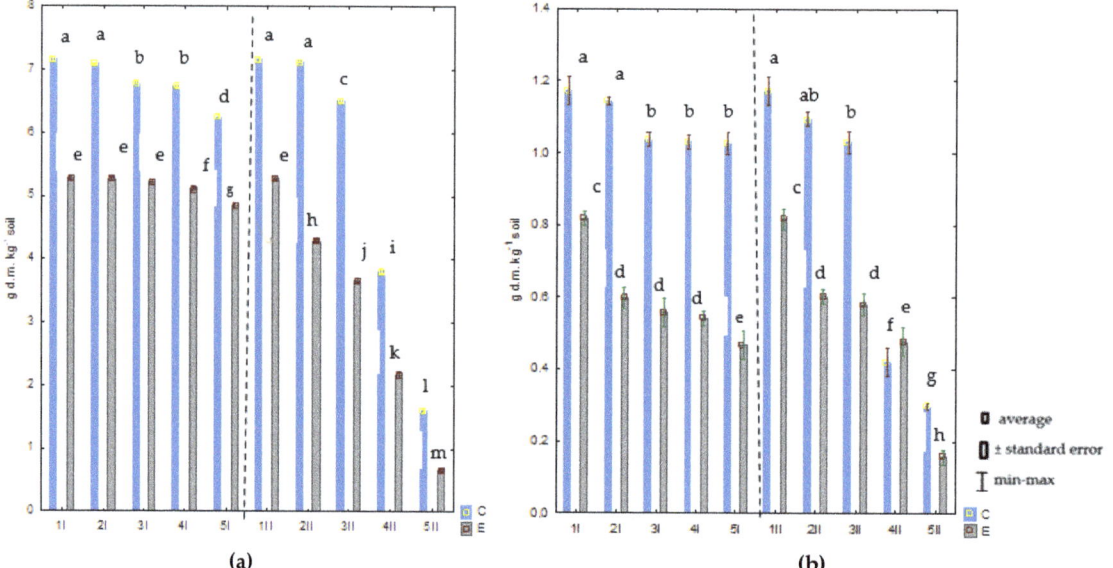

Figure 2. Yield of aboveground parts (**a**) and roots (**b**) of *Avena sativa* L. (g dm kg^{-1} soil) from soil contaminated with chromium (III) and (VI) with Na$_2$EDTA. Explanations: 1–0 mg Cr kg^{-1} of soil; 2–5 mg Cr kg^{-1} of soil; 3–10 mg Cr kg^{-1} of soil; 4–20 mg Cr kg^{-1} of soil; 5–40 mg Cr kg^{-1} of soil; I—Cr(III); II—Cr(VI); C—control, E—Na$_2$EDTA. Homogeneous groups (a–m) were created separately for aboveground parts and roots.

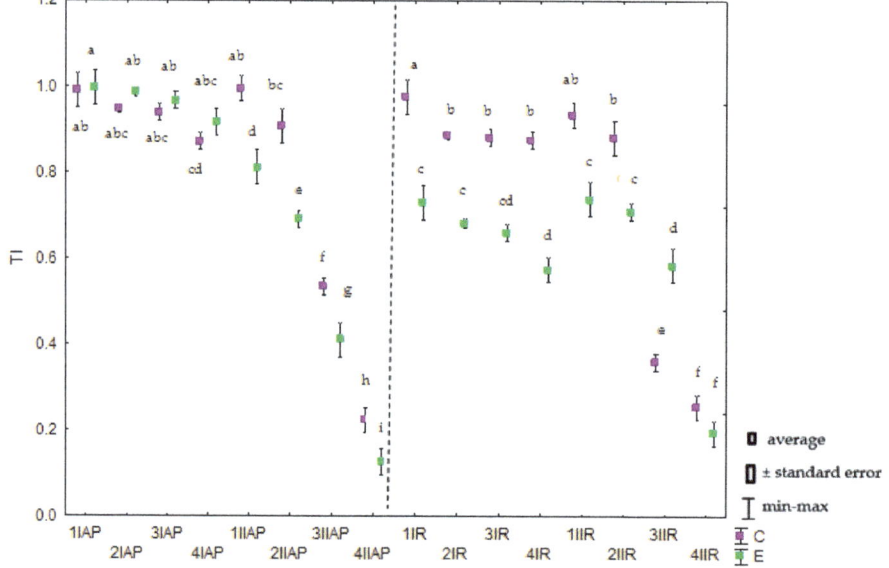

Figure 3. Tolerance index (TI) of *Avena sativa* L. to soil contamination with chromium (III) and (VI). Explanations: 1–0 mg Cr kg^{-1} of soil; 2–5 mg Cr kg^{-1} of soil; 3–10 mg Cr kg^{-1} of soil; 4–20 mg Cr kg^{-1} of soil; 5–40 mg Cr kg^{-1} of soil; I—Cr(III); II—Cr(VI); AP—aboveground parts; R—roots; C—control, E—Na$_2$EDTA. Homogeneous groups (a–i) were created separately for aboveground parts and roots.

Table 1. The content of chromium in *Avena sativa* L. and soil in mg kg^{-1} d.m.

Cr Dose mg kg^{-1} d.m. Soil	Aboveground Parts		Roots		Soil	
	Cr(III)	Cr(VI)	Cr(III)	Cr(VI)	Cr(III)	Cr(VI)
			Control			
0	1.22 [f]	1.34 [e]	12.90 [h]	25.30 [f]	19.30 [g]	20.10 [e]
40	1.66 [d]	6.21 [b]	41.30 [d]	45.40 [c]	43.30 [d]	61.50 [b]
			Na$_2$EDTA			
0	0.88 [g]	0.50 [g]	28.40 [e]	24.20 [g]	17.60 [h]	19.20 [f]
40	2.19 [c]	16.30 [a]	47.90 [b]	86.80 [a]	44.10 [c]	64.90 [a]

Homogeneous groups (a–h) were created separately for aboveground parts, roots, and soil.

There was a higher chromium content in the soil in the experimental series with Cr(VI) than that with Cr(III). This is due to the greater phytotoxic properties of Cr(VI) than Cr(III). This resulted in a lower uptake of chromium by *Avena sativa* L. from soil in the Cr(VI)-contaminated series than the Cr(III)-contaminated series.

Avena sativa L. absorbed more chromium from the soils contaminated with Cr(III) than those contaminated with Cr(VI) (Table 2). Chromium uptake was inhibited by the addition of Na$_2$EDTA. The metal mobility in *Avena sativa* L. was determined using the translocation factor (TF), which was calculated from the chromium levels in the aerial parts and roots (Table 2) The Na$_2$EDTA-amended group had 12% higher TF values for the Cr(III) plants and 27% higher TF values for the Cr(VI) plants compared to the no-Na$_2$EDTA group. The translocation values for Cr(III) were much lower than those for Cr(VI), though they below 1.0 in both cases.

Table 2. Uptake (D) chromium by *Avena sativa* L. and indices of translocation (TF) chromium, accumulation (AF), bioaccumulation index in aboveground parts (BF$_{AG}$), bioaccumulation index in roots (BF$_R$).

Cr dose mg kg^{-1} d.m. Soil	D µg kg^{-1}		TF		AF		BF$_{AG}$		BF$_R$	
	Cr(III)	Cr(VI)	Cr(III)	Cr(VI)	Cr(III)	Cr(VI)	Cr(III)	Cr(VI)	Cr(III)	Cr(VI)
				Control						
0	23.86 [f]	39.25 [b]	0.10 [c]	0.05 [d]	0.73 [f]	1.33 [b]	0.06 [b]	0.07 [b]	0.67 [f]	1.26 [c]
40	52.73 [a]	23.48 [g]	0.04 [e]	0.14 [b]	0.99 [d]	0.84 [e]	0.04 [b]	0.10 [b]	0.95 [e]	0.74 [f]
				Na$_2$EDTA						
0	27.86 [d]	22.42 [h]	0.03 [f]	0.02 [g]	1.66 [a]	1.29 [b]	0.05 [b]	0.03 [b]	1.61 [a]	1.26 [bc]
40	33.06 [c]	24.54 [e]	0.05 [de]	0.19 [a]	1.14 [c]	1.59 [a]	0.05 [b]	0.25 [a]	1.09 [d]	1.34 [b]

Homogeneous groups (a–h) were created separately for each coefficient.

The highest accumulation factor (AF) was observed for *Avena sativa* L. grown with Cr(VI) and Na$_2$EDTA, which reached 1.59 (Table 2). AF > 1 was also noted for plants exposed to Cr(III) and Cr(VI) with Na$_2$EDTA, as well as Cr(VI) without Na$_2$EDTA. Similarly, BF$_R$ > 1 was recorded for Cr(III)- and Cr(VI)-contaminated soil with Na$_2$EDTA, as well as for the no-Cr(VI)/no-Na$_2$EDTA specimens (Table 2). The aerial parts of *Avena sativa* L. showed very low levels of bioaccumulated chromium, whether with or without Na$_2$EDTA (Table 2). The highest BF$_{AG}$ (0.25) was observed for the Cr (VI) + Na2EDTA soil.

2.2. Effect of Chromium on Biochemical and Physicochemical Parameters of Soil

In our experiment, the chromium dose accounted for from 14% (dehydrogenases) to 51% (arylsulfatase) of the effect on the enzyme activity, the Cr oxidation state accounted for from 0% (urease) to 31% (β-glucosidase), and the Na2EDTA amendment accounted for from 4% (β-glucosidase) to 71% (urease) (Figure 1). The effect of soil contamination with chromium (III) and (VI) on soil enzyme activity was interpreted using principal component analysis (PCA) (Figure 4). The combined principal components account for 72.64% of the variation in original variables, of which PCA 1 accounted for 47.61%, and PCA 2 accounted for 25.03% (Figure 3). Two homogeneous groups formed around the principal components. The first group comprised catalase, arylsulfatase, β-glucosidase, and alkaline phosphatase vectors, whereas the second comprised acidic phosphatase, dehydrogenases, and urease. The vectors situated along the axes suggest that chromium (III) and (VI) had an adverse effect on soil enzyme activity. The soils that were uncontaminated with Cr(III) and Cr(VI) had the highest rates of enzyme activity, both in the Na_2EDTA and no-Na_2EDTA groups. The distribution of the data points relative to the vectors seems to indicate that added Na_2EDTA not only did not reduce chromium (III) and (VI)-induced stress, but actually exacerbated the adverse effect of Cr on soil enzyme activity.

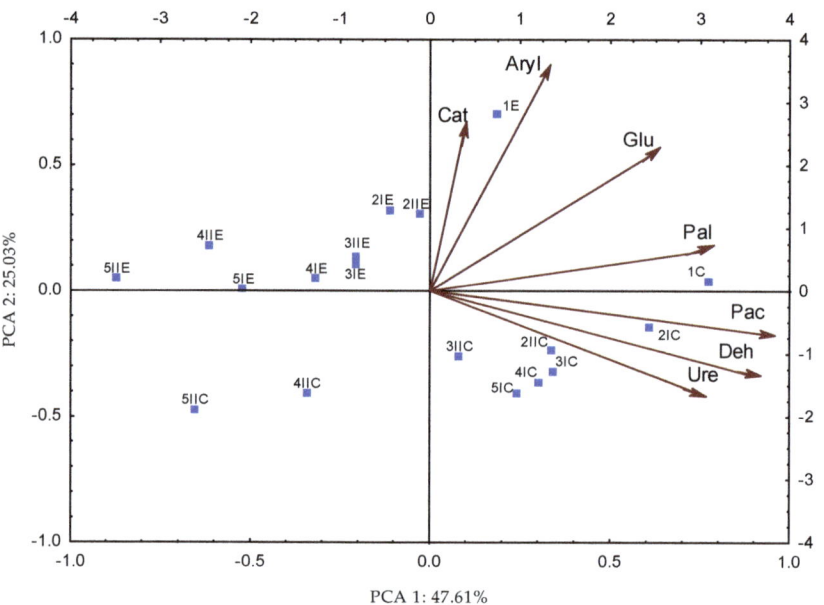

Figure 4. Activity of enzymes in soil contaminated with chromium (III) and (VI) with Na_2EDTA presented by the PCA method. Explanations: 1—0 mg Cr kg^{-1} of soil; 2—5 mg Cr kg^{-1} of soil; 3—10 mg Cr kg^{-1} of soil; 4—20 mg Cr kg^{-1} of soil; 5—40 mg Cr kg^{-1} of soil; I—Cr(III); II—Cr(VI); C—Control, E—soil with Na_2EDTA, Deh—dehydrogenases; Cat—catalase; Ure—urease; Pac—acid phosphatase; Pal—alkaline phosphatase; Glu—β-glucosidase; Aryl—arylsulfatase.

The values of the index of the chromium effects on soil enzyme activity (IF_{Cr}) confirm that chromium had an adverse effect on the biochemical characteristics of soil (Table 3).

Dehydrogenases were found to be the most sensitive to Cr(III) and Cr(VI), whereas catalase proved to be the most resistant. Cr(VI) had more of an inhibitory effect on the tested enzymes than Cr(III). No positive effect of Na2EDTA was observed for the tested enzymes.

Organic carbon content, total nitrogen content, pH, CEC, and BS were mostly unaffected by the chromium species which were tested, remaining fairly stable throughout the study period (Table 4). The hydrolytic acidity increased (except for Cr(VI)-contaminated sites with Na_2EDTA), and the sum of the base exchangeable cations decreased (except for the site with the highest Cr(III) dose in the series without Na_2EDTA) under the influence of applied chromium compounds. Soil amendment with Na2EDTA caused higher values of soil pH, but did not significantly alter the other parameters which were studied. In the Cr(III) specimens, the chromium (III) dose was significantly negatively correlated with the activity of catalase, alkaline phosphatase, β-glucosidase, and arylsulfatase, as well as organic carbon content, total nitrogen content, pH, and base saturation (Table 5). In the case of the chromium (VI) specimens, the Cr(VI) dose was significantly negatively correlated with *Avena sativa* L. yield (aerial parts and roots), the activity of all of the tested enzymes, the contents of C_{org} and N_{Total}, and the EBC and BS values (Table 6).

Table 3. Index of the effect of chromium (III) and (VI) with Na2EDTA on enzyme activity.

(a)

Cr Dose mg kg^{-1} d.m. Soil	Dehydrogenases		Catalase		Urease	
	Cr(III)	Cr(VI)	Cr(III)	Cr(VI)	Cr(III)	Cr(VI)
Control						
5	−0.01 [a]	−0.37 [i]	−0.02 [a]	−0.01 [a]	−0.04 [a]	−0.04 [a]
10	−0.15 [c]	−0.49 [j]	−0.03 [a]	−0.01 [a]	−0.03 [a]	−0.03 [a]
20	−0.16 [d]	−0.88 [o]	−0.04 [a]	−0.02 [a]	−0.03 [a]	−0.03 [a]
40	−0.17 [e]	−0.95 [p]	−0.04 [a]	−0.02 [a]	−0.04 [a]	−0.06 [ab]
$\bar{x}$	−0.12 [A]	−0.67 [C]	−0.03 [D]	−0.01 [B]	−0.04 [A]	−0.04 [A]
Na$_2$EDTA						
5	−0.10 [b]	−0.45 [k]	−0.01 [a]	−0.01 [a]	−0.08 [abc]	−0.07 [ab]
10	−0.23 [f]	−0.75 [l]	−0.02 [a]	−0.01 [a]	−0.08 [abc]	−0.04 [a]
20	−0.25 [g]	−0.78 [m]	−0.03 [a]	−0.01 [a]	−0.123 [bcd]	−0.16 [be]
40	−0.35 [h]	−0.83 [n]	−0.03 [a]	−0.01 [a]	−0.21 [e]	−0.20 [e]
$\bar{x}$	−0.23 [B]	−0.70 [C]	−0.02 [C]	−0.01 [A]	−0.12 [B]	−0.12 [B]

(b)

Cr dose mg kg^{-1} d.m. Soil	Acid Phosphatase		Alkaline Phosphatase		β-glucosidase		Arylsulfatase	
	Cr(III)	Cr(VI)	Cr(III)	Cr(VI)	Cr(III)	Cr(VI)	Cr(III)	Cr(VI)
Control								
5	−0.08 [abc]	−0.02 [ab]	−0.01 [bc]	−0.06 [cd]	−0.01 [a]	−0.06 [abc]	−0.01 [a]	−0.36 [c]
10	−0.11 [cd]	−0.18 [d]	−0.08 [cd]	−0.10 [cd]	−0.02 [a]	−0.06 [abc]	−0.09 [ab]	−0.37 [c]
20	−0.12 [cd]	−0.35 [e]	−0.09 [cd]	−0.11 [d]	−0.03 [a]	−0.16 [cd]	−0.13 [b]	−0.46 [cde]
40	−0.13 [cd]	−0.52 [f]	−0.11 [d]	−0.12 [d]	−0.03 [a]	−0.23 [d]	−0.14 [b]	−0.54 [e]
$\bar{x}$	−0.11 [B]	−0.27 [C]	−0.07 [C]	−0.10 [D]	−0.02 [A]	−0.13 [C]	−0.09 [A]	−0.43 [B]
Na$_2$EDTA								
5	−0.01 [a]	−0.10 [bcd]	−0.04 [bce]	0.07 [a]	−0.02 [a]	−0.06 [abc]	−0.37 [c]	−0.48 [de]
10	−0.02 [a]	−0.13 [cd]	−0.05 [cd]	0.04 [ab]	−0.05 [ab]	−0.15 [bce]	−0.41 [cd]	−0.50 [de]
20	−0.08 [abc]	−0.37 [e]	−0.06 [cd]	−0.02 [bc]	−0.05 [ab]	−0.16 [bce]	−0.45 [cde]	−0.51 [de]
40	−0.10 [bcd]	−0.37 [e]	−0.09 [cd]	−0.07 [cd]	−0.06 [abc]	−0.24 [d]	−0.50 [de]	−0.52 [e]
$\bar{x}$	−0.05 [A]	−0.24 [C]	−0.05 [B]	0.01 [A]	−0.05 [B]	−0.15 [D]	−0.43 [B]	−0.50 [C]

Homogeneous groups (a–n) were generated separately for each enzyme; homogeneous groups for means were calculated separately for each enzyme (A–D).

Table 4. Physicochemical properties of soil contaminated with chromium (III) and (VI) with Na$_2$EDTA.

Cr Dose mg kg^{-1} d.m. Soil	C$_{org}$		N$_{Total}$		pH$_{KCl}$		HAC		EBC (mmol$^{(+)}$ kg^{-1} Soil)		CEC		BS%	
	%													
	Cr(III)	Cr(VI)	Cr(III)	Cr(VI)	Cr(III)	Cr(VI)	Cr(III)	Cr(VI)	Cr(III)	Cr(VI)	Cr(III)	Cr(VI)	Cr(III)	Cr(VI)
Control														
0	0.72 a	0.72 a	0.13 ab	0.13 ab	6.10 c	6.10 c	9.19 g	9.19 g	33.00 e	33.00 e	42.19 e	42.19 e	78.21 ef	78.21 ef
5	0.70 a	0.65 cd	0.13 ab	0.12 bc	5.95 cd	5.95 cd	9.19 g	9.75 def	37.00 d	42.00 a	46.19 d	51.75 a	80.10 bcd	81.16 bc
10	0.66 bc	0.65 cd	0.13 ab	0.10 ef	5.95 cd	5.90 d	9.94 de	10.50 c	33.00 e	41.00 ab	42.94 e	51.50 a	76.85 fg	79.610 cde
20	0.60 e	0.64 cd	0.12 bcd	0.10 ef	5.85 de	5.90 d	10.13 cd	11.25 b	31.00 e	37.00 d	41.13 e	48.25 bcd	75.37 g	76.68 fg
40	0.60 e	0.64 d	0.12 bcd	0.09 f	5.70 e	5.90 d	10.50 c	11.44 b	40.00 abc	32.00 c	50.50 ab	43.44 e	79.21 de	73.67 h
x̄	0.66 A	0.66 A	0.12 A	0.11 B	5.91 B	5.95 B	9.79 C	10.43 B	37.00 B	39.20 A	46.79 B	49.63 A	78.85 B	78.77 B
r	−0.89	−0.63	−0.96	−0.90	−0.97	−0.64	0.91	0.90	−0.12	−0.99	−0.02	−0.99	−0.38	−0.98
Na$_2$EDTA														
0	0.71 a	0.71 a	0.14 a	0.14 a	6.70 a	6.70 a	9.56 efg	9.56 efg	40.00 abc	40.00 abc	49.56 abc	49.56 abc	80.71 bcd	80.71 bcd
5	0.687 b	0.68 b	0.14 a	0.11 cde	6.65 a	6.65 a	9.94 de	8.06 i	33.00 e	40.00 abc	42.94 e	48.06 cd	76.84 fg	83.23 a
10	0.68 b	0.68 b	0.13 ab	0.11 cde	6.60 a	6.65 a	11.63 ab	8.63 h	31.00 e	38.00 cd	42.63 e	46.63 d	72.72 h	81.50 b
20	0.67 b	0.67 b	0.13 ab	0.11 de	6.35 b	6.55 a	12.00 a	9.19 g	31.00 e	39.00 bcd	43.00 e	48.19 bcd	72.08 h	80.93 bc
40	0.67 b	0.66 bc	0.12 bc	0.11 de	5.95 cd	6.55 a	12.00 a	9.38 fg	31.00 e	32.00 e	43.00 e	41.38 e	72.08 h	77.34 f
x̄	0.68 A	0.68 A	0.13 A	0.12 B	6.45 A	6.62 A	11.03 A	8.55 D	33.20 C	37.80 B	44.23 C	46.35 B	74.89 C	81.44 A
r	−0.68	−0.84	−0.88	−0.60	−0.99	−0.88	0.79	0.89	−0.63	−0.92	−0.51	−0.84	−0.75	−0.99

C$_{org}$—total organic carbon; N$_{total}$—total nitrogen; HAC—hydrolytic acidity; EBC—total exchangeable cations; CEC—total exchange capacity of soil; BS—basic cations saturation ratio in soil. Homogeneous groups (a–i) were created separately for each parameter; homogeneous groups for means were calculated separately for each parameter (A–D); r—correlation coefficient significant at $p = 0.05$, $n = 16$.

Table 5. Coefficients of correlation between variables in soil contaminated with chromium (III).

Variable Factors	AP	R	Deh	Cat	Ure	Pac	Pal	Glu	Aryl	C$_{org}$	N$_{Total}$	pH	HAC	EBC	CEC	BS
Dose Cr	-0.26	-0.31	-0.15	-0.85 *	-0.35	-0.32	-0.51 *	-0.63 *	-0.65 *	-0.76 *	-0.72 *	-0.55 *	0.71 *	-0.24	-0.03	-0.56 *
AP	1.00	0.96 *	0.98 *	-0.04	0.88 *	0.96 *	0.58 *	-0.48 *	-0.27	-0.07	-0.21	-0.60 *	-0.72 *	0.18	-0.04	0.54 *
R		1.00	0.96 *	0.05	0.93 *	0.94 *	0.58 *	-0.38 *	-0.06	-0.07	-0.15	-0.55 *	-0.80 *	0.42 *	0.21	0.72 *
Deh			1.00	-0.11	0.88 *	0.96 *	0.55 *	-0.55 *	-0.32	-0.16	-0.31	-0.69 *	-0.69 *	0.27	0.07	0.57 *
Cat				1.00	0.05	0.07	0.17	0.77 *	0.69 *	0.72 *	0.67 *	0.65 *	-0.48 *	0.20	0.07	0.40 *
Ure					1.00	0.89 *	0.68 *	-0.33	-0.00	-0.12	-0.06	-0.42 *	-0.77 *	0.41 *	0.20	0.69 *
Pac						1.00	0.63 *	-0.41 *	-0.236	-0.03	-0.16	-0.54 *	-0.74 *	0.27	0.06	0.60 *
Pal							1.00	-0.04	0.08	0.34	0.19	-0.150	-0.59 *	0.25	0.09	0.49 *
Glu								1.00	0.82 *	0.60 *	0.70 *	0.86 *	-0.15	0.22	0.19	0.20
Aryl									1.00	0.47 *	0.63 *	0.68 *	-0.34	0.49 *	0.44 *	0.46 *
C$_{org}$										1.00	0.63 *	0.53 *	-0.26	0.01	-0.08	0.16
N$_{Total}$											1.00	0.69 *	-0.31	0.14	0.06	0.27
pH												1.00	0.07	-0.05	-0.03	-0.08
HAC													1.00	-0.50 *	-0.23	-0.88 *
EBC														1.00	0.96 *	0.85 *
CEC															1.00	0.67 *

C$_{org}$—total organic carbon, N$_{total}$—total nitrogen, HAC—hydrolytic acidity, EBC—total exchangeable cations, CEC—total exchange capacity of soil, BS—basic cations saturation ratio in soil; AP—yield aboveground parts; R—yield roots; Deh—dehydrogenases; Cat—catalase; Ure—urease; Pac—acid phosphatase; Pal—alkaline phosphatase; Glu—β-glucosidase; Aryl—arylsulfatase; * r—coefficient of correlation significant at: $p = 0.05$, $n = 30$.

Table 6. Coefficients of correlation between variables in soil contaminated with chromium (VI).

Variable Factors	AP	R	Deh	Cat	Ure	Pac	Pal	Glu	Aryl	C$_{org}$	N$_{Total}$	pH	HAC	EBC	CEC	BS
Dose Cr	−0.84 *	−0.82 *	−0.68 *	−0.55 *	−0.43 *	−0.79 *	−0.64 *	−0.94 *	−0.64 *	−0.71 *	−0.84 *	−0.18	0.46 *	−0.47 *	−0.35	−0.59 *
AP	1.00	0.97 *	0.86 *	0.21	0.73 *	0.96 *	0.51 *	0.74 *	0.21	0.33	0.58 *	−0.33	−0.10	0.52 *	0.51 *	0.39 *
R		1.00	0.89 *	0.27	0.65 *	0.94 *	0.51 *	0.75 *	0.24	0.40 *	0.61 *	−0.28	−0.15	0.48 *	0.45 *	0.39 *
Deh			1.00	0.22	0.64 *	0.91 *	0.45 *	0.55 *	0.12	0.43 *	0.54 *	−0.42 *	−0.06	0.09	0.08	0.09
Cat				1.00	−0.25	0.26	0.54 *	0.59 *	0.62 *	0.68 *	0.61 *	0.65 *	−0.69 *	0.14	−0.06	0.54 *
Ure					1.00	0.71 *	0.28	0.26	−0.15	−0.02	0.13	−0.73 *	0.49 *	0.12	0.27	−0.24
Pac						1.00	0.48 *	0.65 *	0.16	0.33	0.60 *	−0.36 *	−0.09	0.37 *	0.35	0.28
Pal							1.00	0.62 *	0.22	0.54 *	0.41 *	0.17	−0.48 *	0.20	0.06	0.44 *
Glu								1.00	0.71 *	0.71 *	0.82 *	0.32	−0.51 *	0.54	0.41 *	0.67 *
Aryl									1.00	0.78 *	0.82 *	0.66 *	−0.46 *	0.19	0.06	0.42 *
C$_{org}$										1.00	0.75 *	0.52 *	−0.60 *	−0.04	−0.22	0.37 *
N$_{Total}$											1.00	0.37 *	−0.47 *	0.29	0.16	0.48 *
pH												1.00	−0.78 *	0.11	−0.12	0.57 *
HAC													1.00	−0.25	0.04	−0.80
EBC														1.00	0.96 *	0.77 *
CEC															1.00	0.56 *

C$_{org}$—total organic carbon, N$_{total}$—total nitrogen, HAC—hydrolytic acidity, EBC—total exchangeable cations, CEC—total exchange capacity of soil, BS—basic cations saturation ratio in soil; AP—yield aboveground parts; R—yield roots; Deh—dehydrogenases; Cat—catalase; Ure—urease; Pac—acid phosphatase; Pal—alkaline phosphatase; Glu—β-glucosidase; Aryl—arylsulfatase; * r—coefficient of correlation significant at: $p = 0.05$, $n = 30$.

3. Discussion

3.1. Effect of Chromium on Avena sativa L. Growth and Development

Our study found that Cr(III) and Cr(VI) did not disrupt *Avena sativa* L. growth and development at doses of 5 mg kg^{-1} soil, but did result in diminished aerial and root biomass at levels from 10 to 40 mg kg^{-1} d.m. soil. The inhibitory effect of chromium on *Avena sativa* L. was determined by the oxidation state. The 40 mg Cr(VI) and Cr(III) kg^{-1} d.m. soil runs showed a diminished aerial biomass of *Avena sativa* L.—that was 78% and 13% lower than in the non-contaminated specimens, respectively. The decrease for the roots was 75% for Cr(VI) and 12% for Cr (III). Cr(III) is less toxic due to its extremely low solubility, which prevents it from entering groundwater or being taken up by plants. Cervantes et al. [51] found that a Cr(III) dose of 100 mg kg^{-1} caused a 40% reduction in the growth of the aerial parts of barley, whereas Cr(VI) reduced growth by 75% (aerial parts) and 90% (roots). Another study, by Wyszkowska et al. [4], showed that Cr(VI), at 60 mg kg^{-1}, reduced the aboveground biomass of *Zea mays* by 90% and the root biomass by 92%. This significant negative effect of Cr (VI) on plants—which also emerged in our study—is caused by disrupted water management, manifested by the wilting and chlorosis of young leaves [47,52]. The reduced biomass is due to the toxic effect of Cr(VI) on photosynthesis and the hindered water/nutrient transport from the soil [53,54]. Stunted root growth may be attributed to the inhibition of root proliferation and elongation, preventing roots from absorbing water and nutrients from the soil [47,55]. Plants take up Cr(III) through a passive mechanism via diffusion at the cell wall cation exchange site [56]. Cr(VI) has structural similarity to phosphate and sulfate, so its uptake occurs through an active process via phosphate and sulfate transporters [57]. The active transport of Cr(VI) results in its immediate conversion to Cr(III) in roots through the action of iron reductase enzymes [58]. This converted Cr(III) binds to the cell wall, thus inhibiting its further transport to the aboveground parts of the plant [59]. In our study, the lower tolerance index (TI) values indicate that Cr(VI) was more toxic to *Avena sativa* L. than Cr(III), and this was further exacerbated by Na$_2$EDTA. This is corroborated by Bareen et al. [60], who demonstrated an intensified phytotoxic effect on *Sorghum bicolor* and *Pennisetum glaucum* specimens treated with both Na$_2$EDTA and Cr(VI). The detrimental effects of Na$_2$EDTA may be caused by an impaired uptake of essential nutrients, such as Zn^{2+} and Ca^{2+}, which in turn decreases cell wall elasticity and viscosity, hampers cell division, disrupts transpiration, and damages cell membranes [61,62]. In the present study, chromium accumulation in the aerial parts and roots was found to be higher in the runs of soil samples contaminated with this metal. According to Rai et al. [53], the chromium concentration in plants will vary depending on the plant species, the metal dose, and the duration of the experiment. For example, *Zea mays* grown on soil contaminated with 10 and 20 mg Cr(VI) kg^{-1} for 30 days contained 15.2 and 16.3 mg Cr kg^{-1} in its aerial parts, respectively [62]. Wyszkowska et al. [4] reported chromium concentrations from 0.66 to 1.04 mg kg^{-1} in the aerial parts of *Zea mays*, and from 1.23 to 17.67 mg kg^{-1} in the roots, when applying doses of 60 mg Cr(VI) kg^{-1}. *Cicer arietinum* L. grown on soil contaminated with from 25 to 75 mg Cr(VI) per kg^{-1} soil has been shown to accumulate from 0.0002 to 0.0001 mg chromium kg^{-1} in its roots and from 0.0009 to 0.0005 mg kg^{-1} in its aerial parts [55]. Similarly, Cr accumulation between 0.01 and 0.03 mgkg^{-1} has been demonstrated in *Oryza sativa* L. that has been exposed to 2.5–200 mg kg^{-1} Cr(VI) [63]. In the present study, the aerial parts of *Avena sativa* L. grown on soil contaminated with 40 mg Cr kg^{-1} contained 1.66 mg Cr(III) and 6.21 mg Cr(VI) per kg^{-1}, whereas the roots contained 41.30 and 45.40 mg kg^{-1}, respectively. The greater accumulation of chromium in *Avena sativa* L. roots, as observed in our study, may be attributed to the reduced transport of chromium from the root to the aerial parts of the plant. The plants may immobilize chromium by compartmentalizing it into the vacuoles, or storing it in the cation exchange sites of the xylem parenchyma cells—a natural defense strategy against metal toxicity [64]. Some smaller proteins act as natural chelates, binding as cations to Cr ions and inhibiting Cr transport [55]. The higher chromium accumulation in the roots may also be explained by the reduction of Cr(VI) to the poorly-soluble Cr(III) [53].

The effect of chromium on *Avena sativa* L. is particularly well demonstrated by the values of the bioaccumulation factors for the aerial parts (BF_{AG}) and the roots (BF_R), as well as by the translocation factor (TF). Bioaccumulation is the ability of plants to neutralize toxic metals into non-toxic or less toxic forms in different plant organs [65]. The BF_{AG} value for *Avena sativa* L. was found to be < 1.0, making the plant a poor candidate for chromium (III) and (VI) phytoextraction. The TF index values were also lower than 1, suggesting that *Avena sativa* L. has a limited capacity to transport chromium from the root to the aerial parts [66,67].

The present study also showed that the addition of Na_2EDTA increased chromium levels and accumulation in *Avena sativa* L. Na_2EDTA can bind to chromium to form a Cr-Na_2EDTA complex or, alternatively, can increase the concentration of the soluble and exchangeable form of Cr by lowering soil pH, thereby increasing bioavailability and promoting transport [68]. In our study, the TF values were higher in Cr(VI)-contaminated samples, and were increased by Na_2EDTA. Han et al. [69] and Ebrahimi et al. [70] also noted increased Cr accumulation and translocation values in their Cr(III)-contaminated samples, with Na_2EDTA amendment leading to their further increases in *Phragmites australis* (Cav.) and *Brassica juncea*.

3.2. Effect of Chromium on Biochemical and Physicochemical Parameters of Soil

Soil enzymes are synthesized by microorganisms and act as biological catalysts which are involved in metabolic processes that break down organic matter. They reflect the microbial activity in the soil and serve as indicators of metabolic capacity trends in an environment [71,72]. Enzyme tests are considered to be one of the cheapest and easiest techniques for quantifying soil contamination [73–75]. The reduction of soluble Cr (VI) to insoluble Cr (III) occurs only in the surface layer of aggregates with higher available organic carbon and higher microbial respiration [76,77]. Therefore, spatial biochemical and microbiological measurements within soil aggregates are needed to characterize and predict the fate of chromium contamination [76]. Soil enzymatic activity is highly sensitive to both natural and anthropogenic disturbances and shows a rapid response to induced changes. Therefore, enzyme activity can be considered an effective indicator of changes in soil quality resulting from environmental stress [78]. The present study found that 5 mg kg^{-1} was the only dose of Cr(III) and Cr(VI) that did not affect enzyme activity to a significant degree. Higher levels of the two metal species inhibited the activity of dehydrogenases, catalase, acidic phosphatase, β-glucosidase, and arylsulfatase, with Cr(VI) being the stronger inhibitor of the two metal species which were tested. Similar findings were reported by Wyszkowska [79], who demonstrated suppressed activity of dehydrogenases, acidic phosphatase, and alkaline phosphatase after exposure to chromium (VI). The results for urease activity were less clear-cut, with Cr(VI) having a stimulating effect at 10 to 40 mg Cr(VI) kg^{-1} soil and an inhibitory effect at the higher doses of 50, 100, and 150 mg Cr(VI) kg^{-1}. Of the enzymes that were analyzed, dehydrogenases were found to be the most sensitive to soil contamination with chromium. Dehydrogenases, being intracellular enzymes, occur exclusively in living cells, and the release of heavy metals (including chromium) into the soil can reduce the abundance and activity of reducing/oxidizing microbes [72,80]. Studies by Huang et al. [72] and Peng et al. [81] also demonstrated this sensitivity of dehydrogenases to chromium pollutants. Conversely, catalase proved to be the least sensitive to Cr(III) and Cr(VI) contamination. This is probably due to a reaction between the metal ion in the soil and the functional group of catalase [82,83]. Our findings are corroborated by the results reported by Samborska et al. [84], Al-Khashman and Shawabkeh [85], and Schulin [86]. The chromium-induced inhibition of enzymes may be due to the interaction with the enzyme substrate, denaturation of the enzyme protein, and interaction with its active components [73]. Na_2EDTA did not eliminate the damaging effects of Cr(III) and Cr(VI) on soil biochemistry—rather, it actually intensified them. The low effectiveness of Na_2EDTA against chromium may stem from the fact that anionic Cr prevents the formation of a stable complex with Na_2EDTA. Na_2EDTA is considered to be the most effective synthetic chelator for the removal of cationic metals, but less so for anionic metals [44,87]. A study by Mahmood-ul-Hassan et al. [88] showed that Cr concentrations were significantly higher in soils enriched with Na_2EDTA than in soils without its addition. Komárek et al. [89] showed a correlation between soluble Cr concentrations with Na_2EDTA addition and indicated that dissolved metals persist in contaminated soil even after crop harvest. The use of Na_2EDTA is problematic, not only because of the excessive mobilization of metals compared to uptake by plants [90], but also because metal complexes with Na_2EDTA persist for a long time [91], hence the risk of excessive leaching of soluble metals to deeper depths, which can result in the contamination of shallow groundwater [90].

4. Materials and Methods

4.1. Soil Preparation

Soil samples were taken from the surface layer (0–20 cm deep) in Tomaszkowo near Olsztyn, Warmińsko-Mazurskie Voivodeship, Poland (53.7161° N, 20.4167° E). The soil was crumbled and air-dried, then passed through a 0.5 mm mesh sieve. The choice of soil was primarily dictated by the fact that Poland—which lies in the Central European zone of the subboreal belt and has a temperate climate with oceanic influence—is dominated by zonal soils. These include brown soils, which account for approximately 52% of the country's area, forming on clay and loam [92]. Prior to the experiment, the soil was analyzed for particle size distribution and basic physicochemical properties (Table 7).

Table 7. Some physicochemical properties of the soil used in the experiment.

Type of Soil	Granulometric Composition (%)			pH_{KCl}	C_{org}	N_{total}	HAC	EBC	CEC	BS%
	Sand	Silt	Clay		g kg^{-1}		mmol$^{(+)}$ kg^{-1} Soil			
ls	69.41	27.71	2.88	6.09	6.18	1.27	8.81	24.00	32.81	73.14

ls—sandy loam, C_{org}—total organic carbon, N_{total}—total nitrogen, HAC—hydrolytic acidity, EBC—total exchangeable cations, CEC—total exchange capacity of soil, BS—basic cations saturation ratio in soil.

4.2. Experimental Procedure

The experiment was conducted in 3.5 kg plastic pots in a greenhouse and consisted of 20 runs in four replications each. The experiment was divided into two groups: non-amended and amended with 1.5 g Na_2EDTA (di-Sodium versenate dihydrate pure p. a., producer POL-AURA, Morąg, Poland) per kg^{-1} soil. For each run, 3.5 kg soil was weighed and contaminated with (depending on the run): Cr(III) as $KCr(SO_4)_2 \cdot 12H_2O$ and Cr(VI) as $K_2Cr_2O_7$ at 5, 10, 20, and 40 mg Cr kg^{-1}. Soils uncontaminated with Cr(III) and Cr(VI) served as the control. In 2015, chromium was classified as one of the six pollutants which are highly dangerous to human health [93]. Na_2EDTA input was set based on Zou et al. [43] and Neugschwentner et al. [45]. To provide optimal conditions for *Avena sativa* L. growth and development, all pots were fertilized with the following macro-nutrients: N—140 mg [$CO(NH_2)_2$], P—60 mg [KH_2PO_4], K—120 mg [KH_2PO_4+KCl], and Mg—20 mg [$MgSO_4 \cdot 7H_2O$]. All components (Cr(III) and Cr(VI), Na_2EDTA, and the fertilizers) were thoroughly mixed with the soil and brought to a moisture content of 50% capillary water capacity. The thus-prepared soil was then potted and sown with the *Avena sativa* L. cultivar 'Bingo' (12 plants per pot). Day time ranged from 13 h, 5 min to 16 h, 51 min. The average air temperature was 16.6 C and air humidity was 77.5%. The experiment lasted for 60 days.

4.3. Assessment of Plant Growth Performance

Once *Avena sativa* L. was harvested (BBCH 61—beginning of flowering), the dry mass yield of aboveground parts and roots was measured. Chromium was quantified in the aerial (aboveground) parts and roots with ICP-OES (N) (inductively coupled plasma optical emission spectrometry) in Thermo Scientific iCAP 7400 Duo with a TELEDYNE CETAC ASX-560 autosampler (Thermo Scientific, Waltham, MA, USA) according to PN-

ISO-11466:2002 [94], after microwave mineralization with 3:1 concentrated nitric acid (V)/hydrogen peroxide.

4.4. Biochemical Determinations

Once *Avena sativa* L. was harvested, the soil samples (passed through a 2 mm mesh sieve) were tested for the activity of dehydrogenases [EC 1.1] (according to the procedure provided by Öhlinger [95]), as well as catalase [EC1.11.1.6], urease [EC 3.5.1.5], acid phosphatase [EC 3.1.3.2], alkaline phosphatase [EC 3.1.3.1], β-glucosidase [EC 3.2.1.21], and arylsulfatase [EC 3.1.6.1] (according to Alef and Nannpieri [96]). Extinction of enzymatic reaction products was measured by a PerkinElmer Lambda 25 spectrophotometer (Peabody, MA, USA). Biochemical determinations were performed in triplicate. The protocol used for the enzyme activity assay is detailed in Zaborowska et al. [97] and Borowik et al. [98].

4.5. Physicochemical and Chemical Tests

The soil samples were tested for soil pH hydrolytic acidity (HAC), sum of exchangeable base cations (EBC), organic carbon (C_{org}), total nitrogen (N_{total}), total cation-exchange capacity (CEC), and base saturation (BS). The test protocol is provided in our previous publications [99,100]. Chromium content of the soil was assayed in non-contaminated pots and those contaminated with 40 mg Cr per kg^{-1} dm soil, after microwave mineralization in an extract of 1:3 concentrated nitric acid (V)/concentrated hydrochloric acid (aqua regia). The assay was done by means of ICP-OES according to PN-ISO 11047:2001(A) [101].

4.6. Calculations and Statistics

Chromium uptake, tolerance index, translocation factor, bioaccumulation factors, and accumulation factor were calculated from *Avena sativa* L. biomass (aboveground parts and roots) and the plant/soil levels of chromium. Index of chromium effect on soil enzyme activity was also calculated. The index computation methods are detailed in our previous papers [4,102].

The results were statistically processed by analysis of variance (ANOVA) at $p \leq 0.05$, using STATISTICA 13.1 [103]. Homogeneous groups were generated using Tukey's test for the following variables: yield of *Avena sativa* L. (aboveground parts and roots), Cr(III) and Cr(VI) in plants and soil, and indices of phytoremediation capacity. Applying multivariate exploratory techniques using Statistica 13.1 software [84], enzyme activity in soil contaminated with Cr(III) and Cr(VI) and with the addition of Na2EDTA was analyzed using principal component analysis—PCA. In turn, the analysis of variance (ANOVA) was used to calculate the coefficient of variation (%) for all considered variables (η^2). The Pearson linear correlation coefficient was also calculated for the variables.

5. Conclusions

Chromium(VI) caused a greater reduction in the aerial and root biomass of *Avena sativa* L. compared with Cr(III). The tolerance indices (TI) showed that *Avena sativa* L. was observed to be more tolerant to Cr(III) contamination than Cr(VI) contamination. The translocation value which was recorded for Cr(III) was much lower than for Cr(VI), though it was at TF < 1 in both cases. Judging by the BF_{AG} < 1, the species does not seem to be suited for chromium (III) and (VI) phytoextraction. Dehydrogenases were found to be the enzymes which were the most sensitive to soil contamination with Cr(III) and Cr(VI). Conversely, catalase was the least sensitive. At 5 mg kg^{-1}, the two chromium species did not affect enzyme activity to a significant degree. However, the higher doses of 10, 20, and 40 mg Cr(III) and Cr(VI) kg^{-1} reduced the yields and soil enzyme activity. Na2EDTA not only did not reduce Cr(III)- and Cr(VI)-induced stress, but actually augmented the adverse effect of Cr on *Avena sativa* L. and soil enzyme activity.

Author Contributions: Conceptualization, E.B.-L., J.W. and J.K.; experimental design and methodology, E.B.-L., J.W. and J.K.; investigation, J.W.; statistical analyses, E.B.-L.; writing original draft, E.B.-L.; review and editing, J.W.; supervision, J.K. All authors have read and agreed to the published version of the manuscript.

Funding: This research was funded by the University of Warmia and Mazury in Olsztyn, Faculty of Agriculture and Forestry, Department of Soil Science and Microbiology (grant No. 30.610.006-110) and was financially supported by the Minister of Education and Science under the program entitled "Regional Initiative of Excellence" for the years 2019–2023, project No. 010/RID/2018/19 (amount of funding: PLN 12,000,000).

Institutional Review Board Statement: Not applicable.

Informed Consent Statement: Not applicable.

Data Availability Statement: Not applicable.

Conflicts of Interest: The authors declare no conflict of interest. The funders had no role in the design of the study; in the collection, analyses, interpretation of data; in the writing of the manuscript; or in the decision to publish the results.

Sample Availability: Samples of the compound are not available from the authors.

References

1. Fibbi, D.; Doumett, S.; Lepri, L.; Checchini, L.; Gonnelli, C.; Coppini, E.; Bubba, M.D. Distribution and mass balance of hexavalent and trivalent chromium in a subsurface, horizontal flow (SF-h) constructed wetland operating as post-treatment of textile wastewater for water reuse. *J. Hazard. Mater.* **2012**, *199–200*, 209–216. [CrossRef]
2. Barrera-Diaz, C.E.; Lugo-Lugo, V.; Bilyeu, B. A review of chemical, electrochemical and biological methods for aqueous Cr(VI) reduction. *J. Hazard. Mater.* **2012**, *223*, 1–12. [CrossRef]
3. Li, G.; Yang, X.; Liang, L.; Guo, S. Evaluation of the potential redistribution of chromium fractionation in contaminated soil by citric acid/sodium citrate washing. *Arab. J. Chem.* **2017**, *10*, 539–545. [CrossRef]
4. Wyszkowska, J.; Borowik, A.; Zaborowska, M.; Kucharski, J. Sensitivity of *Zea mays* and soil microorganisms to the toxic effect of chromium (VI). *Int. J. Mol. Sci.* **2023**, *24*, 178. [CrossRef]
5. Hsu, L.C.; Liu, Y.T.; Tzou, Y.M. Comparison of the spectroscopic speciation and chemical fractionation of chromium in contaminated paddy soils. *J. Hazard. Mater.* **2015**, *296*, 230–238. [CrossRef]
6. Prasad, S.; Yadav, K.K.; Kumar, S.; Gupta, N.; Cabral-Pinto, M.M.S.; Rezania, S.; Radwan, N.; Alam, J. Chromium contamination and effect on environmental health and its remediation: A sustainable approaches. *J. Environ. Manag.* **2021**, *285*, 112174. [CrossRef]
7. Saha, R.; Nandi, R.; Saha, B. Sources and toxicity of hexavalent chromium. *J. Coord. Chem.* **2011**, *64*, 1782–1806. [CrossRef]
8. Rakhunde, R.; Deshpande, L.; Juneja, H.D. Chemical speciation of chromium in water: A review. *Crit. Rev. Environ. Sci. Technol.* **2012**, *42*, 776–810. [CrossRef]
9. Nakkeeran, E.; Patra, C.; Shahnaz, T.; Rangabhashiyam, S.; Selvaraju, N. Continuous biosorption assessment for the removal of hexavalent chromium from aqueous solutions using *Strychnos nux* vomica fruit shell. *Bioresour. Technol. Rep.* **2018**, *3*, 256–260. [CrossRef]
10. Shahid, M.; Shamshad, S.; Rafiq, M.; Khalid, S.; Bibi, I.; Niazi, N.K.; Dumat, C.; Rashid, M.I. Chromium speciation, bioavailability, uptake, toxicity and detoxification in soil-plant system: A review. *Chemosphere* **2017**, *178*, 513–533. [CrossRef]
11. Zaheer, I.E.; Ali, S.; Saleem, M.H.; Imran, M.; Alnusairi, G.S.H.; Alharbi, B.M.; Riaz, M.; Abbas, Z.; Rizwan, M.; Soliman, M.H. Role of iron–lysine on morpho-physiological traits and combating chromium toxicity in rapeseed (*Brassica napus* L.) plants irrigated with different levels of tannery wastewater. *Plant Physiol. Biochem.* **2020**, *155*, 70–84. [CrossRef]
12. Zainab, N.; Amna, K.A.A.; Azeem, M.A.; Ali, B.; Wang, T.; Shi, F.; Alghanem, S.M.; Munis, M.F.H.; Hashem, M.; Alamri, S.; et al. PGPR-mediated plant growth attributes and metal extraction ability of *Sesbania sesban* L. in industrially contaminated soils. *Agronomy* **2021**, *11*, 1820. [CrossRef]
13. Ugwu, E.I.; Agunwamba, J.C. A review on the applicability of activated carbon derived from plant biomass in adsorption of chromium, copper, and zinc from industrial wastewater. *Environ. Monit. Assess.* **2020**, *192*, 240. [CrossRef] [PubMed]
14. Ertani, A.; Mietto, A.; Borin, M.; Nardi, S. Chromium in agricultural soils and crops: A review. *Water Air Soil Pollut.* **2017**, *228*, 190. [CrossRef]
15. Ranieri, E.; Moustakas, K.; Barbafieri, M.; Ranieri, A.C.; Herrera-Melián, J.A.; Petrella, A.; Tommasi, F. Phytoextraction technologies for mercury-and chromium-contaminated soil: A review. *J. Chem. Technol. Biotechnol.* **2020**, *95*, 317–327. [CrossRef]
16. Fu, Z.; Guo, W.; Dang, Z.; Hu, Q.; Wu, F.; Feng, C.; Zhao, X.; Meng, W.; Xing, B.; Giesy, J.P. Refocusing on nonpriority toxic metals in the aquatic environment in China. *Environ. Sci. Technol.* **2017**, *51*, 3117–3118. [CrossRef]
17. Ali, H.; Khan, E.; Ilahi, I. Environmental chemistry and ecotoxicology of hazardous heavy metals: Environmental persistence, toxicity, and tioaccumulation. *J. Chem.* **2019**, *2019*, 6730305. [CrossRef]

18. Amin, H.; Arain, B.A.; Amin, F.; Surhio, M.A. Phytotoxicity of chromium on germination, growth and biochemical at-tributes of *Hibiscus esculentus* L. *Am. J. Plant Sci.* **2013**, *4*, 41293. [CrossRef]
19. Bhalerao, S.A.; Sharma, A.S. Chromium: As an environmental pollutant. *Int. J. Curr. Microbiol. Appl. Sci.* **2015**, *4*, 732–746.
20. Xu, Z.-R.; Cai, M.-L.; Chen, S.-H.; Huang, X.-Y.; Zhao, F.-J.; Wang, P. High-Affinity Sulfate Transporter Sultr1;2 Is a Major Transporter for Cr(VI) Uptake in Plants. *Environ. Sci. Technol.* **2021**, *55*, 1576–1584. [CrossRef]
21. Singh, H.P.; Mahajan, P.; Kaur, S.; Batish, D.R.; Kohli, R.K. Chromium toxicity and tolerance in plants. *Environ. Chem. Lett.* **2013**, *11*, 229–254. [CrossRef]
22. Srivastava, S.; Nigam, R.; Prakash, S.; Srivastava, M.M. Mobilization of trivalent chromium in presence of organic acids: A hydroponic study of wheat plant (*Triticum vulgare*). *Bull. Environ. Contam. Toxicol.* **1999**, *63*, 524–530. [CrossRef] [PubMed]
23. Kota´s, J.; Stasicka, Z. Chromium occurrence in the environment and methods of its speciation. *Environ. Pollut.* **2000**, *107*, 263–283. [CrossRef] [PubMed]
24. Dhal, B.; Thatoi, H.N.; Das, N.N.; Pandey, B.D. Chemical and microbial remediation of hexavalent chromium from contaminated soil and mining/metallurgical solid waste: A review. *J. Hazard. Mater.* **2013**, *15*, 272–291. [CrossRef] [PubMed]
25. Zaheer, I.E.; Ali, S.; Saleem, M.H.; Arslan Ashraf, M.; Ali, Q.; Abbas, Z.; Muhammad Rizwan, M.; El-Sheikh, M.A.; Alyemeni, M.N.; Wijaya, L. Zinc-lysine supplementation mitigates oxidative stress in rapeseed (*Brassica napus* L.) by preventing phytotoxicity of chromium, when irrigated with tannery wastewater. *Plants* **2020**, *9*, 1145. [CrossRef]
26. Hussain, I.; Saleem, M.H.; Mumtaz, S.; Rasheed, R.; Ashraf, M.A.; Maqsood, F.; Rehman, M.; Yasmin, H.; Ahmed, S.; Ishtiaq, C.M.; et al. Choline chloride mediates chromium tolerance in spinach (*Spinacia oleracea* L.) by restricting its uptake in relation to morpho-physio-biochemical attributes. *J. Plant Growth Regul.* **2021**, *41*, 1594–1614. [CrossRef]
27. Narayani, M.; Shetty, K.V. Chromium-resistant bacteria and their environmental condition for hexavalent chromium removal: A review. *Crit. Rev. Environ. Sci. Technol.* **2013**, *43*, 955–1009. [CrossRef]
28. Chen, T.; Chang, Q.R.; Liu, J.; Clevers, J.G.P.W.; Kooistra, L. Identification of soil heavy metal sources and improvement in spatial mapping based on soil spectra information: A case study in northwest China. *Sci. Total Environ.* **2016**, *565*, 155–164. [CrossRef]
29. Available online: https://www.statista.com (accessed on 10 December 2022).
30. Available online: https://www.eea.europa.eu/publications/lrtap-1990-2019 (accessed on 10 December 2022).
31. Clean-Up of Polluted Environment? *Front Plant Sci.* **2018**, *9*, 1476. [CrossRef]
32. Yan, X.; Wang, J.; Song, H.; Peng, Y.; Zuo, S.; Gao, T.; Duan, X.; Qin, D.; Dong, J. Evaluation of the phytoremediation potential of dominant plant species growing in a chromium salt–producing factory wasteland, China. *Environ. Sci. Pollut. Res.* **2020**, *27*, 7657–7671. [CrossRef]
33. Huda, A.K.M.N.; Hossain, M.; Mukta, R.H.; Khatun, M.R.; Haque, M.A. EDTA–enhanced Cr detoxification and its potential toxicity in rice (*Oryza sativa* L.). *Plant Stress* **2021**, *2*, 100014. [CrossRef]
34. Hong, P.K.A.; Banerji, S.K.; Regmi, T. Extraction, recovery, and biostability of EDTA for remediation of lead, copper, zinc and nickel. *Soil Sci. Soc. Am. J.* **1999**, *47*, 47–51. [CrossRef]
35. Grčman, H.; Velikonja-Bolta, Š.; Vodnik, D.; Kos, B.; Leštan, D. EDTA enhanced heavy metal phytoextraction: Metal accumulation, leaching and toxicity. *Plant Soil* **2001**, *235*, 105–114. [CrossRef]
36. Evangelou, M.W.H.; Ebel, M.; Schaeffer, A. Chelate assisted phytoextraction of heavy metals from soil Effect mechanism toxicity and fate of chelating agents. *Chemosphere* **2007**, *68*, 989–1003. [CrossRef]
37. Guo, X.; Wei, Z.; Wu, Q.; Li, C.; Qian, T.; Zheng, W. Effect of soil washing with only chelators or combining with ferric chloride on soil heavy metal removal and phytoavailability: Field experiments. *Chemosphere* **2016**, *147*, 412–419. [CrossRef] [PubMed]
38. Jelusic, M.; Vodnik, D.; Macek, I.; Lestan, D. Effect of EDTA washing of metal polluted garden soils. Part II: Can remediated soil be used as a plant substrate. *Sci. Total. Environ.* **2014**, *475*, 142–152. [CrossRef]
39. Jez, E.; Lestan, D. EDTA retention and emissions from remediated soil. *Chemosphere* **2016**, *151*, 202–209. [CrossRef] [PubMed]
40. Dipu, S.; Kumar, A.A.; Thanga, S.G. Effect of chelating agents in phytoremediation of heavy metals. *Remediat. J.* **2012**, *22*, 133–146. [CrossRef]
41. Cheng, S.; Lin, Q.; Wang, Y.; Luo, H.; Huang, Z.; Fu, H.; Chen, H.; Xiao, R. The removal of Cu, Ni, and Zn in industrial soil by washing with EDTA-organic acids. *Arab. J. Chem.* **2020**, *13*, 5160–5170. [CrossRef]
42. Finžgar, N.; Leštan, D. Multi-step leaching of Pb and Zn contaminated soils with EDTA. *Chemosphere* **2007**, *66*, 824–832. [CrossRef]
43. Zou, Z.; Qiu, R.; Zhang, W.; Dong, H.; Zhao, Z.; Zhang, T.; Wei, X.; Cai, X. The study of operating variables in soil washing with EDTA. *Environ. Pollut.* **2009**, *157*, 229–236. [CrossRef]
44. Udovic, M.; Domen, L. Fractionation and bioavailability of Cu in soil remediated by EDTA leaching and processed by earthworms. *Environ. Sci. Pollut. Res.* **2010**, *17*, 561–570. [CrossRef]
45. Neugschwandtner, R.W.; Tlustos, P.; Komarek, M.; Szakova, J.; Jakoubkova, L. Chemically enhanced phytoextraction of risk elements from a contaminated agricultural soil using *Zea mays* and *Triticum aestivum*: Performance and metal mobilization over a three year period. *Int. J. Phytorem.* **2012**, *14*, 754–771. [CrossRef]
46. Wu, G.; Kanga, H.; Zhang, X.; Shao, H.; Chu, L.; Ruand, C. A critical review on the bio-removal of hazardous heavy metals from contaminated soils: Issues, progress, eco-environmental concerns and opportunities. *J. Hazard. Mater.* **2010**, *174*, 1–8. [CrossRef]
47. Shanker, A.K.; Cervantes, C.; Loza-Tavera, H.; Avudainayagam, S. Chromium toxicity in plants. *Environ. Int.* **2005**, *31*, 739–753. [CrossRef]

48. Abideen, S.N.U.; Abideen, A.A. Protein level and heavy metals (Pb, Cr, and Cd) concentrations in wheat (*Triticum aestivum*) and in oat (*Avena sativa*) plants. *IJIAS* **2013**, *3*, 284–289.
49. Tobiasz-Salach, R.; Pyrek-Bajcar, E.; Bobrecka-Jamro, D. Assessing the possible use of hulled and naked oat grains as energy source. *Econtechmod. Inter. Quart. J.* **2016**, *15*, 35–40.
50. Proszak-Miąsik, D.; Jarecki, W.; Nowak, K. Selected parameters of oat straw as an alternative energy raw material. *Energies* **2022**, *15*, 331. [CrossRef]
51. Cervantes, C.; Campos-Garcia, J.; Devars, S.; Gutiérrez-Corona, F.; Loza-Tavera, H.; Torres-Guzmán, J.C.; Moreno-Sánchez, R. Interactions of chromium with microorganisms and plants. *FEMS Microbiol. Rev.* **2001**, *25*, 335–347. [CrossRef] [PubMed]
52. Akinci, I.E.; Akinci, S. Effect of chromium toxicity on germination and early seedling growth in melon (*Cucumis melo* L.). *Afr. J. Biotechnol.* **2010**, *9*, 4589–4594.
53. Rai, V.; Tandon, P.K.; Khatoon, S. Effect of chromium on antioxidant potential of *Catharanthus roseus* varieties and production of their anticancer alkaloids: Vincristine and vinblastine. *Biomed. Res. Int.* **2014**, *2014*, 934182. [CrossRef] [PubMed]
54. Mathur, S.; Kalaji, H.M.; Jajoo, A. Investigation of deleterious effects of chromium phytotoxicity and photosynthesis in wheat plant. *Photosynthetica* **2016**, *54*, 185–192. [CrossRef]
55. Dey, U.; Mondal, N.K. Ultrastructural deformation of plant cell under heavy metal stress in Gram seedlings. *Cogent Environ. Sci.* **2016**, *2*, 1–12. [CrossRef]
56. Sharma, A.; Kapoor, D.; Wang, J.; Shahzad, B.; Kumar, V.; Bali, A.S.; Zheng, B.; Yuan, H.; Yan, D. Jasrotia, S. Chromium bioaccumulation and its impacts on plants: An overview. *Plants* **2020**, *9*, 100. [CrossRef] [PubMed]
57. de Oliveira, L.M.; Gress, J.; De, J.; Rathinasabapathi, B.; Marchi, G.; Chen, Y.; Ma, L.Q. Sulfate and chromate increased each other's uptake and translocation in As-hyperaccumulat or *Pterisvittata*. *Chemosphere* **2016**, *147*, 36–43. [CrossRef]
58. Zayed, A.; Lytle, C.M.; Qian, J.-H.; Terry, N. Chromium accumulation, translocation and chemical speciation in vegetable crops. *Planta* **1998**, *206*, 293–299. [CrossRef]
59. Shanker, A.K.; Djanaguiraman, M.; Venkateswarlu, B. Chromium interactions in plants: Current status and future strategies. *Metallomics* **2009**, *1*, 375–383. [CrossRef]
60. Bareen, F.; Khadija, R.; Muhammad, S.; Aisha, N. Uptake and leaching of Cu, Cd, and Cr after EDTA application in sand columns using sorghum and pearl millet. *Pol. J. Environ. Stud.* **2019**, *28*, 2065–2077. [CrossRef]
61. Ali, S.Y.; Chaudhury, S. EDTA-enhanced phytoextraction by tagetes sp. and effect on bioconcentration and translocation of heavy metals. *Environ. Proc.* **2016**, *3*, 735. [CrossRef]
62. Naseem, S.; Yasin, M.; Ahmed, A.; Faisal, M. Chromium accumulation and toxicity in corn (*Zea mays* L.) seedlings. *Pol. J. Environ. Stud.* **2015**, *24*, 899–904.
63. Nagarajan, M.; Ganesh, K.S. Effect of chromium on growth, biochemicals and nutrient accumulation of paddy (*Oryza sativa* L.). *Int. Lett. Nat. Sci.* **2014**, *23*, 63–71. [CrossRef]
64. Diwan, H.; Ahmad, A.; Iqbal, M. Chromium-induced alterations in photosynthesis and associated attributes in Indian mustard. *J. Environ. Biol.* **2012**, *33*, 239–244. [PubMed]
65. Saravanan, A.; Jayasree, R.; Hemavathy, R.V.; Jeevanantham, S.; Hamsini, S.; Senthil, K.P.; Yuvaraj, D. Phytoremediation of Cr (VI) ion contaminated soil using Black gram (*Vigna mungo*): Assessment of removal capacity. *J. Environ. Chem. Eng.* **2019**, *7*, 103052.
66. Ramana, S.; Biswas, A.K.; Singh, A.B.; Ahirwar, N.K.; Subba Rao, A. Tolerance of ornamental succulent plant crown of thorns (*Euphorbia milli*) to chromium and its remediation. *Int. J. Phytoremediation* **2014**, *17*, 363–368. [CrossRef] [PubMed]
67. Amin, H.; Arain, B.A.; Abbasi, M.S.; Amin, F.; Jahangir, T.M.; Soomro, N.U. Evaluation of chromium phyto-toxicity, phyto-tolerance, and phyto-accumulation using biofuel plants for effective phytoremediation. *Int. J. Phytoremediation* **2019**, *14*, 1–12. [CrossRef] [PubMed]
68. Bareen, E.F.; Tahira, S.A. Efficiency of seven different cultivated plant species for phytoextraction of toxic metals from tannery effluent contaminated soil using EDTA. *Soil Sediment Contam.* **2010**, *19*, 160–173. [CrossRef]
69. Han, F.X.; Sridhar, B.B.M.; Monts, D.L.; Su, Y. Phytoavailability and toxicity of trivalent and hexavalent chromium to *Brassica juncea*. *New Phytol.* **2004**, *162*, 489–499. [CrossRef]
70. Ebrahimi, M. Effect of EDTA treatment method on leaching of Pb and Cr by *Phragmites australis* (Cav.) Trin. Ex Steudel (common reed). *Caspian J. Environ. Sci.* **2015**, *13*, 153–166.
71. Dick, W.A.; Cheng, L.; Wang, P. Soil acid and alkaline phosphatase activity as pH adjustment indicators. *Soil Biol. Biochem.* **2000**, *32*, 1915–1919. [CrossRef]
72. Huang, Y.; Peng, B.; Yang, Z.; Chai, L.; Zhou, L. Chromium accumulation, microorganism population and enzyme activities in soils around chromium-containing slag heap of steel alloy factory. *Trans. Nonferrous Met. Soc. China* **2009**, *19*, 241–248. [CrossRef]
73. Belyaeva, O.N.; Haynes, R.J.; Birukova, O.A. Barley yield and soil microbial and enzyme activities as affected by contamination of two soils with lead, zinc or copper. *Biol. Fertil. Soils* **2005**, *41*, 85–94. [CrossRef]
74. Lombard, N.; Prestat, E.; van Elsas, J.D.; Simonet, P. Soil-specific limitations for access and analysis of soil microbial communities by metagenomics. *FEMS Microbiology Ecol.* **2011**, *78*, 31–49. [CrossRef]
75. Liao, Y.; Min, X.; Yang, Z.; Chai, L.; Zhang, S.; Wang, Y. Physicochemical and biological quality of soil in hexavalent chromium-contaminated soils as affected by chemical and microbial remediation. *Environ. Sci. Pollut. Res.* **2014**, *21*, 379–388. [CrossRef]
76. Tokunaga, T.K.; Wan, J.; Firestone, M.K.; Hazen, T.C.; Olson, K.R.; Herman, D.J.; Sutton, S.R.; Lanzirotti, A. In situ reduction of chromium(VI) in heavily contaminated soils through organic carbon amendment. *J. Environ. Qual.* **2003**, *32*, 1641–1649. [CrossRef]

77. Dotaniya, M.L.; Rajendiran, S.; Meena, V.D.; Saha, J.K.; Vassanda Coumar, M.; Kundu, S.; Patra, A.K. Influence of chromium contamination on carbon mineralization and enzymatic activities in Vertisol. *Agric. Res.* **2017**, *6*, 91–96. [CrossRef]
78. Quilchano, C.; Maranon, T. Dehydrogenase activity in Mediterranean forest soils. *Biol. Fertil. Soils* **2002**, *35*, 102–107. [CrossRef]
79. Wyszkowska, J. Soil contamination with chromium and its enzymatic activity and yielding. *Polish J. Environ. Stud.* **2002**, *11*, 79–84.
80. Baathe, E. Effects of heavy metals in soil microbial processes and populations (a review). *Water Air Soil Pollut.* **1989**, *47*, 335–379. [CrossRef]
81. Peng, B.; Huang, S.H.; Yang, Z.H.; Chai, L.Y.; Xu, Y.Z.; Su, C.Q. Inhibitory effect of Cr(VI) on activities of soil enzymes. *J. Cent. South. Univ. Technol.* **2009**, *16*, 594–598. [CrossRef]
82. Yang, Z.; Liu, S.; Zheng, D.; Feng, S. Effects of cadmium, zinc and lead on soil enzyme activities. *J. Environ. Sci.* **2006**, *18*, 1135–1141. [CrossRef]
83. Stępniewska, Z.; Wolińska, A.; Ziomek, J. Response of soil catalase activity to chromium contamination. *J. Environ. Sci.* **2009**, *21*, 1142–1147. [CrossRef] [PubMed]
84. Samborska, A.; Stępniewska, Z.; Stępniewski, W. Influence of different oxidation states of chromium (VI, III) on soil urease activity. *Geoderma* **2004**, *122*, 317–322. [CrossRef]
85. Al-Khashman, O.A.; Shawabkeh, R.A. Metals distribution in soils around the cement factory in southern Jordan. *Environ. Pollut.* **2006**, *140*, 387–394. [CrossRef] [PubMed]
86. Schulin, R. Heavy metal contamination along a soil transect in the vicinity of the iron smelter of Kremikovtzi (Bulgaria). *Geoderma* **2007**, *140*, 52–61. [CrossRef]
87. Tome, V.F.; Blanco, R.P.; Lozano, J.C. The ability of *Helianthus annuus* L. and *Brassica juncea* to uptake and translocate natural uranium and 226Ra under different milieu conditions. *Chemosphere* **2009**, *74*, 293–300. [CrossRef]
88. Mahmood-ul-Hassan, M.; Suthar, V.; Ahmad, R.; Yousra, M. Heavy metal phytoextraction—Natural and EDTA-assisted remediation of contaminated calcareous soils by sorghum and oat. *Environ. Monit. Assess.* **2017**, *189*, 591. [CrossRef]
89. Komárek, M.; Tlustoš, P.; Száková, J.; Chrastn, V.; Balík, J. The role of Fe- and Mn-oxides during EDTA enhanced phytoextraction of heavy metals. *Plant Soil Environ.* **2007**, *53*, 216–224. [CrossRef]
90. Römkens, P.; Bouwman, L.; Japenga, J.; Draaisma, C. Potentials and drawbacks of chelate-enhanced phytoremediation of soils. *Environ. Pollut.* **2002**, *116*, 109–121. [CrossRef]
91. Lombi, E.; Zhao, F.J.; Dunham, S.J.; McGrath, S.P. Phytoremediation of heavy metal contaminated soils: Natural hyperaccumulation versus chemically enhanced phytoextraction. *J. Environ. Qual.* **2001**, *30*, 1919–1926. [CrossRef]
92. Available online: https://zpe.gov.pl/a/soils-in-poland (accessed on 10 December 2022).
93. World's Worst Pollution Problems 2015. The New Top Six Toxic Threats: A Priority List for Remediation. Available online: http://www.worstpolluted.org/docs/WWPP_2015_Final.pdf (accessed on 21 November 2020).
94. *PN-ISO-11466:2002*; Polish Committee for Standardization. Soil Quality—Extraction of Trace Elements Soluble in Aqua Regia. Polish Committee for Standardization: Warsaw, Poland, 2002.
95. Öhlinger, R. Dehydrogenase activity with the substrate TTC. In *Methods in Soil Biology*; Schinner, F., Ohlinger, R., Kandler, E., Margesin, R., Eds.; Springer: Berlin/Heidelberg, Germany, 1996; pp. 241–243.
96. Alef, K.; Nannipieri, P. *Methods in Applied Soil Microbiology and Biochemistry*; Alef, K., Nannipieri, P., Eds.; Academic: London, UK, 1998; pp. 316–365.
97. Zaborowska, M.; Wyszkowska, J.; Borowik, A.; Kucharski, J. Bisphenol A—A dangerous pollutant distorting the biological properties of soill. *Int. J. Mol. Sci.* **2021**, *22*, 12753. [CrossRef]
98. Borowik, A.; Wyszkowska, J.; Zaborowska, M.; Kucharski, J. The impact of permethrin and cypermethrin on plants, soil enzyme activity, and microbial communities. *J. Mol. Sci.* **2023**, *24*, 2892. [CrossRef] [PubMed]
99. Borowik, A.; Wyszkowska, J.; Wyszkowski, M. Resistance of aerobic microorganisms and soil enzyme response to soil contamination with Ekodiesel Ultra fuel. *Environ. Sci. Pollut. Res.* **2017**, *24*, 24346–24363. [CrossRef] [PubMed]
100. Boros-Lajszner, E.; Wyszkowska, J.; Kucharski, J. Use of zeolite to neutralise nickel in a soil environment. *Environ. Monit. Assess.* **2018**, *190*, 54. [CrossRef] [PubMed]
101. *PN ISO 11047:2001*; Soil Quality—Determination of Cadmium, Chromium, Cobalt, Copper, Lead, Manganese, Nickel and Zinc in Aqua Regia Extracts of Soil—Flame and Electrothermal Atomic Absorption Spectrometric Methods. Polish Committee for Standardization: Warsaw, Poland, 2013.
102. Boros-Lajszner, E.; Wyszkowska, J.; Kucharski, J. Phytoremediation of soil contaminated with nickel, cadmium and cobalt. *Int. J. Phytoremediation* **2021**, *23*, 252–262. [CrossRef] [PubMed]
103. Dell Inc. *Dell Statistica (Data Analysis Software System)*; Version 13.1; Dell Inc.: Tulsa, OK, USA, 2022.

Disclaimer/Publisher's Note: The statements, opinions and data contained in all publications are solely those of the individual author(s) and contributor(s) and not of MDPI and/or the editor(s). MDPI and/or the editor(s) disclaim responsibility for any injury to people or property resulting from any ideas, methods, instructions or products referred to in the content.

Article

Multivariate Exploratory Analysis of the Bulgarian Soil Quality Monitoring Network

Galina Yotova [1], Mariana Hristova [2], Monika Padareva [1], Vasil Simeonov [1], Nikolai Dinev [2] and Stefan Tsakovski [1,*]

[1] Faculty of Chemistry and Pharmacy, Sofia University "St. Kliment Ohridski", 1 J. Bourchier Blvd., 1164 Sofia, Bulgaria; g.yotova@chem.uni-sofia.bg (G.Y.); m.padareva@abv.bg (M.P.); vsimeonov@chem.uni-sofia.bg (V.S.)

[2] Institute of Soil Science, Agrotechnologies and Plant Protection "N. Poushkarov", Agricultural Academy, 7 Bansko shose Str., 1331 Sofia, Bulgaria; marihristova@hotmail.com (M.H.); ndinev@itp.bg (N.D.)

* Correspondence: stsakovski@chem.uni-sofia.bg; Tel.: +359-2-8161426

Abstract: The goal of the present study is to assess the soil quality in Bulgaria using (i) an appropriate set of soil quality indicators, namely primary nutrients (C, N, P), acidity (pH), physical clay content and potentially toxic elements (PTEs: Cu, Zn, Cd, Pb, Ni, Cr, As, Hg) and (ii) respective data mining and modeling using chemometrical and geostatistical methods. It has been shown that five latent factors are responsible for the explanation of nearly 70% of the total variance of the data set available (principal components analysis) and each factor is identified in terms of its contribution to the formation of the overall soil quality—the mountain soil factor, the geogenic factor, the ore deposit factor, the low nutrition factor, and the mercury-specific factor. The obtained soil quality patterns were additionally confirmed via hierarchical cluster analysis. The spatial distribution of the patterns throughout the whole Bulgarian territory was visualized via the mapping of the factor scores for all identified latent factors. The mapping of identified soil quality patterns was used to outline regions where additional measures for the monitoring of the phytoavailability of PTEs were required. The suggested regions are located near to thermoelectric power plants and mining and metal production facilities and are characterized by intensive agricultural activity.

Keywords: soil analysis; nutrients; potentially toxic elements; multivariate statistics; principal components analysis; cluster analysis; kriging; soil management

Citation: Yotova, G.; Hristova, M.; Padareva, M.; Simeonov, V.; Dinev, N.; Tsakovski, S. Multivariate Exploratory Analysis of the Bulgarian Soil Quality Monitoring Network. *Molecules* **2023**, *28*, 6091. https://doi.org/10.3390/molecules28166091

Academic Editor: Jalal Hawari

Received: 29 June 2023
Revised: 13 August 2023
Accepted: 14 August 2023
Published: 16 August 2023

Copyright: © 2023 by the authors. Licensee MDPI, Basel, Switzerland. This article is an open access article distributed under the terms and conditions of the Creative Commons Attribution (CC BY) license (https://creativecommons.org/licenses/by/4.0/).

1. Introduction

Soil quality can be defined as the ability of the soil to perform functions for its intended use. The capacity of the soil to function involves the balance and integration of sustained biological productivity, environmental quality and plant and animal health [1]. Soil quality cannot be measured directly because of its broad and integrative factors related to different soil uses. Soil quality evaluations are based on soil indicator (attributes) measurements, which reflect the inherent soil properties. Human management and natural disturbances can lead to significant changes in soil properties, which require a soil monitoring network, including the proper selection of an indicator data set related to reliable soil quality assessment.

Some reviews of national soil monitoring networks in Europe reveal serious deficiencies in monitoring the functional capacity of different soils and their changes over time [2,3]. The inspection of national monitoring networks emphasizes clearly unbalanced data sets with a dominance of chemical parameters in soil at the expense of biological and physical indicators [3]. Some of the indicators related to a decline in soil biodiversity and soil erosion are measured very rarely, whereas those related to soil compaction and the decline in soil organic matter and soil contamination are measured at almost all sites [2]. The other important drawback of soil monitoring in Member States of EU is the lack of harmonization of their national networks [4] regarding their design [2,3], soil sampling density [3] and method of analysis [5]. The overcoming of the above-mentioned obstacles and the expansion of the scope of monitored soil functions could be achieved via the establishment of an EU-wide monitoring network, which is the aim of the European-scale LUCAS soil survey (Land Use and Coverage Area Frame Survey) [2].

The Bulgarian soil quality monitoring network is no exception, and it contains well-documented records concerning basic chemical soil indicators and potentially toxic elements. The good sampling density provides the opportunity for the reliable monitoring of some of the basic soil functions like primary productivity, nutrient cycling and soil contamination [3]. Usually, such data sets are used for establishing the geochemical background and threshold of chemical elements [6], but incorporating basic soil chemical indicators in data modeling could provide important information for the phytoavailability of PTEs. The prolonged existence of PTEs in the ecosystem poses a potential risk to human health, caused by the consumption of contaminated plants or the direct inhalation of soil particles [7]. Despite the lower levels of industrial pollution in recent years, it is essential not to underestimate the concerns related to PTEs [8]. Although PTEs like Cu, Mn and Zn are essential micronutrients for plants, their high concentrations could cause toxic effects and serious risks for the food chain [7].

Along with the selection and assessment of an optimal set of soil indicators [3], such a complex and multivariate task as soil quality assessment requires adequate data treatment. The application of multivariate statistical methods, like cluster analysis (CA) and principal components analysis (PCA), can not only reveal hidden interactions between different soil indicators but also identify the factors affecting soil quality, including natural processes and anthropogenic pressures [9–21]. In most of the aforementioned studies, multivariate exploratory approaches are used in order to reveal and estimate the anthropogenic sources of potentially toxic elements.

A solid approach for visualizing the regional distribution of soil indicators or achieved latent factors is geostatistics, namely kriging interpolation using geographical information system (GIS) techniques. This method is often used to illustrate the spatial distribution of soil indicators [11,15,18–26] and, less frequently, to project the spatial distribution of the identified latent factors on a map [10,14,23,27].

The aim of the present study is to reveal the latent factors controlling the soil quality of Bulgarian territories based on the soil indicators collected during the Bulgarian soil quality monitoring program. The GIS-based mapping of soil quality pattern distributions was carried out in order to evaluate the impact assessment of PTEs in agricultural land areas.

2. Results

2.1. Basic Statistics

The 347 topsoil samples of the Bulgarian soil quality monitoring network were analyzed for thirteen soil indicators, namely the pH, the main nutrients (organic carbon—C; total nitrogen—N; total phosphorus—P), physical clay content and eight potentially toxic elements (Cu, Zn, Cd, Pb, Ni, Cr, As, Hg). Thus, the input data matrix consists of 347 rows (objects) and 13 columns (variables). The basic statistics of the input data are provided in Table 1.

Table 1. Basic statistics of the input data set ($n = 347$).

	Dimension	Mean	St. Dev.	Median	Minimum	Maximum
C	g kg^{-1}	18.8	10.7	16.0	0.31	113
N	g kg^{-1}	1.81	0.96	1.60	0.40	9.91
P	mg kg^{-1}	881	585	740	199	4634
Physical clay	%	51.6	19.6	57.5	9.79	83.0
pH	-	6.78	0.98	6.80	3.80	8.80

Table 1. *Cont.*

	Dimension	Mean	St. Dev.	Median	Minimum	Maximum
Cu	mg kg^{-1}	31.4	30.8	23.7	3.60	351
Zn	mg kg^{-1}	63.5	22.3	64.3	1.26	162
Cd	mg kg^{-1}	0.23	0.31	0.16	0.02	4.32
Pb	mg kg^{-1}	20.5	17.8	16.8	3.07	200
Ni	mg kg^{-1}	35.8	20.2	35.4	1.20	208
Cr	mg kg^{-1}	53.0	34.7	45.4	2.30	213
As	mg kg^{-1}	8.21	11.3	6.70	0.04	159
Hg	mg kg^{-1}	0.15	0.13	0.12	0.01	0.97

2.2. Chemometric Data Interpretation

Multivariate statistical methods (CA and PCA) were applied to uncover the relationships between three groups of parameters: basic soil characteristics (pH and physical clay), nutrient elements (C, N and P) and eight potentially toxic elements (Cu, Zn, Cd, Pb, Ni, Cr, As, Hg). First, PCA (Varimax rotation normalized mode) was performed. The PCA results show that the first five latent factors (PCs) explain more than 65% of the total variance. The factor loadings of the selected PCs are presented in Table 2.

Table 2. Factor loadings for five latent factors and their conditional names.

	PC1	PC2	PC3	PC4	PC5
C	0.63 [a]	−0.04	0.00	−0.59	0.20
N	**0.73** [b]	0.00	0.09	−0.52	0.00
P	−0.02	0.00	0.08	**−0.83**	−0.08
Physical Clay	−0.02	0.50	0.05	0.06	−0.39
pH	0.10	0.24	0.04	0.16	**−0.79**
Cu	−0.05	0.24	**0.72**	0.06	0.08
Zn	0.16	0.14	0.70	−0.28	−0.04
Cd	**0.78**	−0.01	−0.03	0.15	−0.08
Pb	0.64	−0.09	0.26	−0.01	0.07
Ni	−0.05	**0.90**	0.05	−0.03	−0.11
Cr	−0.04	**0.87**	0.10	0.00	0.13
As	0.13	−0.12	0.70	0.02	−0.07
Hg	0.15	0.18	0.03	0.30	0.68
Expl.Var.%	15.73	15.37	12.21	11.74	10.25
Conditional name	Mountain soil	Geogenic	Ore deposits	Low nutrition	Hg-specific

[a] Factor loadings with absolute values between 0.4–0.7 are underlined. [b] Factor loadings with absolute values above 0.7 are shown in bold.

The factor score plot presented in Figure 1 illustrates the distribution of sampling points according to the first two principal components. For sampling points with higher PC1 factor scores, the mountain soil pattern prevails (Figure 1a), while for points with higher PC2 factor scores, the geogenic origin dominates.

The relationships between soil parameters obtained in PCA were almost entirely confirmed via the application of hierarchical cluster analysis to the same data set (z-standardized input values, squared Euclidean distances as a measure of similarity and Ward's method of linkage). The grouping of measured soil indicators is presented in Figure 2. Three major clusters are formed: C1 (C, N, Cd, Pb, P); C2 (Cu, Zn, As, Hg); and C3 (physical clay, pH, Ni, Cr).

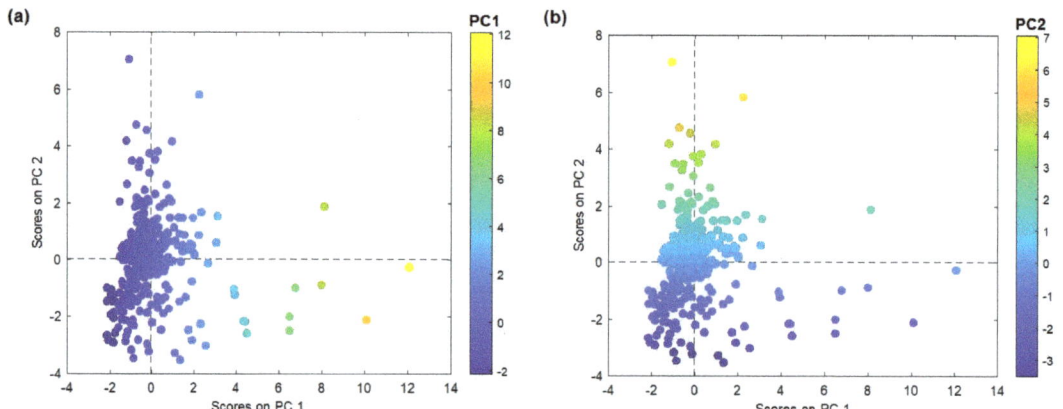

Figure 1. Plot of PC1 vs. PC2 factor scores: (**a**) colored by PC1 factor score values; (**b**) colored by PC2 factor score values.

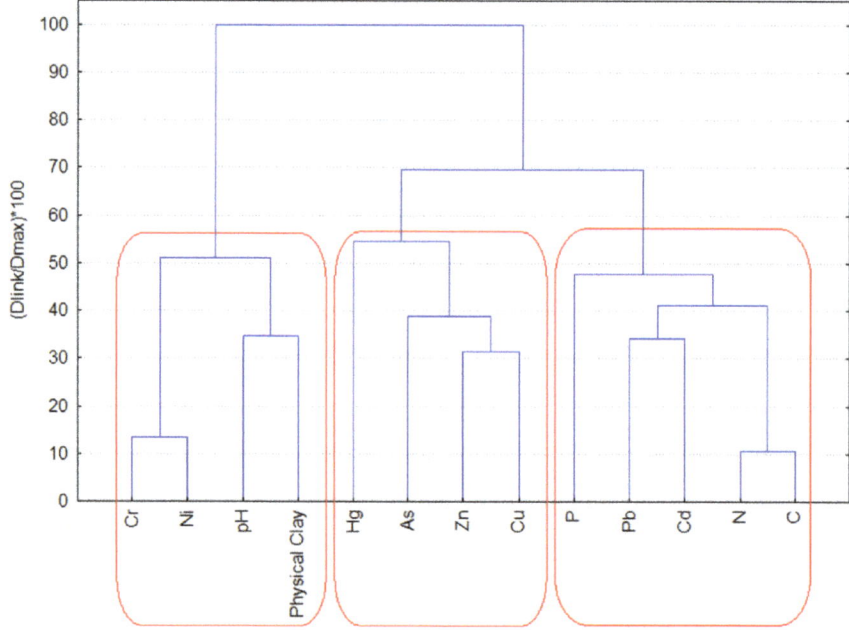

Figure 2. Hierarchical dendrogram of the clustering of the 13 soil indicators.

2.3. Mapping of Principal Components

The next step of the study is to explore the spatial distribution of the principal components using GIS-based maps of the respective sampling sites' factor scores. According to the factor scores of the 347 samples, kriging maps for each one of the five latent factors are presented (Figure 3). The spatial distribution of the principal components is an appropriate method for outlining the regions that are typical for a given factor and for clarifying the origin of a specific component.

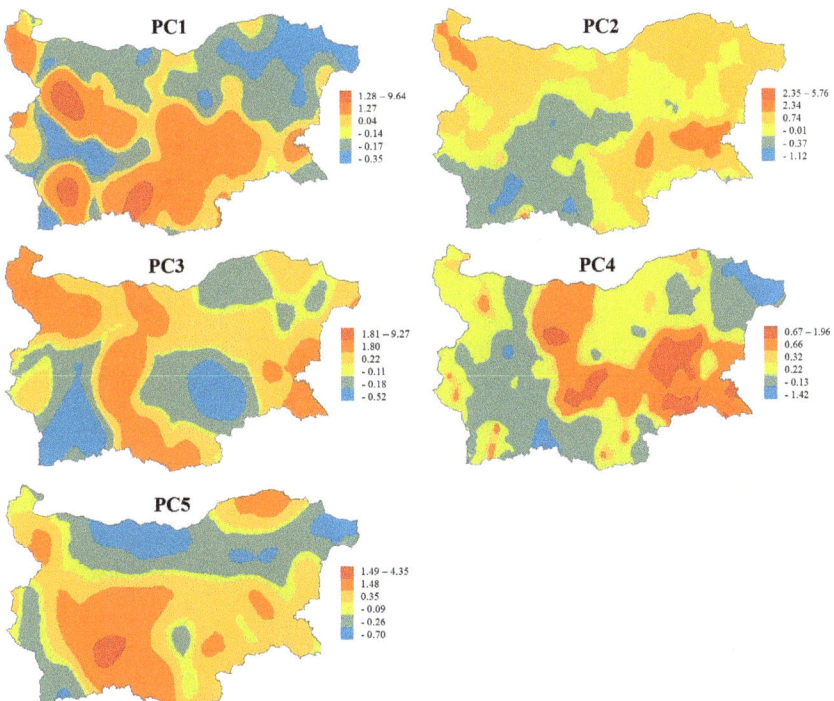

Figure 3. Spatial distribution of the principal components and soil sites determined using the PCA.

3. Discussion

3.1. Basic Statistics

According to the mean values obtained from basic statistics, the soils studied have a medium content of C and N. The C/N ratio shows that most of the soils are characterized by high quality humus, type mul, in which stable organic humates predominate [28]. The confidence interval of the mean pH value characterizes the soils as neutral (from slightly acidic to slightly alkaline). The content of physical clay in most of the soil samples shows that they have good physical and physicomechanical properties, as only 7.8% of the samples were characterized as sands with a physical clay content of less than 20%. The concentration of potentially toxic elements in the samples from the Bulgarian soil quality monitoring network shows low levels of anthropogenic impact. Only 6.1% of the soil samples had PTE concentrations at higher values than the maximum allowable concentrations [29], and there were mainly located in the western and southeastern parts of Bulgaria. The observed exceedances (twenty-one in total) are as follows: five for Cu, one for Cd, three for Pb, three for Ni, three for Cr and six for As.

3.2. Chemometric Data Interpretation

The selected five latent factors (PCs) in the PCA analysis explain more than 65% of the total variance.

The first latent factor (PC1) explains 15.73% of the total variance of the data set. It reveals the relationship between the nutrient elements carbon and nitrogen and the metals Cd and Pb. Such a "pattern" can be related to mountain soils in the Bulgarian network, mainly belonging to the Balkan and Rhodope tectonic zones [30]. It is typical for such soils that the mobility of Cd and Pb is controlled by soil organic matter [21,31].

In PC2, significant loadings possess Ni, Cr and physical clay. This factor is conditionally named "geogenic" [23] and explains 15.37% of the data set variance. Many studies confirm that Ni and Cr content in soil depends on the soil formation and composition of the parent rock material [16,21]. The same conclusion is reported in soil quality studies where only metals are included as indicators [9,18,23,32]. It should be noted that the moderate presence of Cu, Zn and Hg in this factor is an indicator for a specific "natural" content of these metals, which originate in the Earth's crust.

PC3 explains 12.21% of the total variance and can be conditionally named "ore deposits". It resembles the relationship between the significantly contributing Cu, Zn and As and the moderately participating Pb. Usually, these elements are pollutants that originate mainly from industrial, mining and agricultural activity [19–21,32,33], but their significant presence in the Earth's crust may be related to mineral ore deposits (Figure 4). Arsenic is generally recovered from sludge and flue dust in smelters. Furthermore, there are arsenic emissions in coal-burning areas. Its presence in non-ferrous ores is generally regarded to be an environmental problem and not a benefit [34].

Figure 4. Simplified tectonic map of Bulgaria with major exploited ore deposits, metal smelters and thermoelectric power plant.

The behavior of carbon and nitrogen is quite interesting. They participate significantly in the formation of two factors—PC1 and PC4. In the fourth factor, C and N are logically connected with P also being a typical nutrient (all three elements have negative factor loadings in PC4). This factor explains 11.74% of the data set variance and can conditionally be named "low nutrition". The appearance of such a factor in the data structure modeling via PCA could be related to the low humus content in soils, as well as to soils subject to nutrient components exhaustion at intensive agriculture.

The fifth factor (10.25%) reveals a positive relationship between physical clay and pH, which contradicts previous studies [11,35]. The significant factor loadings in PC5 lead to the conclusion that in the Bulgarian soil quality network, high concentrations of Hg are observed in sandy acidic soils. Since anthropogenic sources of Hg are limited within the territory of Bulgaria, the elevated values of Hg can be considered baseline levels, especially in the soils mentioned above. This can be confirmed via the comparison of the mean values of Hg in the FOREGS geochemical database (0.016 mg kg^{-1}) [36] and those in this data set (0.15 mg kg^{-1}).

In general, the results of both multivariate statistical approaches reveal similar patterns of similarity between the soil indicators used—one can identify the same factors responsible for the linkage and correlation between soil indicators, namely geogenic (cluster 3), nutritional (cluster 1) and ore deposits (cluster 2). By using PCA the data interpretation could be additionally improved by commenting on some other features of the data structure, like the specification of the mountain soil type and the unexpected mercury content.

3.3. Mapping of Principal Components

For better description and interpretation of the obtained results a simplified tectonic map of Bulgaria [37] is presented on Figure 4. Moreover, the major exploited ore deposits, as well as the biggest metal smelters and thermoelectric power plant are also indicated.

The first latent factor is related to mountain soils pattern. As expected, sites with the highest factor scores of PC1 are situated mainly in the largest Bulgarian mountains—Rila, Pirin, the Rhodopes, and the Balkans. This spatial distribution confirms that the high content of Pb and Cd is prevalent in mountain soils, such as Cambisols and Luvisols, and is not due to industrial activity in these regions. Moreover, the highest factor scores of PC1 are observed in areas with lead–zinc ore deposits (the central Rhodope zone and the western part of the Balkan zone) and copper mining areas (western part of the Balkan zone and the eastern part of the Srednogorie zone). It should be mentioned that the high factor scores in the central Srednogorie zone may be due to anthropogenic impact caused by contamination with Cd from the biggest coal-fired power plant in Bulgaria (Maritsa Iztok complex) (Figure 4) [38].

The geogenic factor (PC2) has higher factor scores in soils from the South Carpathian orogenic system, western Balkan Zone and eastern part of Srednogorie Zone. Most of the samples are from Chernozems and Vertisols soil types, which are characterized with high physical clay content. The sample sites with low factor scores of PC2 are mainly of Fluvisol, Cambisol and Leptosol soil types with high sand content. The abovementioned soil patterns correspond well with the origin and mapping of the "geogenic" principal component. In previous studies [38], it has been discussed that the presence of ophiolite and diabase–phylitoide complexes in the western Balkan Zone and igneous basic rocks in the western and eastern part of the Srednogorie Zone [39] confirmed the parent rock control of Ni and Cr. Again, high factor scores can be observed in the central Srednogorie zone that could be due to contamination caused by the Maritsa Iztok complex (Figure 4).

The PC3 map presents a spatial distribution of elevated factor scores regions coinciding with the ore deposits in Bulgaria, mainly copper, lead–zinc and polymetallic ores (Figure 4). Dimitrova et al. [40] reported higher concentrations of Cu, Zn, As and Pb in northwestern Bulgarian soils caused by mining activities in the region. It has already been mentioned that the largest mining areas are in the central Rhodope zone, the western part of the Balkan zone and the eastern part of the Srednogorie zone, but unexploited ore deposits are also located in the southern central part of the Moesian platform [41].

Most of the soil samples with the lowest factor scores of PC4, which indicates high nutrient content, are from the most fertile and agricultural (arable) type—Chernozem. These sites are the best for agricultural use, such as Dobrudzha (northeastern Bulgaria) and the valleys of the rivers Iskar, Struma and Vacha. The average concentration of nutrients for the samples with the lowest factor scores (below −1) comprises a very high content of C and P and a high content of N, whereas those with the highest factor scores (above 1) have an average content of C and low content of N and P.

The soil samples, which have high scores in the Hg-specific factor (PC5) are mainly derived from three soil types: Cambisols, Leptosols and Fluvisols. The parent material of the first two types are non-carbonate rocks, which have high acidity and low physical clay content (a high content of sand). The mean mercury concentration in the Bulgarian soil quality monitoring network (0.15 mg kg^{-1}) is quite high and follows the trend reported in other European studies, where soil mercury content increases from northern to southern Europe [42]. The high factor scores in the western Balkan zone and the central Rhodope zone are in a good agreement with the ore deposits, as the elevated Hg soil content in central Srednogorie region is presumably also due to the coal-fired power plant [38]. The northeastern part of Moesian platform (near the town of Silistra) is not an anthropogenically influenced area, which is an indication that such high Hg levels could be considered as baseline ones for the respective region.

The simultaneous consideration of the spatial distribution of soil quality patterns and arable land coverage could outline the regions where additional measures for the monitoring of the phytoavailability of PTEs and their transfer to plants are appropriate. The elevated concentrations of Cd and Pb (associated with PC1) are mainly found in mountain regions with limited agricultural activity. Concerning geogenic PTEs, in PC2, Ni and Cr require special measures for the monitoring of their phytoavailability only in the central Srednogorie region, which is affected by the Maritsa Iztok thermoelectric power plant. The most anthropogenically influenced soil quality pattern (PC3) presenting ore deposits and related mining and metal production activities has to be examined carefully. Excluding the regions with elevated factor scores of geogenic factor (PC2), which could be an indication for the higher regional background concentrations of Cu, Zn, As and Pb, the western Srednogorie zone remains an anthropogenically polluted region with significant agricultural activities. The previous study in this region [7] outlines EDTA soil extraction as a reliable procedure for the phytoavailability estimation of As, Cd, Cr, Cu, Mn and Pb. The soil–plant transfer of the main contaminants of Cu and As around the copper mining and smelter factories is controlled mainly by soil pH, total organic matter and $CaCO_3$. The transfer coefficient of Cu is positively correlated with soil pH and $CaCO_3$ and negatively with total organic matter. These relationships can be explained by the increasing stability of the Cu(II)-EDTA complex at higher pH values and the binding ability of organic soil fractions. The reason for positive correlation between soil–plant transfer and total organic matter can be found in the dissolution of As, which is bound to soil humic substances at the pH of the EDTA leaching procedure. The soil–plant transfers of Zn and Pb are controlled by Al and Fe soil contents, respectively. The only region with a significant presence of Hg-specific soil quality patterns (PC5) and with significant agricultural activity is the aforementioned central Srednogorie region, which is affected by the Maritsa Iztok thermoelectric power plant. It can be concluded that the soil management of the western and central Srednogorie zone needs additional measures for the monitoring of PTEs' phytoavailability.

4. Materials and Methods

4.1. Sampling and Chemical Analysis

In general, the sampling and sample preparation procedures are fully described in national directives and documents, according to international standards (ISO 10381-2:2005, ISO 10381-4:2005 and ISO 11464:2006) [43–45]. The 347 topsoil samples of the Bulgarian soil quality monitoring network were collected at depths of 0–20 cm and in the intersections of an orthogonal 16 km grid across the whole country (Figure 5).

Soil sampling was performed in the 2004–2009 period, and the soil samples were taken from different types of land with different usages. For each sample, a total of 13 soil indicators were measured as follows: pH, main nutrients (organic carbon—C; total nitrogen—N; total phosphorus—P), physical clay content and 8 potentially toxic elements (Cu, Zn, Cd, Pb, Ni, Cr, As, Hg).

The main soil indicators (pH, physical clay) and the nutrients were determined via the following analytical methods: pH (ISO 10390:2005) [46]; organic carbon via sulfochoromic oxidation (ISO 14235:2002) [47]; total nitrogen via the modified Kjeldahl method (ISO 11261:2002) [48]; and total phosphorus via a validated method based on acid mixture microwave digestion and analyzed through ICP-OES. The particle size distribution was determined through the method of sieving and sedimentation (ISO 11277:2009) [49]. Only a fraction of the physical clay (percentage of the soil particles with diameters of less than 0.02 mm) was used for statistical modeling.

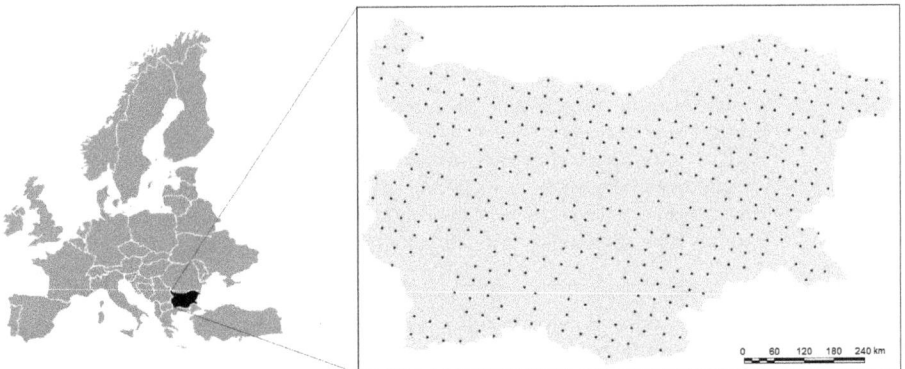

Figure 5. Locations of sampling points of the Bulgarian soil quality monitoring network.

The soil aqua regia extraction of potentially toxic elements was performed according to ISO 11466:1995 [50]. Six of the elements (Cu, Zn, Cd, Pb, Ni, and Cr) were determined via atomic absorption spectrometry (ISO 11047:1998) [51], and As and Hg were determined using ICP-MS, according to the CEN/TS 16171:2012 [52] standard.

4.2. Data Analysis Methods

For the chemometric assessment and interpretation of the obtained data, two multivariate statistical approaches were used: cluster analysis [53] and principal components analysis [54].

Cluster analysis (CA) is a well-known and widely used classification approach in chemometric and environmetric studies, with its hierarchical and non-hierarchical algorithms. The main goal of the hierarchical agglomerative cluster analysis was to spontaneously classify the data into groups of similarity (clusters). Usually, the sampling sites in traditional monitoring surveys are considered objects for classification, but it is also possible to search for links between the variables (different soil quality indicators) that characterize them. A preliminary step of CA is the normalization of the raw input data (e.g., autoscaling or z-transformation) in order to avoid the influence of the different range of chemical dimensions (concentration). As a result, normalized dimensionless numbers replaced the real data values. Then, the distance between the objects (or the variables) of classification was determined, usually by applying squared Euclidean distances as a similarity measure. There are a wide variety of hierarchical algorithms for object linkage—the single linkage, the complete linkage or the average linkage methods—but Ward's method is predominantly used, because it achieved balanced clustering, while taking into account the intra- and inter-cluster distances. The CA results are normally depicted using a tree-like scheme with a hierarchical structure called a dendrogram.

Principal components analysis (PCA) is widely used as a dimension-reducing, -modeling and -display method, which allows the estimation of internal relations in the data set. PCA enables the reduction of the coordinate system of the variables in the direction of the highest variance. The input variables are converted into new ones, which are better descriptors of the data structure. The new variables, called principal components (PCs) or latent factors, are a linear combination of the original variables. Usually, just a few of the latent factors account for a large part of the data set variation. Thus, the data structure in a reduced variable space can be observed and interpreted. As a result of PCA, the autoscaled original data matrix (**D**) was transformed into a product of two matrices as follows:

$$D = UV^T + E$$

where $\mathbf{V}^T$ is factor loadings matrix, revealing the contribution of each one of the original variables (indicators) to the newly formed principal components; $\mathbf{U}$ is factor score matrix, providing the new coordinates of each object (sampling point) in the new space of principal components; and $\mathbf{E}$ is residual matrix.

In this study, the Varimax rotation mode of PCA was applied for the better interpretation of the system (soil quality), as it increases the role of the soil indicators with higher impacts on the formation of principal components and decreases the role of soil indicators with lower impacts.

A GIS-based approach was chosen to present the spatial distribution of principal components achieved via PCA analysis. Principal components maps were plotted using the kriging interpolation of principal component factor scores (row of matrix $\mathbf{U}$) of all sampling points from the Bulgarian soil quality monitoring network.

5. Conclusions

The present study is a pioneer effort to integrate all aspects of a proper soil quality assessment for the territory of Bulgaria. The use of state-of-the-art monitoring and analytical and chemometric approaches made it possible to better understand and classify the factors related to different soil patterns in the country, such as spatial, geogenic, nutritional and anthropogenic. The mapping, based on the chemometric results, offers an additional insight into the soil specificity throughout the country through the specific distribution of the identified soil quality patterns. The content of potentially toxic elements in the soil samples reveals the low level of anthropogenic impact, as higher values of Ni and Cr are of geogenic origin and higher values of Pb and Cd are mainly observed in mountain regions; higher values of Cu, Zn and As are mainly found in regions with ore deposits; and higher values Hg can mainly be located in site-specific sandy acidic soils.

Such an assessment of the soil quality of a whole country requires many different prerequisites, namely a well-organized soil sampling network; well-trained analytical staff able to perform advanced determinations of all specific soil quality indicators—both structural and chemical; and the correct classification, modeling, and interpretation of the collected data. The proposed methodology is able to outline the anthropogenically influenced regions characterized by intensive agricultural activity where additional measures for the impact assessment of PTEs are required.

The results obtained via data mining and modeling ensure better soil quality management and sustainability as well as efficient agricultural decision making.

Author Contributions: Conceptualization, V.S. and S.T.; methodology, S.T.; software, M.P., V.S. and S.T.; validation and formal analysis, G.Y., M.H. and M.P.; investigation, G.Y., M.H., M.P., N.D., V.S. and S.T.; resources, G.Y., M.H., M.P., N.D., V.S. and S.T.; data curation, G.Y., M.H., M.P., N.D., V.S. and S.T.; writing—original draft preparation, G.Y., M.H., M.P., N.D., V.S. and S.T.; writing—review and editing, G.Y., V.S. and S.T.; visualization, G.Y., M.P. and S.T.; supervision, N.D. and S.T.; project administration, N.D. and S.T.; funding acquisition, N.D. and S.T. All authors have read and agreed to the published version of the manuscript.

Funding: This research was funded by the National Science Program "Environmental Protection and Reduction of Risks of Adverse Events and Natural Disasters", approved by the Resolution of the Council of Ministers, No. 577/17 August 2018, and supported by the Ministry of Education and Science (MES) of Bulgaria (agreement No. Д01-271/9 December 2022). The work was partially supported by the National Science Fund (project DFNI E 02/7-12 December 2014).

Institutional Review Board Statement: Not applicable.

Informed Consent Statement: Not applicable.

Data Availability Statement: The data presented in this study are available on request from the authors.

Acknowledgments: This work was carried out within the framework of the National Science Program "Environmental Protection and Reduction of Risks of Adverse Events and Natural Disasters", approved by the Resolution of the Council of Ministers, No. 577/17 August 2018, and supported by the Ministry of Education and Science (MES) of Bulgaria (agreement No. Д01-271/9 December 2022). This work was partially supported by the National Science Fund (project DFNI E 02/7-12 December 2014.

Conflicts of Interest: The authors declare no conflict of interest.

Sample Availability: Samples of the compounds are not available.

References

1. Karlen, D.L.; Mausbach, M.J.; Doran, J.W.; Cline, R.G.; Harris, R.F.; Schuman, G.E. Soil quality: A concept, definition, and framework for evaluation (a guest editorial). *Soil Sci. Soc. Am. J.* **1997**, *61*, 4–10. [CrossRef]
2. Van Leeuwen, J.P.; Saby, N.P.A.; Jones, A.; Louwagie, G.; Micheli, E.; Rutgers, M.; Schulte, R.P.O.; Spiegel, H.; Toth, G.; Creamer, R.E. Gap assessment in current soil monitoring networks across Europe for measuring soil functions. *Environ. Res. Lett.* **2017**, *12*, 124007. [CrossRef]
3. Morvan, X.; Saby, N.P.A.; Arrouays, D.; Le Bas, C.; Jones, R.J.A.; Verheijen, F.G.A.; Bellamy, P.H.; Stephens, M.; Kibblewhite, M.G. Soil monitoring in Europe: A review of existing systems and requirements for harmonization. *Sci. Total Environ.* **2008**, *391*, 1–12. [CrossRef] [PubMed]
4. Batjes, N.H. *Options for Harmonising Soil Data Obtained from Different Sources*; ISRIC—World Soil Information: Wageningen, The Netherlands, 2023.
5. Cornu, S.; Keesstra, S.; Bispo, A.; Fantappie, M.; Smreczak, B.; Wawer, R.; Pavlů, L.; Sobocká, J.; Bakacsi, Z.; Farkas-Iványi, K.; et al. National soil data in EU countries, where do we stand? *Eur. J. Soil Sci.* **2023**, e13398. [CrossRef]
6. Reimann, C.; Fabian, K.; Birke, M.; Filzmoser, P.; Demetriades, A.; Négrel, P.; Sadeghi, M. GEMAS: Establishing geochemical background and threshold for 53 chemical elements in European agricultural soil. *Appl. Geochem.* **2018**, *88*, 302–318. [CrossRef]
7. Yotova, G.; Zlateva, B.; Ganeva, S.; Simeonov, V.; Kudlak, B.; Namiesnik, J.; Tsakovski, S. Phytoavailability of potentially toxic elements from industrially contaminated soils to wild grass. *Ecotoxicol. Environ. Saf.* **2018**, *164*, 317–324. [CrossRef]
8. Antoniadis, V.; Shaheen, S.M.; Boersch, J.; Frohne, T.; Laing, G.D.; Rinklebe, J. Bioavailability and risk assessment of potentially toxic elements in garden edible vegetables and soil around a highly contaminated former mining area in Germany. *J. Environ. Manag.* **2017**, *186 Pt 2*, 192–200. [CrossRef]
9. Andrade, J.M.; Kubista, M.; Carlosena, A.; Prada, D. 3-Way characterization of soils by Procrustes rotation, matrix-augmented principal components analysis and parallel factor analysis. *Anal. Chim. Acta* **2007**, *603*, 20–29. [CrossRef]
10. Bitencourt, D.G.B.; Barros, W.S.; Timm, L.C.; She, D.; Penning, L.H.; Parfitt, J.M.B.; Reichardt, K. Multivariate and geostatistical analyses to evaluate lowland soil levelling effects on physico-chemical properties. *Soil Tillage Res.* **2016**, *156*, 63–73. [CrossRef]
11. Ćujić, M.; Dragović, S.; Đorđević, M.; Dragović, R.; Gajić, B. Environmental assessment of heavy metals around the largest coal fired power plant in Serbia. *Catena* **2016**, *139*, 44–52. [CrossRef]
12. Nosrati, K. Assessing soil quality indicator under different land use and soil erosion using multivariate statistical techniques. *Environ. Monit. Assess.* **2013**, *185*, 2895–2907. [CrossRef] [PubMed]
13. Pandey, B.; Agrawal, M.; Singh, S. Ecological risk assessment of soil contamination by trace elements around coal mining area. *J. Soils Sediments* **2016**, *16*, 159–168. [CrossRef]
14. Qu, M.K.; Li, W.D.; Zhang, C.R.; Wang, S.Q.; Yang, Y.; He, L.Y. Source apportionment of heavy metals in soils using multivariate statistics and geostatistics. *Pedosphere* **2013**, *23*, 437–444. [CrossRef]
15. Sağlam, M.; Dengiz, O.; Saygın, F. Assessment of horizantal and vertical variabilities of soil quality using multivariate statistics and geostatistical methods. *Commun. Soil Sci. Plant Anal.* **2015**, *46*, 1677–1697. [CrossRef]
16. Singh, S.; Raju, N.J.; Nazneen, S. Environmental risk of heavy metal pollution and contamination sources using multivariate analysis in the soils of Varanasi environs, India. *Environ. Monit. Assess.* **2015**, *187*, 345. [CrossRef]
17. Stefanoski, D.C.; De Figueiredo, C.C.; Santos, G.G.; Marchão, R.L. Selecting soil quality indicators for different soil management systems in the Brazilian Cerrado. *Pesq. Agropec. Bras.* **2016**, *51*, 1643–1651. [CrossRef]
18. Wang, G.; Zhang, S.; Xiao, L.; Zhong, Q.; Li, L.; Xu, G.; Deng, O.; Pu, Y. Heavy metals in soils from a typical industrial area in Sichuan, China: Spatial distribution, source identification, and ecological risk assessment. *Environ. Sci. Pollut. Res. Int.* **2017**, *24*, 16618–16630. [CrossRef]
19. Wu, C.; Zhang, L. Heavy metal concentrations and their possible sources in paddy soils of a modern agricultural zone, southeastern China. *Environ. Earth Sci.* **2010**, *60*, 45–56. [CrossRef]
20. Zhang, J.; Wang, Y.; Liu, J.; Liu, Q.; Zhou, Q. Multivariate and geostatistical analyses of the sources and spatial distribution of heavy metals in agricultural soil in Gongzhuling, Northeast China. *J. Soils Sediments* **2016**, *16*, 634–644. [CrossRef]
21. Zhao, L.; Xu, Y.; Hou, H.; Shangguan, Y.; Li, F. Source identification and health risk assessment of metals in urban soils around the Tanggu chemical industrial district, Tianjin, China. *Sci. Total Environ.* **2014**, *468–469*, 654–662. [CrossRef]
22. Diodato, N.; Ceccarelli, M. Multivariate indicator Kriging approach using a GIS to classify soil degradation for Mediterranean agricultural lands. *Ecol. Indic.* **2004**, *4*, 177–187. [CrossRef]

23. Lado, L.R.; Hengl, T.; Reuter, H.I. Heavy metals in European soils: A geostatistical analysis of the FOREGS Geochemical database. *Geoderma* **2008**, *148*, 189–199. [CrossRef]
24. Nazzal, Y.H.; Al-Arif, N.S.N.; Jafri, M.K.; Kishawy, H.A.; Ghrefat, H.; El-Waheidi, M.M.; Batayneh, A.; Zumlot, T. Multivariate statistical analysis of urban soil contamination by heavy metals at selected industrial locations in the Greater Toronto area, Canada. *Geol. Croat.* **2015**, *68*, 147–159. [CrossRef]
25. Rodriguez-Iruretagoiena, A.; De Vallejuelo, S.F.O.; Gredilla, A.; Ramos, C.G.; Oliveira, M.L.S.; Arana, G.; De Diego, A.; Madariaga, J.M.; Silva, L.F.O. Fate of hazardous elements in agricultural soils surrounding a coal power plant complex from Santa Catarina (Brazil). *Sci. Total Environ.* **2015**, *508*, 374–382. [CrossRef]
26. Sun, C.; Zhao, W.; Zhang, Q.; Yu, X.; Zheng, X.; Zhao, J.; Lv, M. Spatial distribution, sources apportionment and health risk of metals in topsoil in Beijing, China. *Int. J. Environ. Res. Public Health* **2016**, *13*, 727. [CrossRef]
27. Ha, H.; Olson, J.R.; Bian, L.; Rogerson, P.A. Analysis of heavy metal sources in soil using kriging interpolation on principal components. *Environ. Sci. Technol.* **2014**, *48*, 4999–5007. [CrossRef]
28. Filcheva, E. *Characteristics of Bulgarian Soils on: Content, Composition and Stocks of Organic Matter. Grouping of Bulgarian Soils*; Minerva: Sofia, Bulgaria, 2007.
29. Regulation No3, 1 August 2008. On the Permissible Limits of Toxic Substances in the Soils. Available online: http://eea.government.bg/bg/legislation/soil/normipochvi.doc/view (accessed on 15 May 2023).
30. Filcheva, E. *Characteristics of Soil Organic Matter of Bulgarian Soils*; LAP Lambert Academic Publishing: Saarbrücken, Germany, 2015.
31. Lee, S.Z.; Chang, L.; Yang, H.H.; Chen, C.M.; Liu, M.C. Adsorption characteristics of lead onto soils. *J. Hazard. Mater.* **1998**, *63*, 37–49. [CrossRef]
32. Qiutong, X.; Mingkui, Z. Source identification and exchangeability of heavy metals accumulated in vegetable soils in the coastal plain of eastern Zhejiang province, China. *Ecotoxicol. Environ. Saf.* **2017**, *142*, 410–416. [CrossRef]
33. Rodríguez, J.A.; Nanos, N.; Grau, J.M.; Gil, L.; López-Arias, M. Multiscale analysis of heavy metal contents in Spanish agricultural topsoils. *Chemosphere* **2008**, *70*, 1085–1096. [CrossRef]
34. Loebenstein, J.R. *The Materials Flow of Arsenic in the United States*; U.S. Bureau of Mines Information: Washington, DC, USA, 1994.
35. Tian, L.; Zhao, L.; Wu, X.; Fang, H.; Zhao, Y.; Yue, G.; Liu, G.; Chen, H. Vertical patterns and controls of soil nutrients in alpine grassland: Implications for nutrient uptake. *Sci. Total Environ.* **2017**, *607–608*, 855–864. [CrossRef]
36. Panagos, P.; Van Liedekerke, M.; Jones, A.; Montanarella, L. European Soil Data Centre: Response to European policy support and public data requirements. *Land Use Policy* **2012**, *29*, 329–338. [CrossRef]
37. Zagorchev, I.; Dabovski, C.; Nikolov, T. *Geology of Bulgaria. Volume 2, Mesozoic Geology*; Prof. M. Drinov Academic Publishing House: Sofia, Bulgaria, 2009.
38. Yotova, G.; Padareva, M.; Hristova, M.; Astel, A.; Georgieva, M.; Dinev, N.; Tsakovski, S. Establishment of geochemical background and threshold values for 8 potential toxic elements in the Bulgarian soil qualitymonitoring network. *Sci. Total Envioron.* **2018**, *643*, 1297–1303. [CrossRef] [PubMed]
39. Kamenov, B. *Magmatic Petrology*; University of Sofia Publishing House: Sofia, Bulgaria, 2003.
40. Dimitrova, D.; Velitchkova, N.; Mladenova, V.; Kotsev, T.; Antonov, D. Heavy metal and metalloid mobilisation and rates of contamination of water, soil and bottom sediments in the Chiprovtsi mining district, Northwestern Bulgaria. *Geol. Balc.* **2016**, *45*, 47–63. [CrossRef]
41. Gerginov, P.; Kerestedjian, T.; Toteva, A.; Mihaylova, B.; Benderev, A. Geological environment, groundwater quality and regulation. *Water Aff.* **2019**, *5–6*, 19–29.
42. Ottesen, R.T.; Birke, M.; Finne, T.E.; Gosar, M.; Locutura, J.; Reimann, C.; Tarvainen, T. Mercury in European agricultural and grazing land soils. *Appl. Geochem.* **2013**, *33*, 1–12. [CrossRef]
43. *ISO 10381-2:2005*; Soil Quality—Sampling—Part 2: Guidance on Sampling Techniques. ISO: Geneva, Switzerland, 2005.
44. *ISO 10381-4:2005*; Soil Quality—Sampling—Part 4: Guidance on the Procedure for Investigation of Natural, Near-Natural and Cultivated Sites. ISO: Geneva, Switzerland, 2005.
45. *ISO 11464:2006*; Soil Quality—Pretreatment of Samples for Physico-Chemical Analysis. ISO: Geneva, Switzerland, 2006.
46. *ISO 10390:2005*; Soil quality—Determination of pH. ISO: Geneva, Switzerland, 2005.
47. *ISO 14235:2002*; Soil Quality—Determination of Organic Carbon by Sulfochromic Oxidation. ISO: Geneva, Switzerland, 2002.
48. *ISO 11261:2002*; Soil Quality—Determination of Total Nitrogen—Modified Kjeldahl Method. ISO: Geneva, Switzerland, 2002.
49. *ISO 11277:2009*; Soil Quality—Determination of Particle Size Distribution in Mineral Soil Material—Method by Sieving and Sedimentation. ISO: Geneva, Switzerland, 2009.
50. *ISO 11466:1995*; Soil Quality—Extraction of Trace Elements Soluble in Aqua Regia. ISO: Geneva, Switzerland, 1995.
51. *ISO 11047:1998*; Soil Quality—Determination of Cadmium, Chromium, Cobalt, Copper, Lead, Manganese, Nickel and Zinc—Flame and Electrothermal Atomic Absorption Spectrometric Methods. ISO: Geneva, Switzerland, 1998.
52. *CEN/TS 16171:2012*; Sludge, Treated Biowaste and Soil—Determination of Elements Using Inductively Coupled Plasma Mass Spectrometry (ICP-MS). CEN/TS: Brussels, Belgium, 2012.

53. Massart, D.L.; Kaufman, L. *The Interpretation of Analytical Chemical Data by the Use of Cluster Analysis*; Wiley Interscience: New York, NY, USA, 1983.
54. Vandeginste, B.; Massart, D.L.; Buydens, L.; De Jong, S.; Lewi, P.; Smeyers-Verbeke, J. *Handbook of Chemometrics and Qualimetrics*; Elsevier: Amsterdam, The Netherlands, 1998.

Disclaimer/Publisher's Note: The statements, opinions and data contained in all publications are solely those of the individual author(s) and contributor(s) and not of MDPI and/or the editor(s). MDPI and/or the editor(s) disclaim responsibility for any injury to people or property resulting from any ideas, methods, instructions or products referred to in the content.

Article

Sediment Assessment of the Pchelina Reservoir, Bulgaria

Tony Venelinov [1], Veronika Mihaylova [2], Rositsa Peycheva [3], Miroslav Todorov [4], Galina Yotova [2], Boyan Todorov [2], Valentina Lyubomirova [2] and Stefan Tsakovski [2,*]

[1] Chair of Water Supply, Sewerage, Water and Wastewater Treatment, Faculty of Hydraulic Engineering, University of Architecture, Civil Engineering and Geodesy, 1 Hr. Smirnenski Blvd., 1046 Sofia, Bulgaria; tvenelinov_fhe@uacg.bg

[2] Chair of Analytical Chemistry, Faculty of Chemistry and Pharmacy, Sofia University "St. Kliment Ohridski", 1 J. Bourchier Blvd., 1164 Sofia, Bulgaria; v.mihaylova@chem.uni-sofia.bg (V.M.); g.yotova@chem.uni-sofia.bg (G.Y.); b.todorov@chem.uni-sofia.bg (B.T.); vlah@chem.uni-sofia.bg (V.L.)

[3] DIAL Ltd., 111 Mina Buhovo Str., 1830 Sofia-Buhovo, Bulgaria; rbarganska11@yahoo.bg

[4] Chair of Hydrotechnics, Faculty of Transportation Engineering, University of Architecture, Civil Engineering and Geodesy, 1 Hr. Smirnenski Blvd., 1046 Sofia, Bulgaria; miro_todorof@yahoo.com

* Correspondence: stsakovski@chem.uni-sofia.bg; Tel.: +359-2-8161426

Abstract: The temporal dynamics of anthropogenic impacts on the Pchelina Reservoir is assessed based on chemical element analysis of three sediment cores at a depth of about 100–130 cm below the surface water. The ^{137}Cs activity is measured to identify the layers corresponding to the 1986 Chernobyl accident. The obtained dating of sediment cores gives an average sedimentation rate of 0.44 cm/year in the Pchelina Reservoir. The elements' depth profiles (Ti, Mn, Fe, Zn, Cr, Ni, Cu, Mo, Sn, Sb, Pb, Co, Cd, Ce, Tl, Bi, Gd, La, Th and U$_{nat}$) outline the Struma River as the main anthropogenic source for Pchelina Reservoir sediments. The principal component analysis reveals two groups of chemical elements connected with the anthropogenic impacts. The first group of chemical elements (Mn, Fe, Cr, Ni, Cu, Mo, Sn, Sb and Co) has increasing time trends in the Struma sediment core and no trend or decreasing ones at the Pchelina sampling core. The behavior of these elements is determined by the change of the profile of the industry in the Pernik town during the 1990s. The second group of elements (Zn, Pb, Cd, Bi and U$_{nat}$) has increasing time trends in Struma and Pchelina sediment cores. The increased concentrations of these elements during the whole investigated period have led to moderate enrichments for Pb and U$_{nat}$, and significant enrichments for Zn and Cd at the Pchelina sampling site. The moderately contaminated, according to the geoaccumulation indexes, Pchelina Reservoir surface sediment samples have low ecotoxicity.

Keywords: Pchelina Reservoir; sediment; principal component analysis; Mann–Kendall test; ecotoxicity

1. Introduction

Chemical elements are among the most widespread of the various pollutants originating from anthropogenic activities, particularly from mining, metallurgy and smelting waste sites. They are one of the most persistent pollutants in the environment, since they do not decompose, nor do they biodegrade into simpler and less harmful substances. Their concentrations in water are quite variable, with the highest concentrations found in the suspended matter (insoluble substances) and the lowest in the liquid phase [1–3]. The distribution of the elements is primarily on the surfaces of sediments, suspended particles, and other solids, decreasing in the order: suspended matter > sediment > water [4,5]. Chemical processes occurring in natural waters, such as oxidation, can cause them to precipitate from the solution [6].

As an integral part of natural water sources, sediments play an important role in the biogeochemical cycle of the elements, as they are the site of deposition and chemical transformation of many compounds entering the water bodies [7]. The elements enter surface waters from many sources, in the form of atmospheric deposits, or are leached

from rocks and soil. They are not biodegradable, but bind to proteins, thus being stored in the bodies of water organisms or excreted in their feces [8], which under certain conditions leads to secondary contamination of the water bodies. In natural waters, chemical elements are found in many forms: as free ions (the most toxic forms for living organisms); in the form of various complexes; as precipitated compounds suspended in the aqueous phase; and adsorbed on the surface of other suspended or colloidal particles [9].

The history of anthropogenic water pollution can be determined by analysis of sediments [10–12]. Reservoir sediments provide fine-scale information on the historical record of metal pollution in a watershed [13–15]. They have recorded the elemental deposition and thus allow establishing a connection between the temporal evolution of the pollution and historical changes in smelting and waste treatment processes [13]. The resulting compositional datasets are usually tested by principal component analyses, self-organizing maps, and cluster analyses, with their pollution load index (PLI), index of geoaccumulation (I_{geo}) and enrichment factors (EFs) being calculated [16–26].

Reservoir sediments accumulate at a high rate (usually > 2 cm/year [27]), in contrast to river sediments (usually < 0.3 cm/year [26]), which is why they can reveal the accumulation of the elements over time. Due to these sedimentation rates, reservoir sediments are considered to be slightly affected by early diagenesis processes and provide preserved historical elemental inputs. Reservoir sediments are also of great concern, since they can turn from a sink to a source of chemical elements for fluvial systems by diffusion at the water/sediment interface, bioturbation or resuspension due to dredging or flooding. Thus, it is important to determine the intensity of pollution by inventorying the elemental concentrations and their spatial distribution in sediments [28].

Pchelina Reservoir was built in 1975 and serves not only as a secondary precipitator of the Struma River but also as a source for irrigation and industrial water supply of Pernik Municipality, Bulgaria. The total volume of the reservoir is 54.8 million m^3 (of which the useful volume is 19.3 million m^3), and its depth reaches 19 m [29]. The main sources of water pollution at Pchelina Reservoir are defined as point and diffuse. Point sources are the sewerage systems from the settlements, discharged without purification in treatment plants, the wastewater treatment plants, the industrial sites, the tailings, and the mines. Diffuse sources of pollution are unregulated landfills for solid waste, settlements without sewerage, landfills, and agricultural activities such as animal husbandry. The main pollutant of the river Struma is the town of Pernik, the heavy industry of which has traditionally been dominated by mining activities and metallurgy. The industrial profile of the town changed in the 1990s, but the wastewaters discharged into the river are still collected in the Pchelina Reservoir. The only available studies of the sediments of Pchelina Reservoir were made by Meuser and co-authors [30,31].

This study aims to propose a methodology for assessing the temporal dynamics of anthropogenic impacts on sediments of Pchelina Reservoir. The proposed methodology includes: (i) analysis of chemical elements and ^{137}Cs radionuclides of three sediment cores taken from three selected sampling points in Pchelina Reservoir, (ii) multivariate and time trend statistical analysis of the sediment cores data, and (iii) calculation of enrichment factors and geochemical indexes of surface sediments followed by ecotoxicity assessment using Phytotoxkit F ™ bioassay.

2. Results

To reveal the relationships between the analyzed chemical elements and/or layers in the sediment cores, a principal component analysis (PCA) was applied. The input data set used for PCA consists of 58 objects (layers in the sedimentary cores) and 20 variables (analyzed chemical elements). PCA of the data from the three sedimentary cores revealed that the first three main components describe almost 80% of the variation of the data. The number of latent variables is determined based on their eigenvalues and the internal model validation error. In the formation of the first principal component, explaining 41.53% of data variance, the following elements have a significant impact: Mn, Fe, Cr, Ni, Cu, Mo, Sn,

Sb and Co (Figure 1). This component separates the elements into two groups according to their time trends in Struma sediment core: (i) Mn, Fe, Zn, Cr, Ni, Cu, Mo, Sn, Sb, Pb, Co, Cd and U have increased concentrations with time; and (ii) Ti, Ce, Tl, Bi, Gd, La and Th have decreased or non-significant time trends.

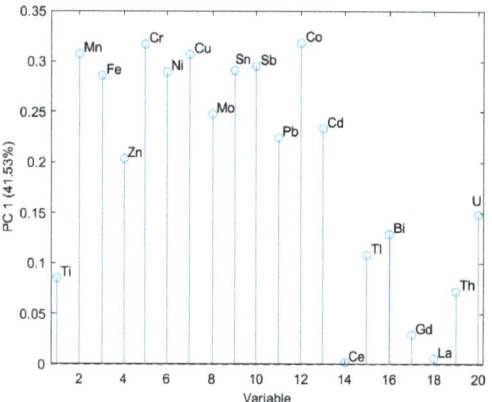
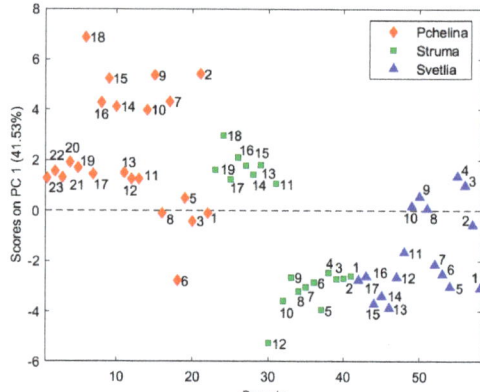

Figure 1. Factor weights and factor scores of the first principal components (the numbers on the graph indicate the historical sequence of layers in sediment core with the first (1) being the oldest).

The second principal component (22.16%) is formed by Ce, Gd, La, Th, and Ti, which significantly decrease over time in the sedimentary cores at Struma and Svetlia rivers (Figure 2). These decreasing trends do not lead to a significant change in the contents of the elements in the Pchelina Reservoir, which are comparable to the contents in the sedimentary cores near the Struma and Svetlia at the beginning of the period.

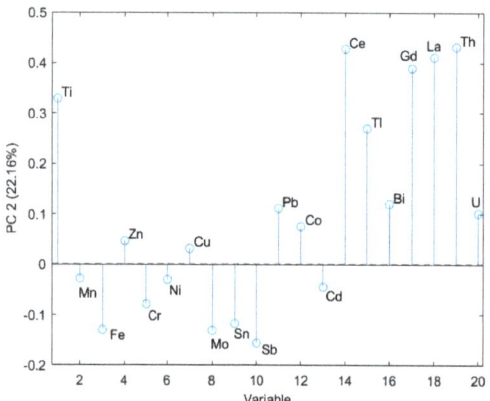
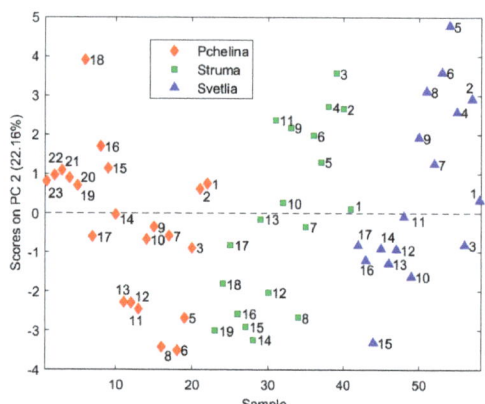

Figure 2. Factor weights and factor scores of the second principal components (the numbers on the graph indicate the historical sequence of layers in sediment core with the first (1) being the oldest).

The third principal component (15.44%) is formed by the elements Zn, Pb, Cd, Bi and U_{nat}. The factor scores of the layers of the Pchelina core show a particularly pronounced positive trend, which leads to the formation of two groups of layers. The first group, covering the beginning of the studied period (1–11), has contents comparable to the sedimentary core of the Svetlia (anthropogenically undisturbed river), while the contents of the elements in the second group (12–23) are the highest for all the three studied sediment cores (Figure 3). The layers in the Struma sediment core have medium factor scores between both groups of Pchelina sediment core.

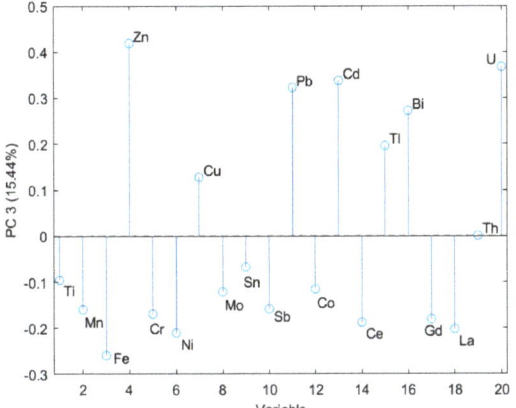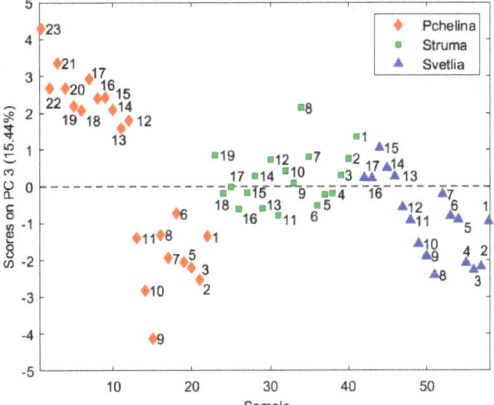

Figure 3. Factor weights and factor scores of the third principal components (the numbers on the graph indicate the historical sequence of layers in sediment core with the first (1) being the oldest).

^{137}Cs has been widely applied as an environmental tracer in the study of sediment recent deposition history (usually within the last 50 years) [32,33]. Each of the 2 cm sediment core fractions was analyzed for its ^{137}Cs content. Only in one of these fractions for each sediment core, radioactivity (γ-activity) was found and the content of ^{137}Cs measured was between 32.8 Bq/kg to 55.3 Bq/kg specific activities. Based on these findings, a conclusion was drawn that, on average, 15 cm of sediment was deposited in the 34 years since the Chernobyl incident (1986). This means that the average sedimentation rate at the sampling points in Pchelina Reservoir is 0.44 cm/year. Such results differ from the literature values—usually > 2 cm/year [27,34]—but are closer to the reported values for the rates of river sediment—usually < 0.3 cm/year) [30].

The element depth profiles of sediments from Pchelina Reservoir at the different sampling points (Pchelina, Struma and Svetlia) are shown in Figure 4. Red points indicate the sample in which the highest radioactivity (γ-activity) has been registered, which corresponds to the Chernobyl pollution of 1986. A similar approach was used by Audry et al. [13] for the sediment core dating of the Lot River reservoirs.

To determine whether significant time trends of elements were observed, the Mann–Kendell test was performed. The results are presented in Table 1, with significant trends ($p < 0.05$) marked with "+" for increasing and "−" for decreasing trends. The significant time tendencies in the sediment cores influenced by the flowing rivers reveal that Struma River (sampling point 1) is the main source of Pchelina Reservoir pollution.

According to the sedimentation rate calculations, it is assumed that any concentration of each element in a layer more than 6 cm below the sample, which corresponds to the Chernobyl pollution (marked in red in Figure 4), is a background concentration. The average result for Fe (used as a conservative element) of all such samples was calculated for each sampling point and used to obtain the enrichment factor for the sample corresponding to 2020 (top layers, Table 2).

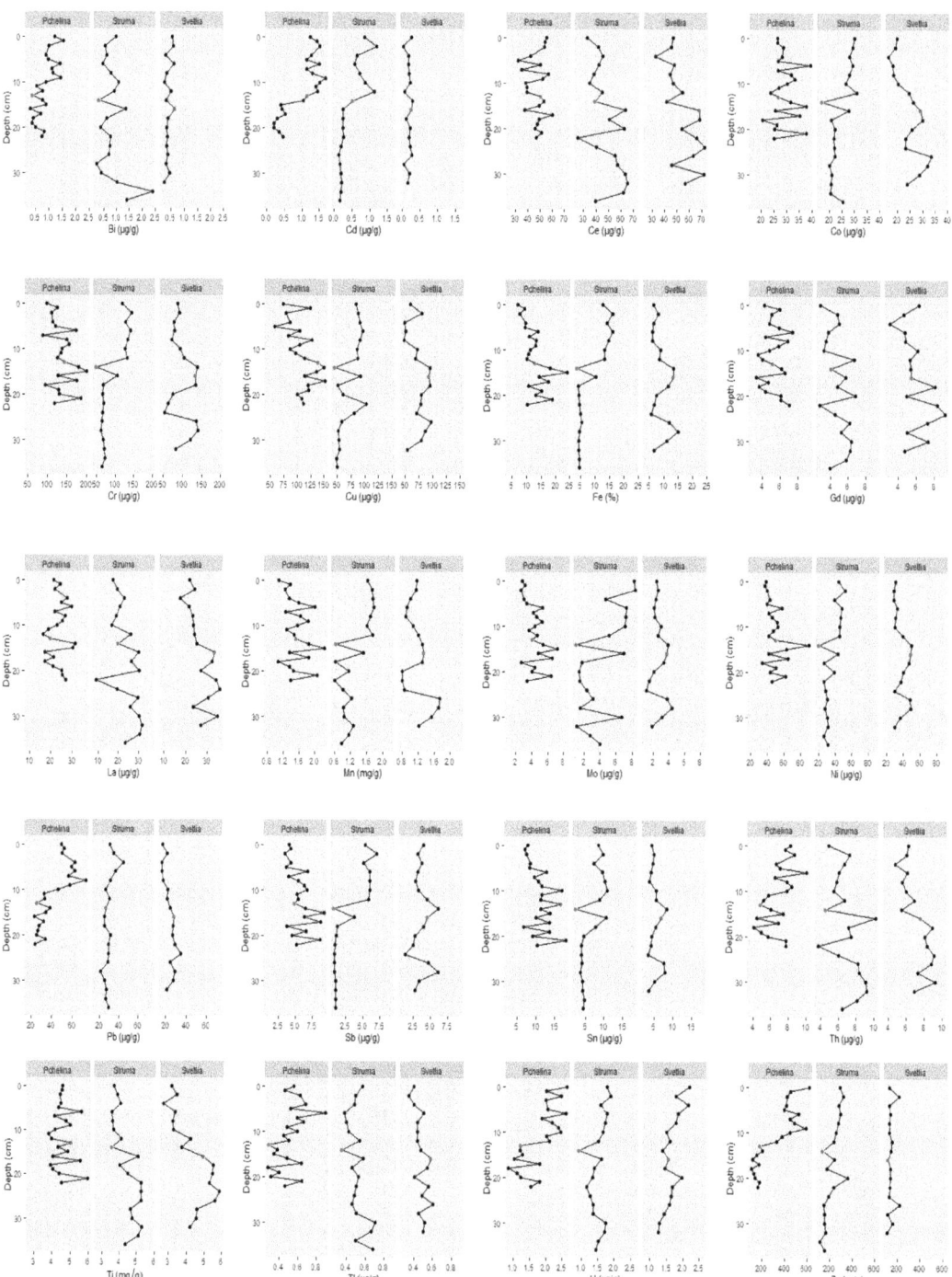

Figure 4. Elements' depth profiles of sediment cores taken at Pchelina, Struma and Svetlia.

Table 1. Trend analysis of sediment cores at three sampling points.

Element	Pchelina	Struma	Svetlia
Bi	+		
Cd	+	+	
Ce		−	−
Co		+	−
Cr		+	
Cu	−	+	
Fe	−	+	
Gd			−
La		−	−
Mn		+	
Mo		+	
Ni	−	+	
Pb	+	+	−
Sb	−	+	
Sn	−	+	
Th	+		−
Ti		−	−
Tl	+		−
U_{nat}	+	+	+
Zn	+	+	

Table 2. Enrichment factor (EF) in the three sampling points (yellow—moderate enrichment, red—significant enrichment).

Element	EF Struma River	EF Svetlia River	EF Pchelina
Ti	0.3	1.2	1.8
Mn	0.5	1.3	1.3
Fe	−	−	−
Zn	0.7	1.8	7.6
Cr	0.7	1.3	1.4
Ni	0.5	1.4	1.5
Cu	0.5	1.4	2.4
Mo	1.0	1.3	1.2
Sn	0.7	1.4	1.3
Sb	2.0	1.7	1.3
Pb	0.4	1.2	3.6
Co	0.5	1.2	1.8
Cd	1.8	3.2	7.5
La	0.2	1.4	1.7
Ce	0.2	1.4	2.4
Tl	0.4	2.7	3.5
Bi	0.4	1.9	1.3
Gd	0.2	1.3	1.7
Th	0.2	1.4	2.3
U_{nat}	0.4	2.8	3.5

Significant enrichment is observed only in the reference point (Pchelina), in terms of Zn and Cd (7.6 and 7.5, respectively). Regarding Cd, the enrichment in Svetlia is moderate (3.2). The enrichment factor is 1.8 in the Struma, which shows significant pollution with this metal in the entire Reservoir. Except for these elements, moderate enrichment is observed for Cu, Pb, Ce and Th only at Pchelina, as well as for Tl and U_{nat} in Svetlia and Pchelina.

The same approach for determination of the reference concentration (average of all results for the layers more than 6 cm below the sample, which corresponds to the Chernobyl pollution—marked in red in Figure 4) for each of the studied elements was used to determine the geoaccumulation index. The results are presented in Table 3.

Table 3. Geoaccumulation index (I_{geo}) in the three sampling points (yellow—uncontaminated to moderately contaminated sample, red—moderately contaminated sample).

Element	I_{geo} Struma	I_{geo} Svetlia	I_{geo} Pchelina
Ti	−1.01	−1.06	−0.70
Mn	−0.03	−0.95	−1.19
Fe	0.85	−1.37	−1.58
Zn	0.32	−0.55	1.35
Cr	0.26	−0.97	−1.13
Ni	−0.01	−0.91	−1.00
Cu	−0.02	−0.90	−0.31
Mo	0.84	−0.98	−1.35
Sn	0.44	−0.93	−1.26
Sb	1.87	−0.61	−1.18
Pb	−0.36	−1.15	0.27
Co	−0.11	−1.10	−0.74
Cd	1.69	0.30	1.33
La	−1.24	−0.90	−0.79
Ce	−1.34	−0.88	−0.34
Tl	−0.58	0.06	0.25
Bi	−0.61	−0.44	−1.19
Gd	−1.51	−0.99	−0.85
Th	−1.28	−0.91	−0.37
U_{nat}	−0.33	0.14	0.21

It is evident from Table 3 that for most elements, the geoaccumulation index corresponds to an uncontaminated sample. This result is especially important for the sediments at the inflow of the Svetlia River (sampling point 2), where only for Cd, Tl and U_{nat} the index corresponds to an unpolluted to moderately polluted sample. The geoaccumulation index in the sediments at the inflow of the Struma River (sampling point 1) corresponds to an unpolluted to moderately contaminated sample for Fe, Cr, Mo and Sn and to a moderately polluted sample for Sb and Cd. The geoaccumulation index for the sediments of Pchelina Reservoir from the village of Radibosh (sampling point 3), which corresponds to an unpolluted to moderately polluted sample in terms of Pb, Tl and U_{nat} (similar to the sediments in Svetlia) and a moderately polluted sample in terms of Zn and Cd (analogous to the sediments in the Struma). The moderate contamination of the sediments of Pchelina Reservoir in all sampling points with Cd is impressive. Similar results were obtained for sediments of Wadi Al-Arab Dam, Jordan [35], where sediments were found uncontaminated with Mn, Fe, and Cu, moderately contaminated with Zn, and strongly to extremely contaminated with Cd. Ye et al. [36] compared deposits of "heavy metals", accumulated in the water-level-fluctuation zone before and after the submergence period and found that Cd is the main pollutant of the sediment.

The results of the conducted biotest Phytotoxkit F ™ are presented in Table 4 and reveal the low ecotoxicity of the surface sediments from Pchelina Reservoir.

Table 4. Ecotoxicological studies using Phytotoxkit F ™ applied to the sediments from the three sampling points.

Sampling Point	Ecotoxicological Effect SG (%)	Ecotoxicological Effect RG (%)
Struma	0	23.26
Svetlia	0	17.56
Pchelina	0	−0.26

The indicator related to seed germination (SG) of Sinapis alba shows the lack of ecotoxicological effect. This means that the number of germinated seeds in the analyzed sediments is equal to their number in the control sample, in this case, all 10 seeds have germinated. The ecotoxicological effect, reflecting the growth of the roots of Sinapis alba, is relatively weak, as it is highest in the sample from Struma River, and even shows a negative value in Pchelina (hormesis). This means that the roots of the test species used are longer than the control sample.

3. Discussion

The background concentrations strongly depend on geological characteristics such as mineral composition, grain size distribution and organic matter content. Thus, establishing geochemical background concentrations of chemical elements is a very important step in environmental pollution assessment [37]. ^{137}Cs as an indicator of sedimentation processes is consistent as it binds almost irreversibly to clay and silt particles and because of its relatively long half-life ($t_{1/2}$ = 30.2 years). Moreover, the Chernobyl nuclear accident of 1986 has been recorded by the European sediments. Thus, ^{137}Cs activity depth profiles are often used for sediment core dating [38].

Based on the measurement of γ-activity and the calculated sedimentation rate in the Pchelina Reservoir of an average of 0.44 cm/year, it can be assumed that three of the points under 1986 (marked in red) are samples of sediment (approximately 6 cm), and the rest are from the natural soil cover, flooded during the construction of the Reservoir (1975). The remaining samples should contain levels of the elements that correspond to the background concentrations of these elements before they are anthropogenically affected. This was the basis for the analysis of the elements' core depth profiles in the present study. It is noteworthy that the concentrations of most of the studied elements increase, except for Bi, La, Ti, Tl, where the concentrations decrease over time, while Ce, Gd, Th practically do not change.

The PCA results divided the analyzed chemical elements into three groups. The elements with significant impact in the formation of the first principal component (Mn, Fe, Cr, Ni, Cu, Mo, Sn, Sb and Co) have an increasing time trend in Struma sediment core (sampling site 1) and no trend (Co, Cr, Mn, Mo) or a decreasing one (Cu, Fe, Ni, Sb, Sn) at the Pchelina sampling site. The factor scores of the layers in the Pchelina sediment core (sampling site 3) increase until the late 1990s and at the end of the investigated period decreased to the levels between 1988 and 1994. The absence of an increasing trend in concentrations of the abovementioned elements at the reference point for Pchelina Reservoir (between the inflows of the two rivers—Struma River—anthropogenically affected and Svetlia River—anthropogenically unaffected) could be explained by the changed profile of the industry in the town of Pernik during the 1990s, since when mining and metallurgy have not been so dominant. These observations are supported by the calculated EFs and geoaccumulation indexes for the top layer at the Pchelina sampling site where only for Cu moderate enrichment is observed.

The chemical elements associated with the second principal component (Ce, Gd, La, Th, and Ti) have decreasing time trends in both sediment cores at the two river inflows and excluding Th in the Pchelina sediment core too. These elements have no anthropogenic origin and the moderate enrichment of Ce and Th at the Pchelina sampling site could be attributed to their geochemical immobility [39]. The third group of chemical elements (Zn, Pb, Cd, Bi and U_{nat}) forming the third principal component have significant increasing trends in Struma and Pchelina sediment cores. This leads to the formation of two distinct groups of layers in the Pchelina sediment core before and after the 1990s (layers 11 and 12). The increased concentrations of these elements at the end of the investigated period have led to moderate enrichments for Pb and U_{nat}, and significant enrichments for Zn and Cd at the Pchelina sampling site. The respected geoaccumulation indexes confirm these observations with uncontaminated to moderately contaminated values for Pb and U_{nat}, and moderately contaminated ones for Zn and Cd. The results from the present study largely confirm the conclusions made by Meuser and co-authors in 2006 [40] where the impact of industry located in the region of Pernik town results in increased concentrations of Pb, Cd, Cr, Cu and Zn.

The results of the conducted biotest Phytotoxkit F ™ reveal the low ecotoxicity of surface sediments, which is an indication that the concentrations of the elements classifying the Pchelina Reservoir samples as moderately contaminated has no significant ecotoxicological effect.

4. Materials and Methods

4.1. Sampling

The sampling of the bottom sediment materials was carried out in the period July–September 2020. Three locations in the Pchelina Reservoir were selected (Figure 5). Sampling point 1 is located at the inflow of the Struma River into the Pchelina Reservoir. Sampling point 2 is located at the inflow of the Svetlia River into the Pchelina Reservoir. The sampling point in Pchelina Reservoir—at the village of Radibosh (sampling point 3)—was chosen between the inflows of the two rivers—Struma River (anthropogenically affected by the industry located in the region of Pernik town) and Svetlia River (anthropogenically unaffected)—as a reference point.

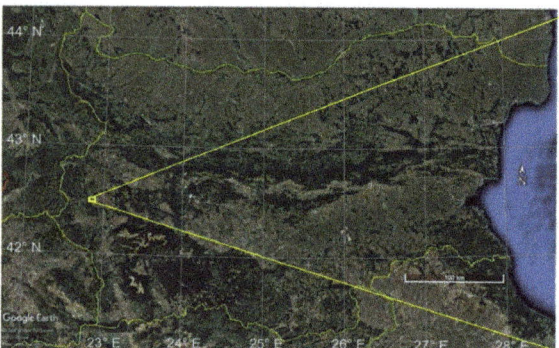

Figure 5. Sampling locations (1-Struma, 2-Svetlia and 3-Pchelina).

Thin-walled tubes with small diameters, which ensures mechanical immersion in the sediment's mass, were used for obtaining semi-intact specimens of fine-grained sediment samples. The basic components of tube-type samplers include steel grade S355 hardened cutting tip, body tube or barrel, and a threaded end. Pipes with a diameter of 48 × 1 mm were used, coupled to a nozzle to reach a depth of approximately 1.20 m (4 ft) below the water surface. The entire Shelby tube system follows the design requirements of ASTM D1587/D1587M–15 [41].

Sampling was carried out with a percussion-swirling motion by hand. Separate sampling tubes were packed and marked on-site and transported to the laboratory at 4 °C.

4.2. Sediment Digestion

Sediment samples were first pre-grinded, sieved through 2-mm sieves and homogenized. Sub-samples of 0.25 g were accurately weighed using an analytical balance, 10 mL of conc. Hydrofluoric acid (HF, 47–51%, Fisher Chemicals, Waltham, MA, USA, Trace Metal Grade) was added and the mixtures were left for 24 h. Subsequent dissolution was performed using a sand bath after adding an additional amount of 10 mL conc. HF (47–51%, Fisher Chemicals, Waltham, MA, USA, Trace Metal Grade) and 5 mL conc. Perchloric acid ($HClO_4$, 70% Fisher Chemicals, Waltham, MA, USA, Trace Metal Grade). The samples were heated until the acid mixture was reduced to 1/3 of the initial volume. Then portions of 10 mL conc. HF were added and heating in a sand bath was performed until complete dissolution of the sediments. Then two portions of 10 mL conc. Nitric acid (HNO_3, 67–69%, Fisher Chemicals, Waltham, MA, USA, Trace Metal Grade) were added and the samples were heated in a sand bath until the volume was reduced to 0.5–1 mL. After cooling, the samples were quantitatively transferred to 50 mL polyethylene tubes by repeated washing with double deionized water. All samples were initially diluted to 50 mL, and immediately before instrumental measurement, an additional dilution of 1 mL to 14 mL was performed.

4.3. Sediment Analysis

4.3.1. Inductively Coupled Plasma Mass Spectrometry (ICP-MS)

The sediment samples were analyzed using Perkin-Elmer SCIEX Elan DRC-e ICP-MS (MDS Inc., Concord, ON, Canada) with cross-flow nebulizer. The spectrometer was optimized (RF power, gas flow, lens voltage) to provide minimal values of the ratios CeO^+/Ce^+ and Ba^{2+}/Ba^+ as well as maximum intensity of the analytes. The concentrations of 20 elements (Bi, Cd, Ce, Co, Cr, Cu, Fe, Gd, La, Mn, Mo, Ni, Pb, Sb, Sn, Th, Ti, Tl, U_{nat} and Zn) were determined. Coefficients for the reduction of the analytical signal (RPa, Dynamic Bandpass Tuning parameter), pre-optimized for sediment matrix, were used for the determination of Fe, Mn, and Ti using ICP-MS. They are presented in Table 5.

Table 5. RPa coefficients used for the determination of Fe, Mn and Ti, in sediment samples.

Isotope	RPa Coefficient
$^{47,49,50}Ti$	0.012
^{54}Fe	0.014
^{56}Fe	0.016
^{57}Fe	0.013
^{55}Mn	0.014

Chemical elements were determined under standard conditions. In the course of the analysis, the appropriate isotopes of the elements in terms of natural distribution and low spectral interference were selected. External calibration by matrix-matched standard solutions was performed. Single element standard solutions of Bi, Cd, Ce, Co, Cr, Cu, Fe, Gd, La, Mn, Mo, Ni, Pb, Sb, Sn, Th, Ti, Tl, U_{nat} and Zn (Fluka, Steinheim, Switzerland) with initial concentration of 1000 mg/L were used to construct the calibration curve after appropriate dilution. The multielement calibration standard solutions were prepared in the concentration range from 0.2 to 20 mg/L for Fe, Mn and Ti, and in the range 0.01 to 100 µg/L for the other elements. All standard solutions were prepared with double deionized water (Millipore purification system Synergy, Molstheim, France). The calibration coefficients for all calibration curves were at least 0.99. The linearity was proven to be four orders of magnitude for Fe, Mn, and Ti, and five orders of magnitude for the other elements. The operating conditions of the ICP-MS are listed in Table 6.

Table 6. ICP-MS instrumental conditions for the determination of chemical elements in sediment samples.

Instrument	Operating Conditions
Cooling Ar gas flow	15 L/min
Auxiliary Ar gas flow	1.20 L/min
Nebulizer gas flow	0.85 L/min
Lens voltage	6.00 V
ICP RF power	1100 W
Integration time	2000 ms
Dwell time	50 ms
Acquisition mode	Peak hop
Sample uptake rate	2 mL/min
Rinse time	180s
Rinsing solution	3% HNO_3

To verify the accuracy of the analysis, stream sediment certified reference materials STSD-1 and STSD-3 of (Canada Center for Mineral and Energy Technology, Geological Survey of Canada) were subjected to analysis using the same sample preparation.

Tables 7 and 8 present the experimentally determined and the certified values of the analyzed elements in the certified reference materials STSD-1 and STSD-3, respectively. If the values of the analytical yields are in the range of 97–108%, the method is considered fit-for-purpose. All measurements were performed in triplicate and the mean value was reported.

Table 7. Measured and certified values of the sediment certified reference material STSD-1.

Element	Isotope	LOQ (µg/g)	Measured Value in STSD-1 (µg/g)	Certified Value (µg/g)	Recovery %
Ti	46	1.28	4632	4600	100.7
Mn	55	0.98	3918	3950	99.2
Fe %	57	1.04	4.73	4.7	100.6
Zn	64	0.17	174	178	97.8
Cr	52	0.17	65.9	67	98.4
Ni	62	0.06	25.3	24	105.4
Cu	65	0.12	37.4	36	103.8
Mo	95	0.07	1.37	<5	–
Sn	116	0.22	4.13	4	103.3
Sb	123	0.07	3.42	3.3	103.6
Pb	208	0.08	35.6	35	101.9
Co	59	0.03	17.4	17	102.6
Cd	112	0.02	1.39	–	–
Ce	140	0.03	50.1	51	98.2
Tl	205	0.02	0.45	–	–
Bi	209	0.02	0.69	–	–
Gd	158	0.03	5.6	–	–
La	139	0.03	29.2	30	97.3
Th	232	0.2	3.83	3.7	103.5
U_{nat}	238	0.2	8.0	8	100.4

Table 8. Measured and certified values of the sediment certified reference material STSD-3.

Element	Isotope	LOQ (µg/g)	Measured Value in STSD-3 (µg/g)	Certified Value (µg/g)	Recovery %
Ti	46	1.28	4370	4400	99.3
Mn	55	0.98	2763	2730	101.2
Fe %	57	1.04	4.49	4.4	102.0
Zn	64	0.17	214	204	105.1
Cr	52	0.17	82.5	80	103.1
Ni	62	0.06	29.2	30	97.3
Cu	65	0.12	38.8	39	99.4
Mo	95	0.07	5.92	6	98.6
Sn	116	0.22	4.31	4	107.6
Sb	123	0.07	4.16	4	104.0
Pb	208	0.08	41.8	40	104.4
Co	59	0.03	16.15	16	100.9
Cd	112	0.02	1.09	–	–
Ce	140	0.03	62	63	98.5
Tl	205	0.02	0.42	–	–
Bi	209	0.02	0.45	–	–
Gd	158	0.03	5.2	–	–
La	139	0.03	38.4	39	98.5
Th	232	0.2	8.6	8.5	101.5
U_{nat}	238	0.2	10.3	10.5	97.6

4.3.2. Gamma-Spectrometry

The individual sediment samples were air-dried and cleaned from plant impurities (roots, leaves, shells, etc.), followed by sieving through 2-mm sieves and homogenization. Samples of about 10 g were packed in standard geometry vessels and measured by gamma-spectrometry for the determination of the total activity of ^{137}Cs. The activity of radionuclides was measured using HPGe detector Canberra 7229 (energy resolution 1.8 and efficiency 16% at 1332.5 keV) coupled to a 4196-channel analyzer Canberra 35Plus. The calibration was achieved using national standard radioactive solutions and standard samples, produced, and standardized at Czech Metrology Institute (Serial No.: 130520-1785043). The accuracy and precision of the analysis have been verified by participation in International Atomic Energy Agency (IAEA) round robin tests. All measurements were performed in triplicate and the mean value was reported.

4.3.3. Ecotoxicological Studies

The ecotoxicological test Phytotoxkit F ™ [42] measures the reduction of seed germination (SG) and the growth of young roots (RG) of selected higher plants (Sorghum saccharatum, Lepidium sativum and Sinapis alba) when seeded in contaminated samples for several days compared to a control sample.

Sample preparation included drying, grinding, and sieving through 2-mm sieves as a preliminary step. Then the Phytotoxkit F ™ tests were performed on water-saturated samples. It was experimentally found that 35 mL of distilled water was required to achieve 100% saturation of control soil with a volume of 90 cm^3. The determination of the water retention capacity of the analyzed sediments was performed on a representative sample (mean of the three analyzed sediments). To 60 mL of distilled water, 90 cm^3 of the sample was added and after equilibrium was reached, the volume of excess water (15 mL) was measured, and the volume of water required to achieve saturation of the sample was calculated (45 mL).

The first step of conducting the Phytotoxkit F ™ test was to place 90 cm^3 of each of the studied sediments, as well as of the control sample, in special plastic test plates. This was followed by hydration with the necessary volume of distilled water to achieve saturation, placement of black filter paper, seeding of 10 seeds of the test plant (Sinapis alba—white mustard) and closing the test plate. Two repetitions for each sediment sample were made, and 3 for the control sample—Reference OECD soil for Phytotoxkit test (Microbiotests, Gent, Belgium). The test plates were placed vertically in an incubator for 72 h at a temperature of 25 ± 1 °C in the dark. As the last step, the number of germinated seeds was counted and the length of the roots was measured, using the free software ImageJ [43].

The ecotoxicological effect (%), reflecting the germination of the seeds is calculated by $100 \times (A - B)/A$, where A is the average number of germinated seeds for the control sample and B is the average number of germinated seeds for the analyzed sample.

The ecotoxicological effect (%), reflecting the growth of the roots, is calculated by $100 \times (A - B)/A$, where A is the mean length (mm) of the plant roots from the control sample and B is the mean length (mm) of the plant roots from the sample.

4.4. Statistical Analysis

4.4.1. Principal Component Analysis (PCA)

The Principal Component Analysis (PCA) is a multivariate approach to data reduction. The aim is to find and interpret the latent interdependencies between the variables (chemical elements) in the data set. Such variables form new ones, called latent factors or principal components. In addition to discovering the data structure, the PCA data set can be modelled, compressed, classified and visualized on a plane. The main task in PCA is the decomposition of the data matrix into two parts—a matrix of factor results and a matrix of factor weights. The factor weights show the participation of each of the old variables in the formation of the main components while the factor results are the coordinates of the objects (layers in the sedimentary cores) in the newly formed variables. The determination of the number of significant principal components is based on their eigenvalues and the percentage of explained variation in the data.

4.4.2. Mann–Kendall Test

The Mann–Kendall test is a non-parametric approach for estimating time trends. The assessment uses all possible discrepancies between the values for a given layer of sediment with those of previous years. Positive differences (increase in the concentration of the element) are marked with +1, negative (decrease in the concentration of the element) with −1, and the lack of difference with 0. The test takes into account only the sign of the differences. The null hypothesis is that there is no time trend, the alternative hypothesis is that there is a positive or negative time trend. The direction of the trend is determined by the sign of the S statistics, which is the difference between the number of positive and negative differences. At the assumed significance level ($\alpha = 0.05$), the null hypothesis is rejected at values of $p < 0.05$.

4.4.3. Enrichment Factor (EF)

To distinguish anthropogenic pollution from the natural content of elements in the sediment, enrichment factors (EF) were calculated by comparing the measured concentrations of chemical elements with the geochemical background values of the study area. To avoid erroneous enrichment results, geochemical normalization based on the concentration of a conservative element is usually used. The purpose of normalization is to correct changes in the nature of the sediment that may affect the distribution of contaminants. Al, Fe, Th, Ti and Zr are usually used as conservative elements [15,44]. The normalized enrichment factor (EF) is determined by the metal/X concentration ratio (X = Fe, Al, Th, Ti, Zr) divided by the background metal concentration/background concentration ratio X:

$$EF = \frac{\left(\frac{Me}{X}\right)_{sample}}{\left(\frac{Me}{X}\right)_{background}}, \quad (1)$$

Five levels of pollution are often identified—EF < 2: low enrichment; EF 2–5: moderate enrichment; EF 5–20: significant enrichment; EF 20–40: strong enrichment; EF > 40: extremely strong enrichment. In addition, values of $0.5 \leq EF \leq 1.5$ suggest that the concentration of the elements may come entirely from natural weathering processes. However, EF > 1.5 shows that a significant part of the microelements did not come from the earth's crust, i.e., their origin is from other sources, such as point and non-point pollution and biota [17].

4.4.4. Geoaccumulation Index (I_{geo})

A similar approach was used for the determination of the Geoaccumulation Index (I_{geo}):

$$I_{geo} = log_2\left(\frac{C_n}{1.5 C_{ref}}\right), \quad (2)$$

where C_n is the concentration of the element in the sample and C_{ref} is the background concentration [45].

The coverage factor of 1.5 used allows the normalization of possible variations in the data for background concentrations, which may also be due to anthropogenic pollution. 7 classes of contamination are known depending on the Geoaccumulation Index ($I_{geo} \leq 0$ —uncontaminated sample; $0 < I_{geo} < 1$ uncontaminated to moderately contaminated sample; $1 < I_{geo} < 2$ moderately contaminated sample; $2 < I_{geo} < 3$ moderately to highly contaminated sample; $3 < I_{geo} < 4$ heavily contaminated sample; $4 < I_{geo} < 5$ heavily to extremely contaminated sample; $I_{geo} > 5$ extremely contaminated sample) [14,46–48]. The highest class corresponds to at least a 100-fold difference with the background concentration.

5. Conclusions

Based on the measurement of γ-activity of the technogenic ^{137}Cs, an accumulation of an average of 15 cm of sediment was established for 34 years, at a sedimentation rate of an average of 0.44 cm/year.

The distribution of the concentrations of chemical elements in the sediment from the three sampling points (1—Pchelina Reservoir at the flow of Struma River, 2—Pchelina Reservoir at the flow of Svetlia River and 3—Pchelina Reservoir near the village of Radibosh) is presented. PCA of the data shows three main components, which describe nearly 80% of the variation of the data. The first main component (41.53% of the data variation) contains Mn, Fe, Cr, Ni, Cu, Mo, Sn, Sb and Co. The factor scores show that the concentrations of these elements decrease in the order Pchelina > Struma > Svetlia. All these elements have a positive time trend in the Struma sediment core, which is an indication that most of the elements in the Reservoir come through the anthropogenically affected river Struma. The second main component (22.16%) is formed by Ce, Gd, La, Th, and Ti, which decrease significantly with time in Svetlia and Struma. The third main component (15.44%) is formed by the elements Zn, Pb, Cd, Bi and U_{nat}. The factor scores of the layers in the sedimentary cores show the anthropogenic origin of most of these elements. There is an increase over time in the sedimentary cores in Struma and Pchelina. The increase in the sedimentary layers of Pchelina is especially pronounced. In the first group (beginning of the studied period), there are contents comparable to the sediment core of Svetlia, while the contents of the elements in the second group are the highest for the three studied sediment cores.

To distinguish anthropogenic pollution from the natural content of elements in the sediment, enrichment factors (EF) have been calculated for which Fe concentrations have been used as a conservative element. Significant enrichment was observed only at the reference point (Pchelina), for Zn and Cd. In terms of Cd, the enrichment in Svetlia is moderate. The enrichment factor is 1.8 in the Struma, which shows significant contamination with this metal in the entire Reservoir. Except for these elements, moderate enrichment is observed for Cu, Pb, Ce and Th only in Pchelina, as well as for Tl and U_{nat} in Svetlia and Pchelina. For most elements, the geoaccumulation index corresponds to an uncontaminated sample. In Svetlia, only for Cd, Tl and U_{nat} the index corresponds to an uncontaminated to moderately contaminated sample. The index of geoaccumulation in the sediments of Struma corresponds to an uncontaminated to moderately contaminated sample for Fe, Cr, Mo and Sn and to a moderately contaminated sample for Sb and Cd. The geoaccumulation index for the sediments of Pchelina corresponds to an unpolluted to moderately contaminated sample for Pb, Tl and U_{nat} (similar to the Svetlia sediments) and a moderately polluted sample for Zn and Cd (similar to the Svetlia sediments). Moderate contamination of the sediments of Pchelina Reservoir in all sampling points is from Cd.

The results of the Phytotoxkit F bioassay revealed the low ecotoxicity of Pchelina Reservoir surface sediments in terms of both seed germination (SG) and root growth (RG) of the plant species Sinapis alba.

Author Contributions: Conceptualization, S.T., T.V. and M.T.; methodology, T.V., V.M., R.P., M.T., G.Y., B.T. and V.L.; software, S.T., T.V. and G.Y.; formal analysis, V.M., R.P., G.Y., B.T. and V.L.; investigation, T.V., V.M., R.P., M.T., G.Y., B.T., V.L. and S.T.; resources, T.V.; data curation, T.V. and S.T.; writing—original draft preparation, T.V. and S.T.; writing—review and editing, T.V., V.M., R.P., M.T., G.Y., B.T. and S.T.; visualization, S.T.; supervision, T.V. and S.T.; project administration, T.V., M.T. and S.T.; funding acquisition, T.V. and S.T. All authors have read and agreed to the published version of the manuscript.

Funding: This research was funded and supported by the UACEG's Research, Consultancy and Design Centre (RCDC), grant number BN 238/20 and by the National Science Program "Environmental Protection and Reduction of Risks of Adverse Events and Natural Disasters", approved by the Resolution of the Council of Ministers № 577/17.08.2018 and supported by the Ministry of Education and Science (MES) of Bulgaria (Agreement № Д01-363/17.12.2020).

Institutional Review Board Statement: Not applicable.

Informed Consent Statement: Not applicable.

Data Availability Statement: The data presented in this study are available on request from the corresponding author.

Acknowledgments: The authors gratefully acknowledge the financial support by the UACEG's RCDC and the National Science Program "Environmental Protection and Reduction of Risks of Adverse Events and Natural Disasters", approved by the Resolution of the Council of Ministers № 577/17.08.2018 and supported by the Ministry of Education and Science (MES) of Bulgaria (Agreement № Д01-363/17.12.2020). The help by the Ministry of the Agriculture, Food and Forestry of the Republic of Bulgaria and Irrigation Systems Ltd. is also acknowledged.

Conflicts of Interest: The authors declare no conflict of interest. The funders had no role in the design of the study; in the collection, analyses, or interpretation of data; in the writing of the manuscript, or in the decision to publish the results.

Sample Availability: Samples of the compounds are not available from the authors.

References

1. Adamiec, E.; Helios-Rybicka, E. Distribution of pollutants in the Odra river system. Part V. Assessment of total and mobile heavy metals content in the suspended matter and sediments of the Odra river system and recommendations for river chemical monitoring. *Pol. J. Environ. Stud.* **2002**, *11*, 675–688.
2. Helios-Rybicka, E.; Adamiec, E.; Aleksander-Kwaterczak, U. Distribution of trace metals in the Odra River system: Water-suspended matter-sediments. *Limnologica* **2005**, *35*, 185–198. [CrossRef]
3. Luck, J.; Workman, S.; Coyne, M.; Higgins, S. Solid material retention and nutrient reduction properties of pervious concrete mixtures. *Biosyst. Eng.* **2008**, *100*, 401–408. [CrossRef]
4. Strakhovenko, V.; Subetto, D.; Ovdina, E.; Belkina, N.; Efremenko, N. Distribution of Elements in Iron-Manganese Formations in Bottom Sediments of Lake Onego (NW Russia) and Small Lakes (Shotozero and Surgubskoe) of Adjacent Territories. *Minerals* **2020**, *10*, 440. [CrossRef]
5. Saeedi, M.; Daneshvar, S.; Karbassi, A. Role of riverine sediment and particulate matter in adsorption of heavy metals. *Int. J. Environ. Sci. Technol.* **2004**, *1*, 135–140. [CrossRef]
6. Miranda, L.; Wijesiri, B.; Ayoko, G.; Egodawatta, P.; Goonetilleke, A. Water-sediment interactions and mobility of heavy metals in aquatic environments. *Water Res.* **2021**, *202*, 117386. [CrossRef]
7. Lecrivain, N.; Clement, B.; Dabrin, A.; Seigle-Ferrand, J.; Bouffard, D.; Naffrechoux, E.; Frossard, V. Water-level fluctuation enhances sediment and trace metal mobility in lake littoral. *Chemosphere* **2021**, *264*, 128451. [CrossRef]
8. Walker, C.; Hopkin, S.; Sibly, R.; Peakall, D. *Principles of Ecotoxicology*, 3rd ed.; CRC Press: Boca Raton, FL, USA, 2005.
9. Manahan, S. *Toxicological Chemistry and Biochemistry*; Lewis Publishers: London, UK; CRC Press Company: Boca Raton, FL, USA; New York, NY, USA; Washington, DC, USA, 2003.
10. Von Gunten, H.; Sturm, M.; Moser, R. 200-year record of metals in lake sediments and natural background concentrations. *Environ. Sci. Technol.* **1997**, *31*, 2193–2197. [CrossRef]
11. Kähkönen, M.; Suominen, K.; Manninen, P.; Salkinoja-Salonen, M. 100 years of sediment accumulation history of organic halogens and heavy metals in recipient and nonrecipient lakes of pulping industry in Finland. *Environ. Sci. Technol.* **1998**, *32*, 1741–1746. [CrossRef]
12. Müller, J.; Ruppert, H.; Muramatsu, Y.; Schneider, J. Reservoir sediments witness of mining and industrial development (Malter Reservoir, eastern Erzgebirge, Germany). *Environ. Geol.* **2000**, *39*, 1341–1351. [CrossRef]
13. Audry, S.; Schafer, J.; Blanc, G.; Jouanneau, J.-M. Fifty-year sedimentary record of heavy metal pollution (Cd, Zn, Cu, Pb) in the Lot River reservoirs (France). *Environ. Pollut.* **2004**, *132*, 413–426. [CrossRef] [PubMed]
14. Zhou, Z.; Wang, Y.; Teng, H.; Yang, H.; Liu, A.; Li, M.; Niu, X. Historical Evolution of Sources and Pollution Levels of Heavy Metals in the Sediment of the Shuanglong Reservoir, China. *Water* **2020**, *12*, 1855. [CrossRef]
15. Buccione, R.; Fortunato, E.; Paternoster, M.; Rizzo, G.; Sinisi, R.; Summa, V.; Mongelli, G. Mineralogy and heavy metal assessment of the Pietra del Pertusillo reservoir sediments (Southern Italy). *Environ. Sci. Pollut. Res.* **2021**, *28*, 4857–4878. [CrossRef] [PubMed]
16. Chatterjee, M.; Silva Filho, E.; Sarkar, S.; Sella, S.; Bhattacharya, A.; Satpathy, K.; Prasad, M.; Chakraborty, S.; Bhattacharya, B. Distribution and possible source of trace elements in the sediment cores of a tropical macrotidal estuary and their ecotoxicological significance. *Environ. Int.* **2007**, *33*, 346–356. [CrossRef]
17. Maanan, M.; Landesman, C.; Maanan, M.; Zourarah, B.; Fattal, P.; Sahabi, M. Evaluation of the anthropogenic influx of metal and metalloid contaminants into the Moulay Bousselham lagoon, Morocco, using chemometric methods coupled to geographical information systems. *Environ. Sci. Pollut. Res. Int.* **2013**, *20*, 4729–4741. [CrossRef]
18. Hargalani, F.; Karbassi, A.; Monavari, S.; Azar, P. A novel pollution index based on the bioavailability of elements: A study on Anzali wetland bed sediments. *Environ. Monit. Assess.* **2014**, *186*, 2329–2348. [CrossRef]

19. Zahra, A.; Hashmi, M.; Ahmed, R. Enrichment and geo-accumulation of heavy metals and risk assessment of sediments of the Kurang Nallah—Feeding tributary of the Rawal Lake Reservoir, Pakistan. *Sci. Total Environ.* **2014**, *470*, 925–933. [CrossRef] [PubMed]
20. Mwanamoki, P.; Devarajan, N.; Niane, B.; Ngelinkoto, P.; Thevenon, F.; Nlandu, J.; Mpiana, P.; Prabakar, K.; Mubedi, J.; Kabele, C.; et al. Trace metal distributions in the sediments from river-reservoir systems: Case of the Congo River and Lake Ma Vallée, Kinshasa (Democratic Republic of Congo). *Environ. Sci. Pollut. Res.* **2015**, *22*, 586–597. [CrossRef] [PubMed]
21. de Paula Filho, F.; Marins, R.; de Lacerda, L.; Aguiar, J.; Peres, T. Background values for evaluation of heavy metal contamination in sediments in the Parnaíba River Delta estuary, NE/Brazil. *Marine Pollut. Bull.* **2015**, *91*, 424–428. [CrossRef]
22. Wang, G.; Yinglan, A.; Jiang, H.; Fu, Q.; Zheng, B. Modeling the source contribution of heavy metals in surficial sediment and analysis of their historical changes in the vertical sediments of a drinking water reservoir. *J. Hydrol.* **2015**, *520*, 37–51. [CrossRef]
23. García-Ordiales, E.; Esbrí, J.; Covelli, S.; López-Berdonces, M.; Higueras, P.; Loredo, J. Heavy metal contamination in sediments of an artificial reservoir impacted by long-term mining activity in the Almadén mercury district (Spain). *Environ. Sci. Pollut. Res.* **2016**, *23*, 6024–6038. [CrossRef]
24. Ladwig, R.; Heinrich, L.; Singer, G.; Hupfer, M. Sediment core data reconstruct the management history and usage of a heavily modified urban lake in Berlin, Germany. *Environ. Sci. Pollut. Res. Int.* **2017**, *24*, 25166–25178. [CrossRef]
25. Fural, S.; Kükrerb, S.; Cürebal, I. Geographical information systems based ecological risk analysis of metal accumulation in sediments of İkizcetepeler Dam Lake (Turkey). *Ecol. Indic.* **2020**, *119*, 106784. [CrossRef]
26. Fonseca, M.; Ferreira, F.; Choueri, R.; Fonseca, G. M-Triad: An improvement of the sediment quality triad. *Sci. Tot. Environ.* **2021**, *770*, 145245. [CrossRef]
27. Arnason, J.; Fletcher, B. A 40-year record of Cd, Hg, Pb, and U deposition in sediments of Patroon Reservoir, Albany County, NY, USA. *Environ. Pollut.* **2003**, *123*, 383–391. [CrossRef]
28. Callender, E. Geochemical effects of rapid sedimentation in aquatic systems: Minimal diagenesis and the preservation of historical metal signatures. *J. Paleolimn.* **2000**, *23*, 243–260. [CrossRef]
29. Gartsiyanova, K. Assessment of the water quality of the "Pchelina" Reservoir. *Probl. Geogr.* **2017**, *1*, 62–71.
30. Meuser, H. *Contaminated Urban Soils*; Springer: Dordrecht, The Netherlands; Berlin/Heidelberg, Germany; London, UK; New York, NY, USA, 2010.
31. Meuser, H.; Hartling, A. *Pollutant Dispersion in the Industrially Influenced Soils and Sediments of the River Struma and the Pchelina Water Reservoir*; Applied Science University: Sofia, Bulgaria, 2008; p. 65.
32. Alvarez-Iglesias, P.; Quintana, B.; Rubio, B.; Perez-Arlucea, M. Sedimentation rates and trace metal input history in intertidal sediments from San Simon Bay (Ría de Vigo, NW Spain) derived from ^{210}Pb and ^{137}Cs chronology. *J. Environ. Rad.* **2007**, *98*, 229–250. [CrossRef] [PubMed]
33. Guo, W.; Yue, J.; Zhao, Q.; Li, J.; Yu, X.; Mao, Y. A 110 Year Sediment Record of Polycyclic Aromatic Hydrocarbons Related to Economic Development and Energy Consumption in Dongping Lake, North China. *Molecules* **2021**, *26*, 6828. [CrossRef]
34. Johnson, T. Sedimentation in large lakes. *Annu. Rev. Earth Planet. Sci.* **1984**, *12*, 179–204. [CrossRef]
35. Ghrefat, H.; Yusuf, N. Assessing Mn, Fe, Cu, Zn, and Cd pollution in bottom sediments of Wadi Al-Arab Dam, Jordan. *Chemosphere* **2006**, *65*, 2114–2121. [CrossRef]
36. Ye, C.; Li, S.; Zhang, Y.; Zhang, Q. Assessing soil heavy metal pollution in the water-level-fluctuation zone of the Three Gorges Reservoir, China. *J. Hazard. Mat.* **2011**, *191*, 366–372. [CrossRef]
37. Dung, T.; Cappuyns, V.; Swennen, R.; Phung, N. From geochemical background determination to pollution assessment of heavy metals in sediments and soils. *Rev. Environ. Sci. Biotechnol.* **2013**, *12*, 335–353. [CrossRef]
38. Abril, J. Constraints on the use of ^{137}Cs as a time-marker to support CRS and SIT chronologies. *Environ. Pollut.* **2004**, *129*, 31–37. [CrossRef]
39. Perelman, A.; Kasimov, N. *Landscape Geochemistry*; Chapters 29.5 and 31.1; Publ, H., Ed.; Astrea-200: Moscow, Russia, 1999. (In Russian)
40. Meuser, H.; Hartling, A.; Dupke, G. Pollutant dispersion of alluvial soils in the industrialized region of Pernik, Bulgaria. In *EUROSOIL 2008—Book of Abstracts*; Blum, W., Gerzabek, M., Vodrazka, M., Eds.; University of Natural Resources and Applied Life Sciences (BOKU): Vienna, Austria, 2008; p. 111.
41. *ASTM D1587/D1587M—15. Standard Practice for Thin-Walled Tube Sampling of Fine-Grained Soils for Geotechnical Purposes*; ASTM International: West Conshohocken, PA, USA, 2015.
42. *ISO 11269-1:2012. Soil Quality—Determination of the Effects of Pollutants on Soil Flora—Part 1: Method for the Measurement of Inhibition of Root Growth*; ISO: Geneva, Switzerland, 2012.
43. Schneider, C.; Rasband, W.; Eliceiri, K. NIH Image to ImageJ: 25 years of image analysis. *Nat. Methods* **2012**, *9*, 671–675. [CrossRef]
44. Tessier, E.; Garnier, C.; Mullot, J.; Lenoble, V.; Arnaud, M.; Raynaud, M.; Mounier, S. Study of the spatial and historical distribution of sediment inorganic contamination in the Toulon bay (France). *Marine Pollut. Bull.* **2011**, *62*, 2075–2086. [CrossRef] [PubMed]
45. Müller, G. Index of geoaccumulation in sediments of the Rhine River. *Geo. J.* **1969**, *2*, 108–118.

46. Marziali, L.; Guzzella, L.; Salerno, F.; Marchetto, A.; Valsecchi, L.; Tasselli, S.; Roscioli, C.; Schiavon, A. Twenty-year sediment contamination trends in some tributaries of Lake Maggiore (Northern Italy): Relation with anthropogenic factors. *Environ. Sci. Pollut. Res.* **2021**, *28*, 38193–38208. [CrossRef] [PubMed]
47. Nkinda, M.; Rwiza, M.; Ijumba, J.; Njau, K. Heavy metals risk assessment of water and sediments collected from selected river tributaries of the Mara River in Tanzania. *Discov. Water* **2021**, *1*, 3. [CrossRef]
48. Navarrete-Rodríguez, G.; Castañeda-Chávez, M.d.R.; Lango-Reynoso, F. Geoacumulation of Heavy Metals in Sediment of the Fluvial–Lagoon–Deltaic System of the Palizada River, Campeche, Mexico. *Int. J. Environ. Res. Public Health* **2020**, *17*, 969. [CrossRef] [PubMed]

Article

Rare Earth Elements and Bioavailability in Northern and Southern Central Red Sea Mangroves, Saudi Arabia

Mohammed Othman Aljahdali [1,*] and Abdullahi Bala Alhassan [1,2,*]

[1] Department of Biological Sciences, Faculty of Science, King Abdulaziz University, P.O. Box 80203, Jeddah 21589, Saudi Arabia
[2] Department of Biology, Faculty of Life Sciences, Ahmadu Bello University, Zaria 810001, Nigeria
* Correspondence: moaljahdali@kau.edu.sa (M.O.A.); balahassan80@gmail.com (A.B.A.)

Abstract: Different hypotheses have been tested about the fractionation and bioavailability of rare earth elements (REE) in mangrove ecosystems. Rare earth elements and bioavailability in the mangrove ecosystem have been of significant concern and are recognized globally as emerging pollutants. Bioavailability and fractionation of rare earth elements were assessed in Jazan and AlWajah mangrove ecosystems. Comparisons between rare earth elements, multi-elemental ratios, geo-accumulation index (Igeo), and bio-concentration factor (BCF) for the two mangroves and the influence of sediment grain size types on concentrations of rare earth elements were carried out. A substantial difference in mean concentrations (mg/kg) of REE (La, Ce, Pr, Nd, Sm, Eu, Gd, Tb, Dy, Ho, Er, Tm, Yb, and Lu) was established, except for mean concentrations of Eu, Gd, Tb, Tm, and Lu. In addition, concentrations of REEs were higher in the Jazan mangrove ecosystem. However, REE composition in the two mangroves was dominated by the lighter REE (LREE and MREE), and formed the major contribution to the total sum of REE at 10.2–78.4%, which was greater than the HREE contribution of 11.3–12.9%. The Post Archean Australian Shale (PAAS) normalized values revealed that lighter REE (LREE and MREE) were steadily enriched above heavy REE. More so, low and negative values of $R_{(H/M)}$ were recorded in the Al Wajah mangrove, indicating higher HREE depletion there. The values of BCF for REEs were less than 1 for all the REEs determined; the recorded BCF for Lu (0.33) and Tm (0.32) were the highest, while the lowest BCF recorded was for Nd (0.09). There is a need for periodic monitoring of REE concentrations in the mangroves to keep track of the sources of this metal contamination and develop conservation and control strategies for these important ecosystems.

Keywords: mangrove; rare earth elements; distribution; bioaccumulation; *Avicennia marina*; Red Sea

1. Introduction

The Red Sea is a channel that forms a linkage between the Mediterranean Sea (north) and the Indian Ocean (south). The sea is a marine biodiversity hotspot with a high abundance of coral reefs, mangroves, and sea grass [1,2]. In aquatic ecosystems such as the Red Sea, suspended sediments and particulate matter may account for almost 90% of metal burden [3].

Rare earth elements (REEs) are a collection of seventeen chemical elements in the periodic table and are generally trivalent elements, excluding Ce and Eu, which tend to exist as Ce (IV) and Eu (II). REEs starting from La and ending with Sm are considered light rare earth elements (LREEs), while those ranging between Gd and Lu are considered heavy rare earth elements (HREEs) [4]. The light rare earth elements (LREEs) and heavy rare earth elements (HREEs) have analogous geochemical behaviors. They give a better understanding of complex procedures of a geochemical nature that single proxies cannot readily discriminate due to their coherent and expectable characteristics [4,5].

REEs do not occur in pure metal form even though they occur in nature, although Promethium, the rarest, only occurs in trace quantities in natural materials as it has no stable or long-term isotopes [6]. Globally, REEs are recognized as emerging micro-pollutants in aquatic ecosystems [7,8]. Modern technologies, on the other hand, utilize REE for its unique physicochemical properties in high-tech applications [9]. For example, AgInSe$_2$ (AIS) is one of the most attractive materials in thin film solar cell applications because of its high optical absorption coefficient [10]. As a result, it is unlikely that REE would spread naturally in most coastal habitats such as mangroves because they have already been impacted by anthropogenic activities [8,11,12].

Mangroves are important intertidal coastal systems that provide multiple ecological functions. They regulate material exchange at the interfaces between land, atmosphere, and marine ecosystems [13,14]. Mangrove ecosystems are dynamic in nature, often subjected to rapid changes in physicochemical properties such as water content, pH, salinity, texture, and redox conditions due to tidal flushing, and the associated flooding could influence metal contamination. Flooding may develop redox cycles in the aquatic environment, with alternating periods of oxidizing and reducing conditions [12,15–17]. Therefore, the application of REE can be useful in tracing the channels and processes in which these elements are involved, particularly in contaminated environments such as those found in the Jazan and AlWajah mangrove ecosystems and their biota.

Notably, there are few or no studies providing a comprehensive investigation of the bioavailability of rare earth elements in the mangrove ecosystems of Jazan and AlWajah in the northern and southern central Red Sea. As a result, this study will open the way for periodic monitoring of REE concentrations in mangrove ecosystems, as well as tracking the sources of metal contamination, allowing for the development of policies for the control and conservation of these important ecosystems.

2. Results

2.1. REE Composition in Sediment and Influence of Grain Sizes

In this study, the results for REE composition in sediment showed significant variation (t-test, $p < 0.05$) between the two mangrove ecosystems investigated (Table 1). A substantial difference in mean concentrations (mg/kg) of REEs (La, Ce, Pr, Nd, Sm, Eu, Gd, Tb, Dy, Ho, Er, Tm, Yb, and Lu) was also recorded; however, except for the mean concentrations of Eu, Gd, Tb, Tm, and Lu, significantly higher concentrations of REEs were recorded in sediment samples collected from the Jazan mangrove ecosystem (Table 1). Generally, the concentrations of REEs were lowest in the Al Wajah mangrove ecosystem. The sum of REEs ($\sum$REE = 112.54 ± 10.48 mg/kg) recorded at the Jazan mangrove was about double that of Al Wajah ($\sum$REE = 78.47 ± 7.89 mg/kg). Nevertheless, the REE composition in the two mangroves was dominated by the lighter REEs (LREE and MREE) and formed the major contribution to the total sum of REEs at 10.2–78.4%, which was greater than the HREE contribution of 11.3–12.9%. In addition, the sum of LREEs (La, Ce, Pr, Nd) was about seven- and eight-fold that of the composition of MREEs (Sm, Eu, Gd) and HREEs (Tb, Dy, Ho, Er, Tm, Yb, Lu), respectively (Table 1).

Table 1. Rear earth elements' composition (mg/kg) and fractions in northern and southern central Red Sea mangroves.

Sediment	AlWajah	Jazan	Average	p-Value	Leaves	AlWajah	Jazan	Average	p-Value
La	12.67 ± 1.73	17.95 ± 2.32	15.31 ± 2.64	0.007	La	0.70 ± 0.11	0.86 ± 0.12	0.78 ± 0.08	0.887
Ce	28.59 ± 2.68	38.35 ± 4.46	33.47 ± 4.88	0.004	Ce	1.48 ± 0.13	1.64 ± 0.20	1.56 ± 0.06	0.776
Pr	3.61 ± 0.23	5.73 ± 0.12	4.67 ± 1.06	0.001	Pr	0.18 ± 0.02	0.34 ± 0.01	0.26 ± 0.07	0.766
Nd	16.62 ± 2.14	22.60 ± 3.59	19.61 ± 2.99	0.002	Nd	0.63 ± 0.09	0.79 ± 0.03	0.71 ± 0.06	0.775
Sm	3.62 ± 0.29	5.65 ± 0.19	4.64 ± 1.02	0.020	Sm	0.17 ± 0.01	0.32 ± 0.04	0.25 ± 0.05	0.895
Eu	0.96 ± 0.05	2.34 ± 0.06	1.65 ± 0.69	0.318	Eu	0.06 ± 0.01	0.21 ± 0.01	0.14 ± 0.05	0.448
Gd	3.43 ± 0.30	5.45 ± 0.21	4.44 ± 1.01	0.449	Gd	0.15 ± 0.02	0.32 ± 0.04	0.24 ± 0.07	0.669
Tb	0.58 ± 0.04	0.89 ± 0.03	0.74 ± 0.16	0.496	Tb	0.05 ± 0.01	0.21 ± 0.01	0.13 ± 0.07	0.768
Dy	3.32 ± 0.31	5.09 ± 0.06	4.21 ± 0.89	0.001	Dy	0.12 ± 0.02	0.28 ± 0.02	0.20 ± 0.06	0.557
Ho	0.66 ± 0.07	0.94 ± 0.02	0.80 ± 0.14	0.004	Ho	0.05 ± 0.01	0.22 ± 0.01	0.14 ± 0.09	0.975

Table 1. Cont.

	AlWajah	Jazan				AlWajah	Jazan		
Sediment			Average	p-Value	Leaves			Average	p-Value
Er	1.92 ± 0.20	3.27 ± 0.02	2.60 ± 0.68	0.001	Er	0.09 ± 0.02	0.24 ± 0.01	0.17 ± 0.04	0.856
Tm	0.32 ± 0.03	0.52 ± 0.01	0.42 ± 0.10	0.128	Tm	0.05 ± 0.01	0.24 ± 0.01	0.15 ± 0.03	0.368
Yb	1.89 ± 0.20	3.25 ± 0.02	2.57 ± 0.68	0.001	Yb	0.07 ± 0.02	0.22 ± 0.01	0.15 ± 0.08	0.697
Lu	0.29 ± 0.02	0.51 ± 0.01	0.40 ± 0.11	0.136	Lu	0.05 ± 0.01	0.21 ± 0.01	0.13 ± 0.04	0.849
ΣREE	78.47 ± 7.89	112.54 ± 10.48	95.51 ± 17.04	0.001	-	3.84 ± 0.76	5.94 ± 0.91	4.98 ± 0.22	0.568
(La/Yb)n	0.49 ± 0.02	0.41 ± 0.01	0.45 ± 0.04	0.008	-	-	-	-	-
(Sm/La)n	1.96 ± 0.04	2.17 ± 0.02	2.07 ± 0.12	0.005	-	-	-	-	-
(Yb/Sm)n	1.03 ± 0.01	1.13 ± 0.01	1.08 ± 0.05	0.006	-	-	-	-	-
R(M/L)	0.25 ± 0.01	0.28 ± 0.03	0.27 ± 0.12	0.043	-	-	-	-	-
R(H/M)	−0.02 ± 0.01	0.030 ± 0.001	0.005 ± 0.030	0.004	-	-	-	-	-
δCe	0.97 ± 0.11	0.86 ± 0.09	0.92 ± 0.06	0.023	-	-	-	-	-
δEu	1.27 ± 0.46	1.98 ± 0.39	1.63 ± 0.36	0.015	-	-	-	-	-

The values in the table are mean ± standard error.

The principal component analysis biplot revealed the influence of sediment grain size types on REE concentrations in sediment and the site's contribution to the total variation (Figure 1A). The relationship revealed by the PCA was based on component 1 (52.3%) and component 2 (16.7%), accounting for the total variation of 69.0% (Figure 1A). The coarse sediment correlation grain size and clay silt grain size showed a positive correlation with REEs (Figure 1A,B). The PCA loadings further confirm the positive relationship between the coarse sediment (r = 0.04) and clay silt sediment (r = 0.58) with the REEs to be true (Figure 1B). In addition, clay silt sediment (r = 0.58) has more influence on the concentrations of REEs in the sediment than the other sediment grain sizes; this is revealed to be true in Figure 1B. Relationship analysis based on the site revealed that the Jazan mangrove ecosystem is more influenced by rare earth elements than the Al Wajah mangrove ecosystem.

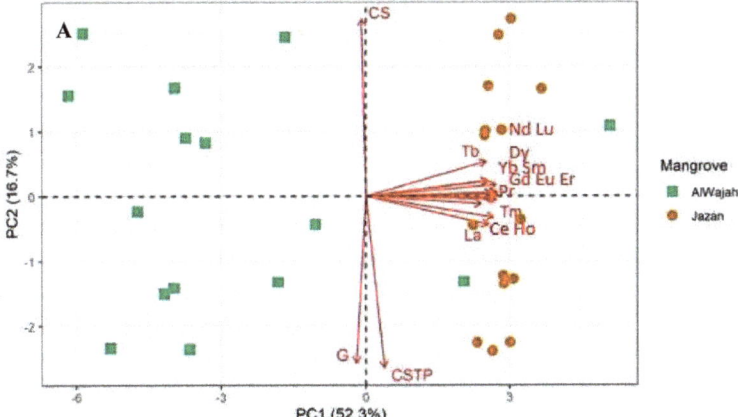

Figure 1. Cont.

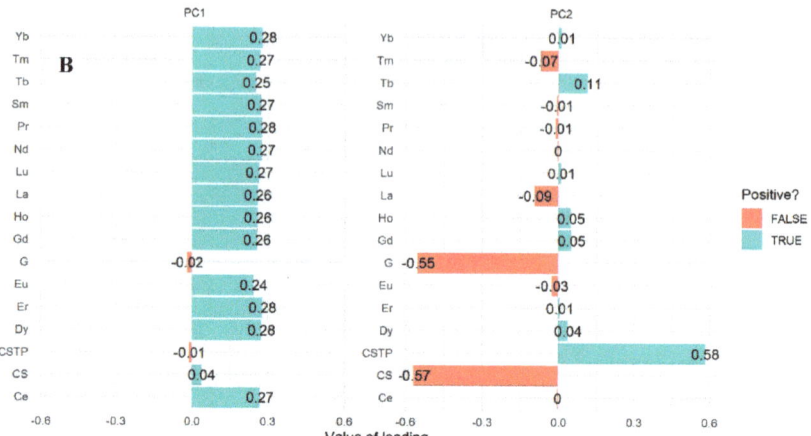

Figure 1. Principal component analysis biplot (**A**) and loadings (**B**) for the relationship between rare earth elements in AlWajah and Jazan mangrove ecosystems. G—Gravel, CSTP—Clay and Silt particles, CS—Coarse sandy.

2.2. Fractionation of REE and Sediment Quality Index

The Post-Archean Australian Shale (PASS) (Taylor and McLennan, 1985) normalized REE patterns of the sediments for the two mangrove ecosystems plotted provide a better understanding of the pattern of accumulation of REE in this study (Figure 2). The results reveal $\sum$REE relative enrichment and comparative trends of fractionation for REE. The fraction $(La/Yb)_n$ was higher (0.49) in the Al Wajah mangrove than the Jazan mangrove (0.41), with an average of 0.45 ± 0.04 for the two mangroves (Table 1). For the fractions using $(Sm/La)_n$, a significantly higher value (2.17) was revealed at Jazan, while the lowest value was recorded in Al Wajah sediment samples. The average value for the $(Sm/La)_n$ fraction was 2.07, which is the highest median proportion when compared to other fractions, revealing a significant LREE and MREE accumulation (Figure 2; Table 1). The Al Wajah mangrove had the lowest value (1.03) of $(Yb/Sm)_n$, while a significantly higher value (1.13) was recorded for sediment sampled from the Jazan mangrove ecosystem. The average for $(Yb/Sm)_n$ in the two mangrove ecosystems was 1.08.

The multi-elemental ratios R(M/L) and R(H/M) indicate positive values corresponding to patterns of fractionation with average MREE enrichment and average HREE depletion. This was supported by the positive range values for R(H/M), the very low positive value for R(H/M) in Jazan, and the negative value at the Al Wajah mangrove, and also a range value from negative to positive (Table 1). There was more enrichment of MREEs at Jazan, with the highest value of R(M/L) (0.28). The low values of R(H/M) and the negative value for Al Wajah indicate HREE depletion, with even more depletion at the Al Wajah mangroves. There exists a significant difference (*t*-test; $p < 0.05$) in the multi-elemental ratios (R(M/L) and R(H/M)) between the two mangrove ecosystems. Ce and Eu anomalies in the two mangroves were revealed by computing the anomalies during the normal and expected shale-normalized REE concentrations, to enable quantification of the probable anomalous concentration. The result revealed a small negative anomaly for Ce, as the average value was almost equal to one. Congruent with the average value, the values of Ce in Al Wajah (0.97) and Jazan (0.86) are still small and negative, with that of Al Wajah being slightly lower. In contrast, the Eu anomaly is small and positive (1.63), with slightly lower (1.27) and higher (1.98) Eu anomaly values at the Al Wajah and Jazan mangroves, respectively (Figure 2; Table 1).

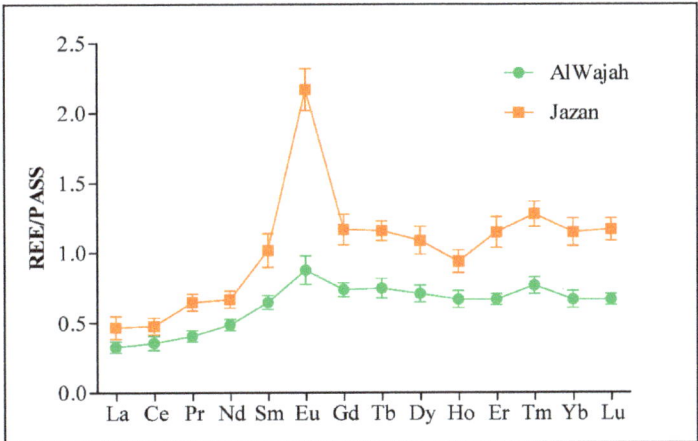

Figure 2. Rare earth elements/Post-Archean Australian Shale (REE/PAAS) patterns for sediments of northern and southern central Red Sea mangrove ecosystems. Values are mean ± standard error.

The sediment quality index used in this study was the geo-accumulation index (Igeo), which has seven classes of enrichment as described by Muller (1969). Using the Igeo, sediments from the two mangroves were strong to extremely contaminated ($4 \leq Igeo \geq 5$) with La, Ce, Pr, Nd, Sm, and Gd, and moderately to strongly contaminated ($1 \leq Igeo \geq 3$) with Dy, Er, and Yb (Figure 3). In addition, Igeo revealed that the sediment was not contaminated (<0) with Eu, Tb, Ho, Tm, or Lu, with negative Igeo values (Figure 3).

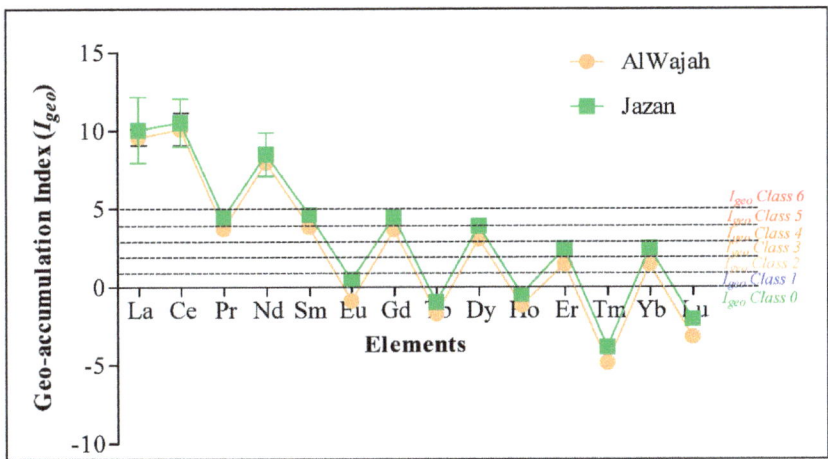

Figure 3. Geo-accumulation index of rare earth elements at Al Wajah and Jazan mangrove ecosystems. Values are mean ± standard error.

2.3. REEs in Mangrove Avicennia Marina and Bio-Concentration Factor (BCF)

The REE concentration in *A. marina* leaves was not significant (t-test; $p > 0.05$) between the two mangrove ecosystems assessed (Table 1). However, a similar pattern of REE distribution in *A. marina* leaves and mangrove sediment was established, with higher concentrations in Jazan mangrove leaves. The higher sum total of REEs in *A. marina* leaves from the Jazan mangrove was about 1/18 of the total sum in sediment, while the lowest value for *A. marina* leaves in the Al Wajah mangrove was about 1/20 of the sum total in sediment (Table 1). This indicates that the ∑REE in the Al Wajah and Jazan mangrove sediments were 20- and 18-fold that of the ∑REE in their *A. marina* leaves, respectively. This is supported by the distribution of REEs in sediment and leaves presented in Figure 4, with higher distributions in sediment than leaves and in the Jazan and Al Wajah mangrove ecosystems. For specific REEs in *A. marina* leaves and sediment, the heat map shows that all the concentrations of the 14 REEs determined in *A. marina* leaves are associated (0.0–0.5) with the concentrations in the sediment (Figure 5). However, the BCF values are less than 1 for all the REEs determined; the recorded BCF values for Lu (0.33) and Tm (0.32) were the highest, while the lowest BCF recorded was for Nd (0.09) (Figure 6).

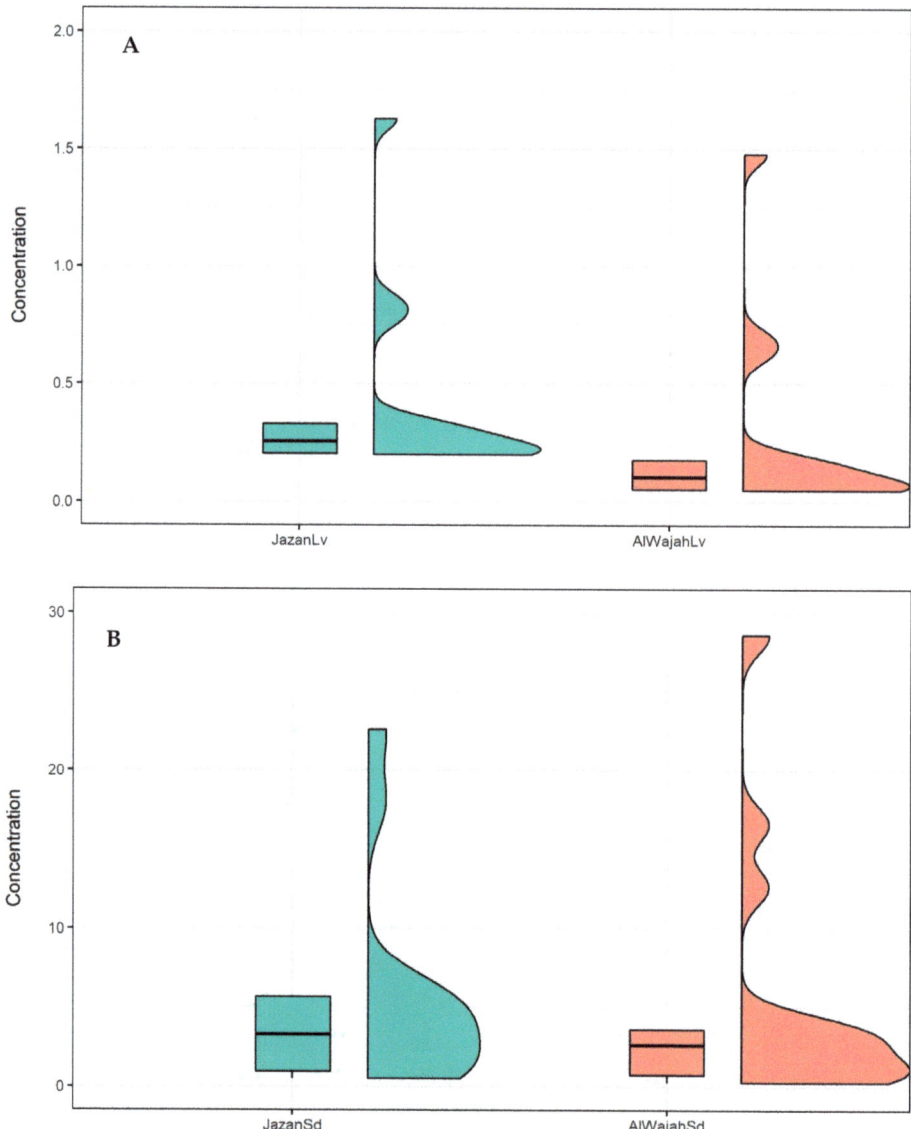

Figure 4. Box and violin plot for distribution of REE in mangrove sediment (**A**) and *A. marina* (**B**). The boxplot for each mangrove displayed the distribution based on the minimum, median, and maximum value, while the violin plot depicts the distribution and density of variables for each mangrove.

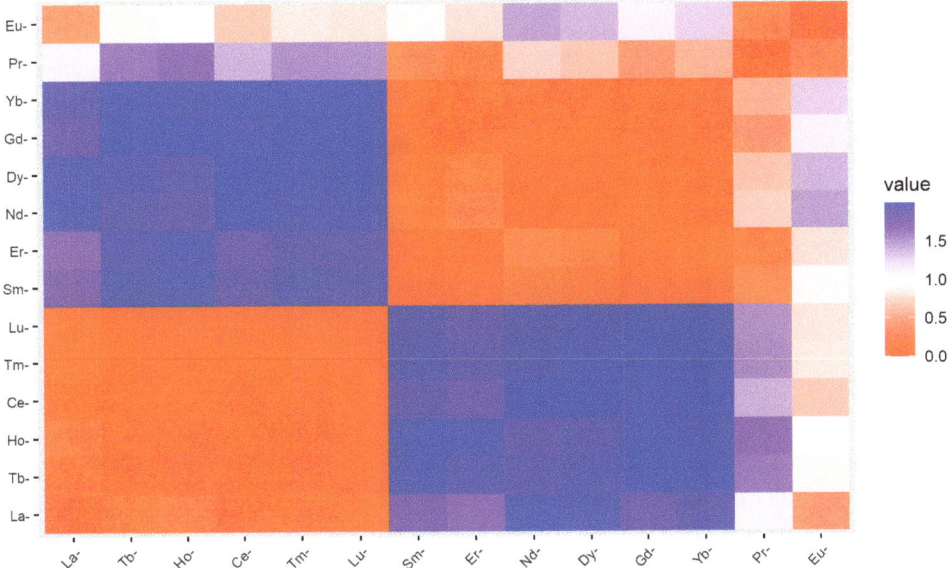

Figure 5. Heat map for the relationship between rare earth elements in sediment and *A. marina*.

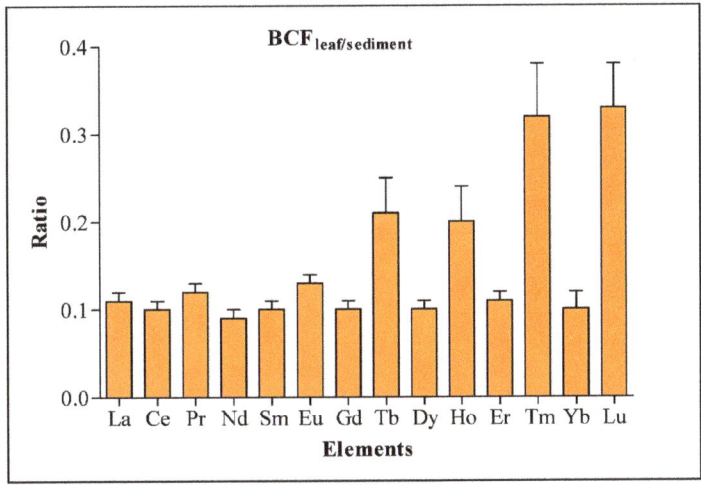

Figure 6. Bio-concentration factor (BCF) for REE composition in northern and southern central Red Sea mangrove ecosystems. Values are mean ± standard error.

3. Discussion

3.1. Influence of Sediment Grain Size on REE Concentrations and Fractionation

Studies indicate that sediment grain size type is a vital factor for REE accumulation in sediments from the Al Wajah and Jazan mangrove ecosystems; this was supported by results from the two mangroves subjected to data analysis. Aquatic ecosystems encompass a combination of diverse products of physicochemical processes functioning in various aspects of drainage basins [18]. This could be linked to or support the influence of grain size types on elements concentrations and distribution, which has also been reported previously to partly reflect the impacts of the hydrodynamic environment [19]. It is important to note that transportation, resuspension, and deposition of allochthonous sediments into an aquatic environment such as the mangrove ecosystem could be influenced by hydrodynamics [19,20]. However, the fractionation of REEs is largely due to mineralogical controls and the diversity of sediment detrital minerals composed of different grain size types triggered by the complex nature of patterns for drainage weathering and hydrodynamic sorting [21,22]. In other studies, different minerals were reported to compose specific REE characteristics due to different grain size types, and differentiation in sizes during transportation and sediment deposition caused some level of differentiation in mineralogy [18,21].

Fine-grain particles have large surface area as one of their properties that can influence an increase in element sorption [23,24]. As such, it is reasonable to hypothesize that there exists a positive correlation between clay silt sediment particles and concentrations of REEs. This could constitute a reason for or enable a better understanding of the correlation between the sediment grain size types and concentrations of REEs in this study [25,26]. Variations in metal concentrations and distribution and their relationship with sediment grain size types have been reported elsewhere, and notably, these studies demonstrated that fine-grain sediment in mangroves possess the capacity for adsorption of REEs [27,28].

The Jazan mangrove is a coastal area open to the public with lots of industrial activities and other anthropogenic activities, and this together with changes in sediment nature (grain size) could form a major reason for higher concentrations of REEs [12,29]. Notably, $\sum$REE in Jazan mangrove sediment was almost two-fold that of the $\sum$REE determined in Al Wajah mangrove sediment, which is a natural reserve area with rare or few anthropogenic activities. Elsewhere, in the assessment of REEs in six mangrove ecosystems in the central Red Sea [19], the average of $\sum$REE (42.56 mg/kg) was lower than (about 1/3 of) the value reported from Jazan mangrove (112.54 mg/kg) in this study was established. In another study on the Egyptian coast of the Red Sea [30], the $\sum$REE (47.55 mg/kg) reported was about 1/2 of the $\sum$REE (112.54 mg/kg) in the Jazan mangrove ecosystem. The depletion of HREEs relative to the lighter ones observed in this study is similar to previous observations elsewhere on the assessment of REEs in mangroves and mangrove soil profiles [11,24,31]. HREE depletion in benthic sediment could be as a result of their greater tendency towards the formation of soluble carbonates in a stable state and complexes of organic forms with ligands than the lighter REEs (LREE and MREE) [32]. HREEs have a high tendency towards being less reactive than LREEs and MREEs, and are well adapted or linked with phases in solid states because they have more pronounced complexation together with ligands located on surfaces of colloids and particles [19,33]. The removal and/or preference of LREEs could also be based on this phenomenon [11].

3.2. REE Fractionation and Sediment Quality Index (Igeo)

The utilization of Post-Archean Australian Shale (PAAS) [34] in the normalization of REE concentrations in sediments of marine ecosystems is vast and quite vital in revealing the comparative fractionation of REEs and the source of REEs, and enables an easy comparison of various findings in ecosystems. The results of REE fractionation are widely utilized as tracers for the determination of the effect of contamination sources and mangrove sediment on flora and fauna diversity, and determination of the chemistry of the environment [8,19].

The use of REE fractionation in the determination of detailed LREE, MREE, and HREE enrichment is vital in the assessment of REEs in an ecosystem, even though direct computation using the sum of the concentrations is measured initially. It is important to note that (La/Yb)n, (Sm/La)n, and (Yb/Sm)n fractions of Post-Archean Australian Shale (PAAS) normalized values were used as a model for REE fractionation and represent LREEs, MREEs, and HREEs, respectively [12,24]. Higher fractions of PAAS normalized (Sm/La)n were recorded in Al Wajah (1.96) and Jazan (2.17) compared to (Yb/Sm)n, and the average (Sm/La)n (2.07) being higher than the average (Yb/Sm)n (1.08) is an indication of the predominance of lighter REEs (LREEs and MREEs) over HREEs in sediments, and more precisely, the levels of lighter REEs (LREEs and MREEs) in Jazan are higher than those of the Al Wajah mangrove. However, the higher fraction (La/Yb)n (0.49) at Al Wajah than (La/Yb)n (0.41) is an indication of the predominance of LREEs at Al Wajah, while MREEs and HREEs were predominant at the Jazan mangrove, as reflected by the higher values of (Sm/La)n (2.17) and (Yb/Sm)n (1.13).

Fractionation of REEs has also been defined by some authors using a scale of 1; a ratio equal to 1 was stated to be an indication of no fraction, whereas a ratio < 1 indicates depletion, and a ratio > 1 implies enrichment of REEs [12,35]. In support of the aforementioned, multi-elemental ratios R(M/L) and R(H/M) with positive and negative values signify fractionation patterns enriched with lighter rare earth elements (MREEs and LREEs), and depletion of HREEs [36,37]. In another study, in the Pichavaram mangrove ecosystem, similar findings were reported on REE fractionation, with an insight into higher concentrations of lighter REEs being linked to clay-silt sediment composition in the mangrove ecosystem [37,38]. This was supported by the positive correlation (r = 0.58) between the clay silt grain size type and the lighter REEs. Nonetheless, it is critical to note that deposition of sediment in a high-sediment regime and rapid burial could lead to a reduced time frame of exposure to REEs in dissolved form with sediment. This causes a restriction in the capacity of adsorption of the sediment and the possibility of REE depletion, leading to dissimilarities in concentrations of REEs in mangrove sediment [11,37].

Geo-accumulation index (Igeo) is an important sediment quality index used for the determination of the extent of contamination of metal and the role of human activities in sediment metal accumulation [29,39,40]. Anthropogenic sources such as industrialization and chemicals or substances from anthropogenic activities such as pesticides and fertilizers contained in agricultural waste could form the key reason for strong to extreme contamination (4 $\leq$ Igeo $\geq$ 5) of certain REEs determined in this study [41–44]. Notably, REEs are often used as fertilizers, which is a direct approach or application to plant/sediment interphase for the purposes of growth, yield, and quality improvement. However, this process or usage of REEs could increase their concentration and sediment or soil contamination [9].

The use of the various applications in several technologies for different materials' production and finished products involves the exploitation of REEs; this has led to pristine ecosystem contamination as a result of poor methods of industrial effluent disposal [9,19]. Dy, Er, and Yb might have originated primarily from anthropogenic activities and crustal material due to their moderate to strong (1 $\leq$ Igeo $\geq$ 3) contamination of the sediment. Conversely, the negative Igeo (Igeo < 0) values for Eu, Tb, Ho, Tm, and Lu suggest possible sources of these REEs to be local and natural [19,45]. Similarly, in China, high REE contamination was linked to anthropogenic activities such as industrial activities, with iron and steel smelting as the major activities [29,46].

3.3. Bio-Concentration Factor (BCF) and REE in Avicennia Marina

Bioaccumulation of metals in plants due to the interaction between plants and sediment is commonly evaluated using Bio-Concentration Factor (BCF) [19,24]. The values for BCFs for all the REES determined in this study were less than 1 (BCFs < 1); this either signifies hypo-accumulation of the REEs by *A. marina* mangrove or the role of an effective mechanism that is involved in detoxification or exclusion of chemicals in *A. marina* [19,47]. Elsewhere, in islands of Indian Sundarban and some mangroves stands in the central Red Sea, results of BCFs reported from these mangroves were in line with our findings with BCFs for REEs less than 1 (BCFs < 1). However, the highest BCF (0.33) determined in this present study was almost three-fold that of the highest value of BCF (0.10) established in Indian Sundarban [12], and 0.01 times greater than the highest BCFs previously determined in six mangroves from the central Red Sea [19].

The composition of bioavailability forms of sediment REEs and the presence of an efficient mechanism or capacity for REE uptake in the plant can significantly affect the phytoextraction of REEs [19,48]. The major reason for the substantial dissimilarity in REE composition in *A. marina* leaves could be due to the bioavailable REE content in the two mangroves investigated. Previous studies have established a positive correlation between increased REE phytoextraction, the concentration of REE in sediment, and environmental variations due to anthropogenic influence or weathering of chemical nature, with the tendency of affecting REE sequestration [19,49].

4. Materials and Methods

4.1. Study Area

The Red Sea of Saudi Arabia encompasses an area of mangroves of approximately 135 Km^2, and the mangroves are disseminated to the northern boundary at 28.207302° N [50]. The arid environment with high temperatures and sparse rainfall is associated with the central Red Sea. In the central Red Sea of the kingdom of Saudi Arabia, some mangrove habitats appear as a narrow fringe supporting halophytes. These mangroves could at times be flooded [26,51]. The abundance of mangroves and levels of anthropogenic activities were used as criteria in the selection of the mangrove ecosystems investigated. Two (2) mangrove ecosystems in Jazan and AlWajah (Figure 7) were selected to accomplish our objectives. In the Jazan mangrove (16°56′38.3″ N, 42°32′31.9″ E), there are various anthropogenic activities, with less control in the catchment, while the Al Wajah mangrove (25°18′25.9″ N, 37°06′51.1″ E) is a reserved area with fewer anthropogenic activities and control of such activities. These two mangrove ecosystems are dominated by monospecific stands of the *Avicennia marina*.

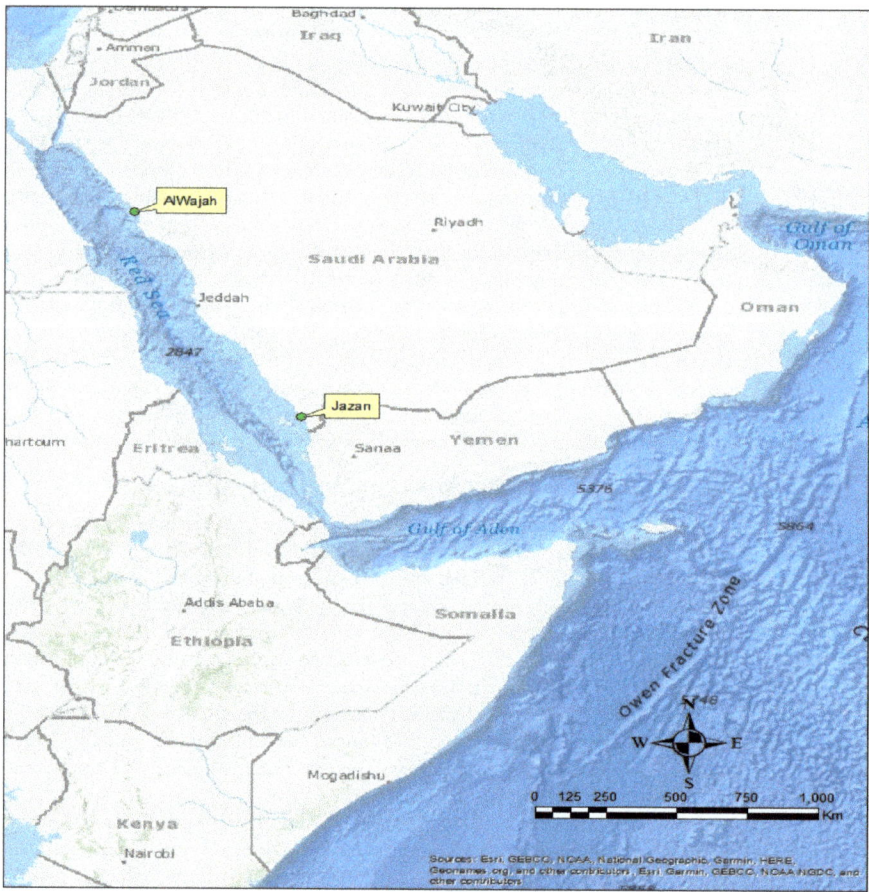

Figure 7. Map showing Al Wajah and Jazan mangrove ecosystems in northern and southern Central Red Sea.

4.2. Sampling and Determination of REE

Thirty (30) samples in each mangrove ecosystem (a total of 60 samples) were collected from two mangrove ecosystems in the northern and southern central Red Sea, Saudi Arabia. *A. marina* leaves and mangrove surface sediments between 0–20 cm were sampled twice monthly from May 2020 to April 2021 from Jazan and Alwajah mangrove ecosystems. At the time of sample collection, variation in the water depth was from 1 to 11 m. For each ecological unit, 15 sediment samples and leaf samples each from 15 mangrove trees were collected in replicate. Van Veen grab-250 cm^2 was utilized in the collection of sediment samples, which were kept in zip lock bags inside an icebox to be transferred to the laboratory for further analyses. For sediment samples, 0.4 g was weighed into a digestion vessel of 50 mL capacity, and 8 mL HNO_3:HCl (1:1) for acid digestion was added. An Anton-Paar PE Multiwave 3000 microwave oven was then used for the digestion of samples at 200 °C for 1 h [52]. Digested samples were kept in a volumetric flask at room temperature, and topped up to 50 mL with Ultrapure Millipore Q water, shaken, and allowed to sit overnight. Filtration of the solution was done using a GF/F filter (Whatman), which was then analyzed for concentrations of rare earth elements (La, Ce, Pr, Nd, Sm, Eu, Gd, Tb, Dy, Ho, Er, Tm, Yb, and Lu) using an Agilent 7700× dual pump Inductively Coupled Plasma-Mass Spectrometer (ICP-MS) [53].

For leaf samples, samples were cleaned using deionized water. Both leaf and sediment samples were dried in an oven at 40–45 °C for 48 and then crushed into powder form with agate mortar and pestle and sieved through 53 μm nylon mesh. The leaf sample was acid digested in HNO_3:H_2O_2 (3:1) at 180 °C for 45 min using 0.2 g of the sieved samples. The formation of calibration curves was achieved by analyzing standard mixture solutions comprising 14 elements at concentrations of 0.5, 1, 5, 10, 20, 50, and 100 μg/L, with 0.999 linear fitting rates. The analytical method for quality control was assessed using standard reference materials GSS-1 and GSV-2 for sediments and leaves, respectively. To confirm repeatability and sensitivity, the solutions of known concentrations used as standard solutions were placed into the sequence of samples for every eight samples. The repossessions of REEs in percentage from the accuracy of the analytical method ranged from 92.68–103.21% and 81.82–116.67% for sediment and leaves, respectively (Table S1, Supplementary Materials). The acceptance of analytical precision and accuracy was based on when the standard deviation was <5% for the rare earth elements from the results of the replications of measurements of the samples and standards.

4.3. Grain Size Analysis in Sediment

The total weight of the oven-dried sediment was measured. Fragmentation of solidified aggregates was done by soaking the dried sediments in distilled water for 24 h. The sediments were washed and separated into fractions of gravel (>2 mm), coarse grain (0.063~2 mm), and mud (clay and silt, <0.063 mm) after passing through 0.063 mm and 2 mm sieves. The computation of percentages of sediment grain sizes relative to the total weight was achieved after the fractions from the residue were dried at 40 °C and weighed [26,38,54].

4.4. Sediment Quality Index and Bio-Concentration Factor (BCF)

The geo-accumulation index based on seven enrichment classes (Table S2, Supplementary Materials) [55] was utilized as the sediment quality index, and was used to measure REE contamination levels in the sediment of the two mangroves investigated using the following formula:

$$I_{geo} = log_2 \left[\frac{C_n}{1.5 \times B_n} \right] \quad (1)$$

where C_n = Concentration of a particular REE in the sediment, B_n = Geochemical background level of that REE obtained for sedimentary rocks (Shales) as described by Turekian and Wedepohl [56], and 1.5 = Correction factor [57] to reduce the effect of variations as a result of sediment lithology.

The bioavailability of REE in *A. marina* was determined using the bio-concentration factor (BCF) to reveal the efficiency of the mangrove to accumulate REEs using the following formula:

$$BCF = \frac{C_{leaves}}{C_{sediment}} \quad (2)$$

where C_{leaf} and $C_{sediment}$ = Concentrations of a given REE in leaves and sediment, respectively.

4.5. Data Analyses

The Student's *t*-test was used for comparison between mean REE concentrations in sediments, leaves, BCF, Igeo, and sediment grain sizes of the two mangrove ecosystems. Principal component analysis (PCA) biplot and loadings were used to determine the relationship between sediment grain sizes and REE concentrations in sediments, while a heat map was used to determine the relationship between REEs in sediment and *A. marina* leaves. The data were analyzed using R for Windows (v. 4.0.3).

The pattern of distribution and REE bioavailability were categorized using fraction ratios (La/Yb, Sm/La, and Yb/Sm) after the concentrations of REE were normalized (*n*) to the Post-Archean Australian Shale (PAAS) [34]. The calculation of multi-elemental ratios was done as described by Duvert et al. [36] and Noack et al. [58] using the formulas below:

$$R_{(M/L)} = log\frac{MREE_n}{LREE_n} = log\left[\frac{(Gd_n + Tb_n + Dy_n)/3}{(La_n + Pr_n + Nd_n)/3}\right] \quad (3)$$

$$R_{(H/M)} = log\frac{HREE_n}{MREE_n} = log\left[\frac{(Tm_n + Yb_n + Lu_n)/3}{(Gd_n + Tb_n + Dy_n)/3}\right] \quad (4)$$

where $R_{(M/L)}$ = Ratio between medium and light REEs, $R_{(H/M)}$ = Ratio between heavy and medium REEs, and *n* refers to PAAS-normalized concentrations.

The non-inclusion of Ce and Eu in the formulas is because of their potential to exhibit an oxidation state. However, a geometric method was used to compute the anomalies of Ce and Eu; this was achieved by assuming that the closest neighboring elements act linearly on log-linear plots [36,59]. The formulas below were used to compute the anomalies:

$$\delta Ce = \frac{2Ce_n}{La_n + Pr_n} \quad (5)$$

$$\delta Eu = \frac{2Eu_n}{Sm_n + Gd_n} \quad (6)$$

where δCe and δEu are the measure of the anomalies for Ce and Eu, and *n* refers to PAAS normalized concentrations.

5. Conclusions

Fractionating causes a significant enrichment of lighter REEs over HREEs in Al Wajah and Jazan mangroves. This is supported by positive and negative multi-elemental ratios R(M/L) and R(H/M), and the enrichment of lighter REEs over the HREEs is attributed to HREEs having a high tendency towards being less reactive than LREEs and MREEs, and the preference for removal of the lighter REEs. However, a higher fraction (La/Yb)n (0.49) at Al Wajah than (La/Yb)n (0.41) is an indication of the predominance of LREEs at Al Wajah, while MREEs and HREEs were predominant at the Jazan mangrove, as reflected by the higher values of (Sm/La)n (2.17) and (Yb/Sm)n (1.13). In addition, the anomalies of Eu were negative for the two mangroves investigated, possibly as a result of dominant reducing conditions in mangrove sediments.

The REE concentrations in Al Wajah and Jazan mangrove ecosystems were significant, with higher concentrations in the Jazan mangrove ecosystem. Different anthropogenic impacts in these two mangroves could form the key reasons for the differences recorded. Clay silt sediment grain size type influences increase REE concentration; however, BCF reveals hypo-accumulation potential or capacity of REE by *A. marina*, with similarity in REE distribution patterns in sediment and *A. marina*. In addition, using the scale of Igeo, based on six classes of classification of the level of contamination, the mangrove sediments were not significantly different and were strong to extremely contaminated with La, Ce, Pr, Nd, Sm, and Gd, and moderately to strongly contaminated with Dy, Er, and Yb, but not contaminated (<0) with Eu, Tb, Ho, Tm, or Lu, showing negative Igeo values. In addition, BCFs for all the REEs determined in this study signify hypo-accumulation of the REEs by *A. marina* mangrove or the role of an effective mechanism that is involved in detoxification or exclusion of chemicals in *A. marina*. There is a dire need for periodic monitoring of REE concentrations in the mangroves investigated, especially the Jazan mangrove ecosystem. This is to keep track of sources of this metal contamination and to develop strategies for control and conservation of these important ecosystems.

Supplementary Materials: The following supporting information can be downloaded at https://www.mdpi.com/article/10.3390/molecules27144335/s1. Table S1: Analytical results achieved on certified reference materials for sediment and leaves; Table S2: Classification of Sediment quality (Geo-Accumulation Index).

Author Contributions: M.O.A. and A.B.A. designed the study and carried out the field surveys and laboratory analysis. A.B.A. performed statistical analyses. M.O.A. and A.B.A. contributed to the initial and final drafting of the manuscript, improvement, and approval for submission. All authors have read and agreed to the published version of the manuscript.

Funding: This research was funded by Deanship of Scientific Research at King Abdulaziz University, Kingdom of Saudi Arabia (grant number D-1443-986-130), and The APC was also funded by Deanship of Scientific Research at King Abdulaziz University, Kingdom of Saudi Arabia.

Institutional Review Board Statement: Not applicable.

Informed Consent Statement: Not applicable.

Data Availability Statement: Data can be shared upon personal request to the authors.

Acknowledgments: The authors acknowledge with thanks the Deanship of Scientific Research at King Abdulaziz University for their technical and financial support. Our appreciation is extended to the editor for handling the process and to the reviewers for their review and valuable comments.

Conflicts of Interest: The authors declare no conflict of interest.

Sample Availability: Samples of the compounds studied are available from the authors.

References

1. Pan, K.; Lee, O.O.; Qian, P.-Y.; Wang, W.-X. Sponges and sediments as monitoring tools of metal contamination in the eastern coast of the Red Sea, Saudi Arabia. *Mar. Pollut. Bull.* **2011**, *62*, 1140–1146. [CrossRef] [PubMed]
2. Roberts, C.M.; McClean, C.J.; Veron, J.E.N.; Hawkins, J.P.; Allen, G.R.; McAllister, D.E.; Mittermeier, C.G.; Schueler, F.W.; Spalding, M.; Wells, F. Marine biodiversity hotspots and conservation priorities for tropical reefs. *Science* **2002**, *295*, 1280–1284. [CrossRef] [PubMed]
3. Amin, B.; Ismail, A.; Arshad, A.; Yap, C.K.; Kamarudin, M.S. Anthropogenic impacts on heavy metal concentrations in the coastal sediments of Dumai, Indonesia. *Environ. Monit. Assess.* **2009**, *148*, 291–305. [CrossRef] [PubMed]
4. Sukumaran, P.V. Elements that rule the world: Impending REE metal crisis. *J. Geol. Soc. India* **2012**, *80*, 295.
5. Johannesson, K.H.; Cortés, A.; Leal, J.A.R.; Ramírez, A.G.; Durazo, J. Geochemistry of rare earth elements in groundwaters from a rhyolite aquifer, central México. In *Rare Earrth Elements in Groundwater Flow Systems*; Springer: Cham, Switzerland, 2005; pp. 187–222.
6. Castor, S.B.; Hedrick, J.B. Rare earth elements. *Ind. Miner. Rocks* **2006**, *1*, 769–792.
7. Gonzalez, V.; Vignati, D.A.L.; Leyval, C.; Giamberini, L. Environmental fate and ecotoxicity of lanthanides: Are they a uniform group beyond chemistry? *Environ. Int.* **2014**, *71*, 148–157. [CrossRef]
8. Bosco-Santos, A.; Luiz-Silva, W.; da Silva-Filho, E.V.; de Souza, M.D.C.; Dantas, E.L.; Navarro, M.S. Fractionation of rare earth and other trace elements in crabs, Ucides cordatus, from a subtropical mangrove affected by fertilizer industry. *J. Environ. Sci.* **2017**, *54*, 69–76. [CrossRef]
9. Balaram, V. Rare earth elements: A review of applications, occurrence, exploration, analysis, recycling, and environmental impact. *Geosci. Front.* **2019**, *10*, 1285–1303. [CrossRef]

10. Pathak, D.; Bedi, R.K.; Kaur, D. Characterization of laser ablated AgInSe2 films. *Mater. Sci.* **2010**, *28*, 199.
11. de Freitas, T.O.P.; Pedreira, R.M.A.; Hatje, V. Distribution and fractionation of rare earth elements in sediments and mangrove soil profiles across an estuarine gradient. *Chemosphere* **2021**, *264*, 128431. [CrossRef]
12. Mandal, S.K.; Ray, R.; González, A.G.; Mavromatis, V.; Pokrovsky, O.S.; Jana, T.K. State of rare earth elements in the sediment and their bioaccumulation by mangroves: A case study in pristine islands of Indian Sundarban. *Environ. Sci. Pollut. Res.* **2019**, *26*, 9146–9160. [CrossRef] [PubMed]
13. Sholkovitz, E.; Szymczak, R. The estuarine chemistry of rare earth elements: Comparison of the Amazon, Fly, Sepik and the Gulf of Papua systems. *Earth Planet. Sci. Lett.* **2000**, *179*, 299–309. [CrossRef]
14. Hoyle, J.; Elderfield, H.; Gledhill, A.; Greaves, M. The behaviour of the rare earth elements during mixing of river and sea waters. *Geochim. Cosmochim. Acta* **1984**, *48*, 143–149. [CrossRef]
15. Wasserman, J.C.; Figueiredo, A.M.G.; Pellegatti, F.; Silva-Filho, E.V. Elemental composition of sediment cores from a mangrove environment using neutron activation analysis. *J. Geochem. Explor.* **2001**, *72*, 129–146. [CrossRef]
16. Censi, P.; Sprovieri, M.; Saiano, F.; Di Geronimo, S.I.; Larocca, D.; Placenti, F. The behaviour of REEs in Thailand's Mae Klong estuary: Suggestions from the Y/Ho ratios and lanthanide tetrad effects. *Estuar. Coast. Shelf Sci.* **2007**, *71*, 569–579. [CrossRef]
17. Silva-Filho, E.V.; Sanders, C.J.; Bernat, M.; Figueiredo, A.M.G.; Sella, S.M.; Wasserman, J. Origin of rare earth element anomalies in mangrove sediments, Sepetiba Bay, SE Brazil: Used as geochemical tracers of sediment sources. *Environ. Earth Sci.* **2011**, *64*, 1257–1267. [CrossRef]
18. Garzanti, E.; Andó, S.; France-Lanord, C.; Censi, P.; Vignola, P.; Galy, V.; Lupker, M. Mineralogical and chemical variability of fluvial sediments 2. Suspended-load silt (Ganga–Brahmaputra, Bangladesh). *Earth Planet. Sci. Lett.* **2011**, *302*, 107–120. [CrossRef]
19. Alhassan, A.B.; Aljahdali, M.O. Sediment Metal Contamination, Bioavailability and Oxidative Stress Response in Mangrove Avicennia marina in Central Red Sea. *Front. Environ. Sci.* **2021**, *9*, 185. [CrossRef]
20. Liu, S.; Zhang, H.; Zhu, A.; Wang, K.; Chen, M.-T.; Khokiattiwong, S.; Kornkanitnan, N.; Shi, X. Distribution of rare earth elements in surface sediments of the western Gulf of Thailand: Constraints from sedimentology and mineralogy. *Quat. Int.* **2019**, *527*, 52–63. [CrossRef]
21. Wu, K.; Liu, S.; Kandasamy, S.; Jin, A.; Lou, Z.; Li, J.; Wu, B.; Wang, X.; Mohamed, C.A.; Shi, X. Grain-size effect on rare earth elements in Pahang River and Kelantan River, Peninsular Malaysia: Implications for sediment provenance in the southern South China Sea. *Cont. Shelf Res.* **2019**, *189*, 103977. [CrossRef]
22. Zhang, X.; Zhang, H.; Chang, F.; Xie, P.; Li, H.; Wu, H.; Ouyang, C.; Liu, F.; Peng, W.; Zhang, Y. Long-range transport of aeolian deposits during the last 32 kyr inferred from rare earth elements and grain-size analysis of sediments from Lake Lugu, Southwestern China. *Palaeogeogr. Palaeoclim. Palaeoecol.* **2021**, *567*, 110248. [CrossRef]
23. Prasad, M.B.K.; Ramanathan, A.L. Sedimentary nutrient dynamics in a tropical estuarine mangrove ecosystem. *Estuar. Coast. Shelf Sci.* **2008**, *80*, 60–66. [CrossRef]
24. Alhassan, A.B.; Aljahdali, M.O. Fractionation and Distribution of Rare Earth Elements in Marine Sediment and Bioavailability in Avicennia marina in Central Red Sea Mangrove Ecosystems. *Plants* **2021**, *10*, 1233. [CrossRef] [PubMed]
25. Shi, X.; Liu, S.; Fang, X.; Qiao, S.; Khokiattiwong, S.; Kornkanitnan, N. Distribution of clay minerals in surface sediments of the western Gulf of Thailand: Sources and transport patterns. *J. Asian Earth Sci.* **2015**, *105*, 390–398. [CrossRef]
26. Aljahdali, M.; Alhassan, A.B.; Zhang, Z. Environmental Factors Causing Stress in Avicennia marina Mangrove in Rabigh Lagoon along the Red Sea: Based on a Multi-Approach Study. *Front. Mar. Sci.* **2021**, *8*, 328. [CrossRef]
27. Caetano, M.; Prego, R.; Vale, C.; de Pablo, H.; Marmolejo-Rodríguez, J. Record of diagenesis of rare earth elements and other metals in a transitional sedimentary environment. *Mar. Chem.* **2009**, *116*, 36–46. [CrossRef]
28. Chaudhuri, P.; Nath, B.; Birch, G. Accumulation of trace metals in grey mangrove Avicennia marina fine nutritive roots: The role of rhizosphere processes. *Mar. Pollut. Bull.* **2014**, *79*, 284–292. [CrossRef]
29. Aljahdali, M.O.; Alhassan, A.B. Saudi Journal of Biological Sciences Ecological risk assessment of heavy metal contamination in mangrove habitats, using biochemical markers and pollution indices: A case study of *Avicennia marina* L. in the Rabigh lagoon, Red Sea. *Saudi J. Biol. Sci.* **2020**, *27*, 1174–1184. [CrossRef]
30. El-Taher, A.; Badawy, W.M.; Khater, A.E.M.; Madkour, H.A. Distribution patterns of natural radionuclides and rare earth elements in marine sediments from the Red Sea, Egypt. *Appl. Radiat. Isot.* **2019**, *151*, 171–181. [CrossRef]
31. Brito, P.; Malvar, M.; Galinha, C.; Caçador, I.; Canário, J.; Araújo, M.F.; Raimundo, J. Yttrium and rare earth elements fractionation in salt marsh halophyte plants. *Sci. Total Environ.* **2018**, *643*, 1117–1126. [CrossRef]
32. Kuss, J.; Garbe-Schönberg, C.-D.; Kremling, K. Rare earth elements in suspended particulate material of North Atlantic surface waters. *Geochim. Cosmochim. Acta* **2001**, *65*, 187–199. [CrossRef]
33. Sholkovitz, E.R. The geochemistry of rare earth elements in the Amazon River estuary. *Geochim. Cosmochim. Acta* **1993**, *57*, 2181–2190. [CrossRef]
34. Taylor, S.R.; McLennan, S.M. The Continental Crust: Its Composition and Evolution. United States. 1985. Available online: https://www.osti.gov/biblio/6582885 (accessed on 23 January 2022).
35. Sow, M.A.; Payre-Suc, V.; Julien, F.; Camara, M.; Baque, D.; Probst, A.; Sidibe, K.; Probst, J.L. Geochemical composition of fluvial sediments in the Milo River basin (Guinea): Is there any impact of artisanal mining and of a big African city, Kankan? *J. Afr. Earth Sci.* **2018**, *145*, 102–114. [CrossRef]

36. Duvert, C.; Cendón, D.I.; Raiber, M.; Seidel, J.-L.; Cox, M.E. Seasonal and spatial variations in rare earth elements to identify inter-aquifer linkages and recharge processes in an Australian catchment. *Chem. Geol.* **2015**, *396*, 83–97. [CrossRef]
37. Andrade, G.R.P.; Cuadros, J.; Barbosa, J.M.P.; Vidal-Torrado, P. Clay minerals control rare earth elements (REE) fractionation in Brazilian mangrove soils. *CATENA* **2022**, *209*, 105855. [CrossRef]
38. Prasad, M.B.K.; Ramanathan, A.L. Distribution of rare earth elements in the Pichavaram mangrove sediments of the southeast coast of India. *J. Coast. Res.* **2008**, *24*, 126–134. [CrossRef]
39. Aljahdali, M.O.; Alhassan, A.B. Spatial Variation of Metallic Contamination and Its Ecological Risk in Sediment and Freshwater Mollusk: Melanoides tuberculata (Müller, 1774)(Gastropoda: Thiaridae). *Water* **2020**, *12*, 206. [CrossRef]
40. Godwyn-Paulson, P.; Jonathan, M.P.; Rodríguez-Espinosa, P.F.; Rodríguez-Figueroa, G.M. Rare earth element enrichments in beach sediments from Santa Rosalia mining region, Mexico: An index-based environmental approach. *Mar. Pollut. Bull.* **2022**, *174*, 113271. [CrossRef]
41. Ghosh, S.; Ram, S.S.; Bakshi, M.; Chakraborty, A.; Sudarshan, M.; Chaudhuri, P. Vertical and horizontal variation of elemental contamination in sediments of Hooghly Estuary, India. *Mar. Pollut. Bull.* **2016**, *109*, 539–549. [CrossRef]
42. Aljahdali, M.O.; Alhassan, A.B. Metallic pollution and the use of antioxidant enzymes as biomarkers in Bellamya unicolor (Olivier, 1804) (Gastropoda: Bellamyinae). *Water* **2020**, *12*, 202. [CrossRef]
43. Alhassan, A.; Balarabe, M.; Gadzama, I. Assessment of some heavy metals in macobenthic invertebrate and water samples collected from Kubanni reservoir Zaria, Nigeria. *FUW Trends Sci. Technol. J.* **2016**, *1*, 55–60.
44. Aljahdali, M.O.; Alhassan, A.B. Heavy metal accumulation and anti-oxidative feedback as a biomarker in seagrass Cymodocea serrulata. *Sustainability* **2020**, *12*, 2841. [CrossRef]
45. Balaram, V. Recent advances in the determination of elemental impurities in pharmaceuticals–Status, challenges and moving frontiers. *TrAC Trends Anal. Chem.* **2016**, *80*, 83–95. [CrossRef]
46. Zhou, H.; Chun, X.; Lü, C.; He, J.; Du, D. Geochemical characteristics of rare earth elements in windowsill dust in Baotou, China: Influence of the smelting industry on levels and composition. *Environ. Sci. Process. Impacts* **2020**, *22*, 2398–2405. [CrossRef] [PubMed]
47. Cluis, C. Junk-greedy greens: Phytoremediation as a new option for soil decontamination. *BioTeach J.* **2004**, *2*, 1–67.
48. Cao, X.; Chen, Y.; Wang, X.; Deng, X. Effects of redox potential and pH value on the release of rare earth elements from soil. *Chemosphere* **2001**, *44*, 655–661. [CrossRef]
49. Mleczek, P.; Borowiak, K.; Budka, A.; Niedzielski, P. Relationship between concentration of rare earth elements in soil and their distribution in plants growing near a frequented road. *Environ. Sci. Pollut. Res.* **2018**, *25*, 23695–23711. [CrossRef] [PubMed]
50. Rasul, N.M.A.; Stewart, I.C.F. *Oceanographic and Biological Aspects of the Red Sea*; Springer: Cham, Switzerland, 2018; ISBN 3319994174.
51. Almahasheer, H.; Duarte, C.M.; Irigoien, X. Nutrient limitation in central Red Sea mangroves. *Front. Mar. Sci.* **2016**, *3*, 271. [CrossRef]
52. The U.S. Environmental Protection Agency. Method 3051A: Microwave assisted acid digestion of sediments, sludges, soils, and oils. In *SW-846 Methods A*; The U.S. Environmental Protection Agency: Washington, DC, USA, 1994; Volume 3051, p. 1997.
53. Kawabata, K.; Kishi, Y.; Kawaguchi, O.; Watanabe, Y.; Inoue, Y. Determination of rare-earth elements by inductively coupled plasma mass spectrometry with ion chromatography. *Anal. Chem.* **1991**, *63*, 2137–2140. [CrossRef]
54. Aljahdali, M.O.; Munawar, S.; Khan, W.R. Monitoring Mangrove Forest Degradation and Regeneration: Landsat Time Series Analysis of Moisture and Vegetation Indices at Rabigh Lagoon, Red Sea. *Forests* **2021**, *12*, 52. [CrossRef]
55. Muller, G. Index of geoaccumulation in sediments of the Rhine River. *Geojournal* **1969**, *2*, 108–118.
56. Turekian, K.K.; Wedepohl, K.H. Distribution of the elements in some major units of the earth's crust. *Geol. Soc. Am. Bull.* **1961**, *72*, 175–192. [CrossRef]
57. Stoffers, P.; Glasby, G.P.; Wilson, C.J.; Davis, K.R.; Walter, P. Heavy metal pollution in Wellington Harbour. *N. Z. J. Mar. Freshw. Res.* **1986**, *20*, 495–512. [CrossRef]
58. Noack, C.W.; Dzombak, D.A.; Karamalidis, A.K. Rare earth element distributions and trends in natural waters with a focus on groundwater. *Environ. Sci. Technol.* **2014**, *48*, 4317–4326. [CrossRef] [PubMed]
59. Lawrence, M.G.; Greig, A.; Collerson, K.D.; Kamber, B.S. Rare earth element and yttrium variability in South East Queensland waterways. *Aquat. Geochem.* **2006**, *12*, 39–72. [CrossRef]

Article

Enhanced Toxicity of Bisphenols Together with UV Filters in Water: Identification of Synergy and Antagonism in Three-Component Mixtures

Błażej Kudłak [1], Natalia Jatkowska [1,*], Wen Liu [2], Michael J. Williams [2], Damia Barcelo [3] and Helgi B. Schiöth [2]

[1] Department of Analytical Chemistry, Faculty of Chemistry, Gdańsk University of Technology, 11/12 Narutowicza Str., 80-233 Gdańsk, Poland; blakudla@pg.edu.pl
[2] Functional Pharmacology, Department of Neuroscience, Uppsala University, 751 24 Uppsala, Sweden; wen.liu@neuro.uu.se (W.L.); michael.williams@neuro.uu.se (M.J.W.); helgi.schioth@neuro.uu.se (H.B.S.)
[3] Catalan Institute for Water Research (ICRA), Parc Científic i Tecnològic de la Universitat de Girona, C/Emili Grahit, 101 Edifici H2O, E-17003 Girona, Spain; dbcqam@cid.csic.es
* Correspondence: natjatko@pg.edu.pl

Abstract: Contaminants of emerging concern (CEC) localize in the biome in variable combinations of complex mixtures that are often environmentally persistent, bioaccumulate and biomagnify, prompting a need for extensive monitoring. Many cosmetics include UV filters that are listed as CECs, such as benzophenone derivatives (oxybenzone, OXYB), cinnamates (2-ethylhexyl 4-methoxycinnamate, EMC) and camphor derivatives (4-methylbenzylidene-camphor, 4MBC). Furthermore, in numerous water sources, these UV filters have been detected together with Bisphenols (BPs), which are commonly used in plastics and can be physiologically detrimental. We utilized bioluminescent bacteria (Microtox assay) to monitor these CEC mixtures at environmentally relevant doses, and performed the first systematic study involving three sunscreen components (OXYB, 4MBC and EMC) and three BPs (BPA, BPS or BPF). Moreover, a breast cell line and cell viability assay were employed to determine the possible effect of these mixtures on human cells. Toxicity modeling, with concentration addition (CA) and independent action (IA) approaches, was performed, followed by data interpretation using Model Deviation Ratio (MDR) evaluation. The results show that UV filter sunscreen constituents and BPs interact at environmentally relevant concentrations. Of notable interest, mixtures containing any pair of three BPs (e.g., BPA + BPS, BPA + BPF and BPS + BPF), together with one sunscreen component (OXYB, 4MBC or EMC), showed strong synergy or overadditive effects. On the other hand, mixtures containing two UV filters (any pair of OXYB, 4MBC and EMC) and one BP (BPA, BPS or BPF) had a strong propensity towards concentration dependent underestimation. The three-component mixtures of UV filters (4MBC, EMC and OXYB) acted in an antagonistic manner toward each other, which was confirmed using a human cell line model. This study is one of the most comprehensive involving sunscreen constituents and BPs in complex mixtures, and provides new insights into potentially important interactions between these compounds.

Keywords: bisphenol A; bisphenol A analogues; sunscreens; toxicity of mixtures; acute toxicity; environmental pollution

1. Introduction

For years, the increase in anthropopressure on the natural environment, resulting from economic, industrial and agricultural activity, has caused significant changes to both abiotic and biotic systems. Numerous reports provide information about contaminant facilitated degradation and disintegration of the natural environment. For several years, a key interest has been to understand the environmental and biological effects of compounds generally referred to as CECs (contaminants of emerging concern) [1]. CECs typically exhibit a high level of environmental persistence and are not easily biodegradable, often demonstrating the ability to bioaccumulate and biomagnify. Moreover, most of these contaminants are considered to be bioavailable and capable of disrupting endocrine systems [2].

The cosmetic industry is flooding the market with numerous products in different formats. A wide range of new cosmetics enter the market each season, including products that protect the skin from harmful UV radiation. Many of these products contain highly diverse chemicals, including UV filters, which are considered CECs. These chemicals encompass organic compounds that are classified into different families, such as benzophenone derivatives (e.g., oxybenzone (OXYB)), salicylates, cinnamates (e.g., 2-ethylhexyl 4-methoxycinnamate (EMC)), camphor derivatives (e.g., 4-methylbenzylidene-camphor (4MBC)), p-aminobenzoic acid and its derivatives. Furthermore, CEC listed inorganic compounds are also used as UV filters, such as titanium dioxide (TiO_2) and zinc oxide (ZnO), which create a physical barrier to excessive solar radiation [3].

Oxybenzon (OXYB) is a popular ingredient in UV-protective products, owing to its low production price, ease of synthesis and high solubility in organics. Along with its threats to animal species, OXYB also threatens plant species, due to the fact that at relatively low concentrations (e.g., 5 μM), it can cause negative reactions in plant photosynthesis systems [4], excessive production of reactive oxygen species (ROS) and disassembly of membranes in algae [5]. Furthermore, several studies performed on fish (*Danio rerio* and *Oryzias latipes*) demonstrate that exposure to OXYB (at 749 μg/L median measured levels) greatly impacts fecundity, lowering the number of eggs laid and hatched [6].

2-Ethylhexyl 4-methoxycinnamate (EMC) is used as a UV filter to help in the treatment of sunburns and scars. EMC is detectable in samples of almost all everyday consumer products, in dust collected in private and public buildings, sewage sediments, surface waters and even mammalian excrements at levels of up to several hundred nM [7]. Considered a CEC, in some species it has been shown to be a possible factor in reproduction impairments at very low levels, including 0.4 mg/kg (*Potamopyrgus antipodarum*) and 10 mg/kg (*Melanoides tuberculata*) [8].

4-Methylbenzylidene-camphor (4MBC) is also commonly used in multi-ingredient sunscreen compositions. 4MBC is also detectable in dust and sewage waters at elevated levels (1780 μg/kg d.w.) [9], resulting from the flushing of household waters, surface outflows and industrial discharges [10]. Moreover, like other UV filters, 4MBC may be present in plastics and other everyday items, posing an additional risk to marine ecosystems.

Much more is known about the environmental impact of bisphenol analogues, including BPA, BPS and BPF. These compounds are commonly used in the production of polycarbonate plastics, epoxy resins, food storage containers, plastic helmets, toys, as well as many other products. BPF-based resins have very good viscous properties and are resistant to organic solvents; therefore, it is more and more commonly used in industry. Furthermore, public acceptance that BPA exposure induces hormonal impairment has led to a considerable increase in the use of BPS. Unfortunately, this means that "BPA-free" products are not free of BPS. The presence of bisphenol analogues is ubiquitous, having been reported in indoor dust, foodstuffs, surface water, sewage sludge, sediment samples, human urine, blood and in maternal and cord serum, generally with concentration levels lower than BPA, but in the same order of magnitude [11–14].

Toxicological studies have been carried out for most of the UV filter and bisphenol compounds, which revealed that they cause adverse changes in living organisms, including affecting survival, behavior, growth, metabolism, development and reproduction, in addition to showing hormonal-like activity [15]. Given the properties of such xenobiotics, it is essential to carry out research aimed at determining the ecological impact resulting from their common presence in the environment. Unfortunately, most research focuses on the quantification of contaminants in the samples. Such information is obviously very useful, yet it may not be sufficient for a complete assessment of the environmental condition, as it does not allow for the determination of an actual biological impact of given pollutant, especially when considering the fact that contaminants do not occur individually within the environment but are found as mixtures.

The co-occurrence and the resulting interactions of contaminants make it extremely difficult to foresee the environmental and physiological effects of such exposure. Therefore, it becomes necessary to identify and determine the type of interactions that occur between contaminants (first in model mixtures then in relation to environmental concentration levels). To date, the research carried out in this area has primarily focused on determining the impact of mixtures comprising UV filters [16]. However, it is apparent in the environment that such contaminants co-occur in mixtures with other xenobiotic compounds, which can significantly alter the level and action of their toxicity. Our study is the first systematic attempt to understand the combined impact of these emerging pollutants using bioluminescent bacteria and then testing if these responses can be recapitulated in human cancer cells. In this way, we not only test the effects of various pollutant mixtures, but try to confirm the relevance of the results using human cells.

2. Results

In the following subsections, the results of mixture toxicity studies with bioluminescent bacteria are given.

2.1. Impact of Three-Component Mixtures on BPA Toxicity

2.1.1. BPA + OXYB + Second Bisphenol

Results of the impact of a third component on the toxicity of BPA and OXYB—in the form of MDR values—are given in Supplementary Table S1 and Figure 1. With the CA model, the impact of BPS on the BPA + OXYB mixture had a clear concentration dependence trend (Figure 1A), whereby increasing BPS concentrations moved the mixture's toxicological potential into strong synergy. On the other hand, in the IA model, at the lowest concentration of BPS, the impact clearly went in the direction of antagonism. A similar trend was observable when the third component was BPF, but the magnitude of this action was clearly weaker.

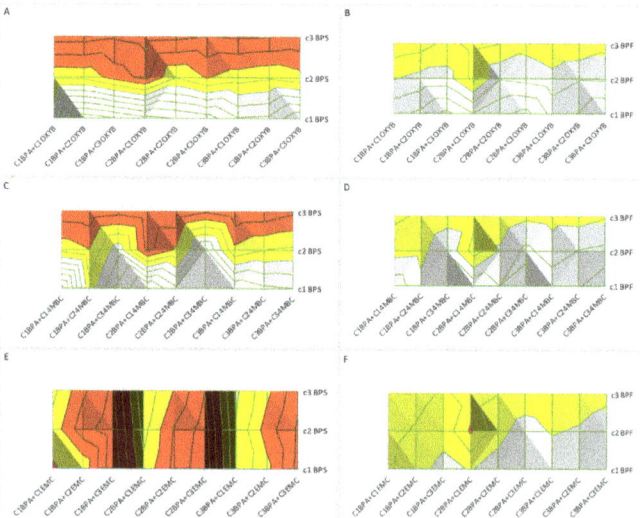

Figure 1. MDR values of *Aliivibrio fischeri* bioluminescent bacteria results for: (**A**) CA modeling of the BPA, OXYB and BPS mixture, (**B**) CA modeling of the BPA, OXYB and BPF mixture, (**C**) CA modeling of the BPA, 4MBC and BPS mixture, (**D**) CA modeling of the BPA, 4MBC and BPF mixture, (**E**) CA modeling of the BPA, EMC and BPS mixture, (**F**) CA modeling of the BPA, EMC and BPF mixture (n = 2). Red color indicates confirmed synergy, blue indicates antagonism, while yellow and green refer to under- and overestimation, respectively.

2.1.2. BPA + OXYB + Second UV Filter

Underestimation was observed in nine cases in our studies to understand the impact of EMC on a BPA + OXYB mixture. At the lowest concentrations, this three-component mixture had a trend towards synergy (for CA modeling, refer to Supplementary Table S1 and Figure 1B). This may suggest and justify continuing studies in the direction of low-content mixtures, which more precisely reflect environmentally relevant mixture compositions. Nevertheless, both IA and CA models adequately showed variation of toxicity with three-component mixtures in our study with bioluminescent bacteria.

2.1.3. BPA + 4MBC + Second Bisphenol Analogue

Results of a third component on the toxicity of a BPA and 4MBC mixture are presented in Supplementary Table S2 and Figure 1C, in the form of MDR values. Interestingly, a trend was observed when the concentration of a second bisphenol analogue (BPS) was increased, where the mixture components became synergistic in their behavior (such as in the case of BPA + OXYB + second bisphenol analogue). In the case of BPF, with CA modeling, this trend was also observable, but the magnitude was weaker. IA modeling resulted in a tendency towards overestimation.

2.1.4. BPA + 4MBC + Second UV Filter

The impact of EMC on the toxicity of a BPA + 4MBC mixture was very well forecasted by the IA model (refer to Supplementary Table S2); The CA model predicted several instances of underestimation, and this behavior was observed in only six cases of EMC C1 and C2.

2.1.5. BPA + EMC + Second Bisphenol Analogue

A CA model of a BPA, EMC and BPS mixture showed synergistic effects in all cases and underestimated the impact (Figure 1E and electronic Supplementary Table S3). Some concentration dependence was also visible in mixtures containing the lowest BPS content, as well as with BPA C2 and C3. With increasing EMC concentration, the mixture became more synergistic. Similar behavior was noticeable for mixtures where BPF was present as the third component (Figure 1F), but there was only one confirmed case of synergy. On the other hand, both for BPS and BPF, in almost all cases, the IA model showed no significant interactions (only one case of overestimation was observed—C1 BPS + C1 BPA + C2 EMC).

2.2. Impact of Three-Component Mixtures on BPS Toxicity

2.2.1. BPS + OXYB + Second Bisphenol

MDR results of the BPF impact on a BPS + OXYB mixture are presented in Supplementary Table S4 and Supplementary Figure S2A. For all cases, for BPS C1 no significant discrepancies between observed and calculated toxicity values were present using either the CA or the IA model. The OXYB and BPF mixture showed underestimation (for C1 BPF) and synergy (C2 and C3 of BPF) with BPS C2, having a clear concentration-dependence trend; interestingly, BPS C3 made all mixtures synergistic.

2.2.2. BPS + OXYB + Second UV Filter

The impact of 4MBC and EMC on BPS + OXYB toxicity are presented in Supplementary Figure S2B,C, respectively, as well as in Supplementary Table S4. At the lowest BPS concentration studied, both models indicated no interactions between any chemical found in the mixture. The BPS C2 and C3 concentration dependence trend was similar to one observed with the BPS + OXYB + BPF cocktail; however, the IA model showed a trend towards overestimation. Interestingly, EMC C2 and C3 exhibited a strong undeniable synergistic impact on all BPS + OXYB mixtures, while the impact of EMC C1 was underestimated, with a trend towards synergy at the lowest concentrations of all analytes present in the cocktail.

2.2.3. BPS + 4MBC + Second UV Filter

Results of MDR for mixtures containing BPS, 4MBC and EMC are presented in Supplementary Figure S2D, as well as in Supplementary Table S5. EMS had a tendency to underestimate the impact on the BPS and 4MBC mixture in almost all cases studied. Of note, at environmentally relevant levels, almost all other mixtures confirmed the synergistic impact (with CA model) of these pollutant cocktails.

2.2.4. BPS + EMC + Second Bisphenol

The CA model of BPS + EMC mixtures with BPF again showed interesting trends (Supplementary Figure S2E and Supplementary Table S6). The lowest BPS concentrations, with increasing concentrations of BPF and EMC, had a trend towards synergy. The magnitude of this trend was not strong but was noticeable. The BPS C2 and C3 concentration levels trended towards a slight weakening of synergy, the only exception being the mixture of BPS C2 + EMC C3 + BPF C2, which showed signs of underestimation. The IA model was again resistant to concentration variations and was not suitable for predicting plausible synergy/antagonism of chemicals acting in a similar manner.

2.3. Impact of Three-Component Mixtures on BPF Toxicity

2.3.1. BPF + 4MBC + Second Bisphenol

In the case of CA modeling, BPS had a clear synergistic impact on the BPF + 4MBC toxicological output of the bioluminescent bacteria. Interestingly, a trend of going from underestimation to synergy was clearly correlated with increasing BPS content (Supplementary Figure S3A and Supplementary Table S7). From an aqueous ecosystems point of view, this may demonstrate the detrimental effects of replacing BPA with BPS in everyday products/plastics production and uncontrolled waste disposal.

2.3.2. BPF + 4MBC + Second UV Filter

The impact of EMC and OXYB on the toxicity of BPF + 4MBC is presented in a graphic manner in Supplementary Figure S3B,C, respectively (as well as in Supplementary Table S7). No cases of synergy were present (except one result of the mixture with OXYB), and underestimation was confirmed in all cases of BPF C3.

2.3.3. BPF + OXYB + Second UV Filter

The results of MDR modeling for BPF + OXYB + EMC are presented in Figure S3D (and in Supplementary Table S8). The behavior of these three-component mixtures was similar to the ones shown in Supplementary Figure S2B,C, with many underestimated cases confirmed. Only one case of synergy was indicated.

2.4. Mixture of Three UV Filters Studied

The CA model for three compounds theoretically acting with the same MOA (Mode of Action) reflects the impact of such mixtures on bioluminescent bacteria (data presented in Supplementary Table S9). The results of MDR for the IA modeling are shown in Figure 2, where a clear trend towards overestimation was visible, indicating that sunscreen components compete at the concentrations studied.

Figure 2. MDR values for IA modeling of the 4MBC, OXYB and EMC mixture of bioluminescent bacteria results (*n* = 2). Red color indicates confirmed synergy, blue indicates antagonism, while yellow and green refer to under- and overestimation, respectively.

2.5. Human Breast Cell Toxicity with MCF10A—A Non-Tumoral, Epithelial Cell Line

In order to study the possible impact of selected UV filters (EMC and 4MBC) with BPA and DBP (dibutyl phthalate) on human breast cells (with a non-tumoral, epithelial cell line named MCF10A), we first performed dose-response studies on each of the single substances (data not shown), which was followed by experiments using mixtures, performed for 24 and 72 h. The concentrations studied are given in Section 4.6. (for CA and IA models). Both UV filters studied had an antagonistic impact on the BPA and DBP mixture (see Table 1). The IA model was the more reliable one, as the chemicals belonged to different classes. The studies showed a small decrement of magnitude for antagonism with increasing concentrations of DBP in the mixtures with prolonged exposure time, while 4MBC appeared to be a stronger antagonist when compared to EMC.

Table 1. MDR results of CA and IA toxicity modeling with MCF10A human breast cancer cells (MDR values > 2.0 exhibit antagonism, MDRs < 0.5 show synergism and MDR values of 0.50–0.71 and 1.40–2.00 mean, respectively, under- and overestimation of the presented models *, *n* = 3).

	S3: EMC				S3: 4MBC			
	24 h		72 h		24 h		72 h	
	CA	IA	CA	IA	CA	IA	CA	IA
C1 BPA + C1 DBP + S3 C1	0.923	2.728	0.852	2.521	0.980	2.897	1.003	2.964
C1 BPA + C1 DBP + S3 C3	0.921	2.708	0.836	2.553	0.970	2.879	1.033	3.018
C1 BPA + C2 DBP + S3 C1	0.869	2.543	0.835	2.486	0.931	2.722	0.856	2.573
C1 BPA + C2 DBP + S3 C3	0.851	2.500	0.774	2.355	0.932	2.756	0.869	2.576
C1 BPA + C3 DBP + S3 C1	0.857	2.492	0.780	2.327	0.907	2.632	0.829	2.506
C1 BPA + C3 DBP + S3 C3	0.840	2.464	0.761	2.309	0.895	2.634	0.856	2.563

* red color indicates confirmed synergy, blue indicates antagonism, while yellow and green refer to under- and overestimation, respectively.

3. Discussion

UV filter compounds are commonly used as active ingredients in many personal care products (PCPs) to protect against sunburns, premature aging and skin cancer caused by the UV light irradiation [17]. In order to ensure PCPs provide adequate protection, a mixture of individual substances, which usually contain three to eight UV filters, is common [3]. Growing concern among consumers about the harmful effects of solar radiation has significantly increased the use of products containing sunscreens, which is directly contributing to their presence in the biome. Currently, there are no regulations inhibiting such combinations [18]. As a consequence, UV filters have been detected in different environmental samples in the ng/L to low µg/L range. For example, OXYB has been detected in seawater samples in the range 13.2 ± 0.4–31.7 ± 0.25 ng/L [19] and <5 to

19.2 µg/L [17]. Moreover, OXYB has been detected in freshwater with quantitative levels ranging from 5 to 79 ng/L, as well as in bottom sediments ranging from <7–82.1 ng/g d.w. [19] and <0.03–65.7 ng/kg d.w. [20]. Some organic UV filters (such as 2-hydroxy-4-methoxybenzophenone, 4-methylbenzylidene camphor and isoamyl 4-methoxycinnamate, among others) have been shown to accumulate in mussels, corals, crabs and fish, with concentrations ranging from few to hundreds of ng/g d.w. (e.g., OXYB and EMC have been detected in cod liver tissue in the range <20–1037 ng/g and <30–36.9 ng/g, respectively [19] (Langford et al., 2015)) and in the livers of dolphins [21].

Our study shows, for the first time, how given selected mixtures of BPs and sunscreen constituents interact in three-component mixtures, at environmentally relevant concentrations. Interestingly, mixtures containing two BPs (BPA + BPS, BPA + BPF, BPS + BPF) and one sunscreen component (OXYB, 4MBC or EMC), show strong synergy in a prevailing number of tests. Moreover, mixtures containing two UV filters (any pair of OXYB, 4MBC and EMC) and one BP (BPA, BPS or BPF) have a strong underestimation potential, which is clearly concentration dependent. The three-component mixtures of UV filters (4MBC, EMC and OXYB) act in an antagonistic manner for each other. Interestingly, the antagonist effects of BPs and these UV filters are conserved in a model human cell line, indicating that these results may be relevant for mammals. We also show that BPF and BPS have as significant impact on the toxicity of these mixtures, which is in agreement with the effects of BPA.

The results show important synergistic interactions between BPs and the three sunscreen constituents, which belong to different chemical classes. The strongest synergy relate to the synergy within three-component mixtures that hold two different types of bisphenols and each of the three sunscreens (see Figure 1A,C,E). The results point out that BPS seems to be important for the synergy action for both OXYB and 4MBC, while the highest doses of EMC are important for the synergy action with the two other BPs. The consistent dose dependency driven by both the BPS and EMC suggest that further studies should be addressed to understand better the mechanisms behind these three-component mixture effects. Overall, our new data suggest that closer attention should be paid to the potential of sunscreens to be much more detrimental in environmentally occurring pollutant mixtures, when compared to their individual effects. The synergy of these UV filters with BPs, which are known to leak out of plastic containers, should prompt investigations into the coexistence of these products in several different kinds of situations. This includes when sunscreens are stored in plastic containers for a long time, sometimes at relatively high temperatures, and then applied to the human skin in relatively high doses [22]. It is relevant, in this context, that sunscreens have the ability to penetrate deep into skin tissues, where they are likely to enable the transfer of other compounds [23–25]. For example, in comprehensive skin tests, EMC was shown to penetrate deep into skin tissue. While the penetration was less significant, similar observations have been reported for OXYB and its metabolites [26]. It was also reported that exposure of human macrophages to EMC led to reduced immunity, increasing the risk of asthma and allergy-related complications, due to elevated excretion of cytokines [27]. Limited studies can be found on the interactions between sunscreen constituents and other pollutants [28]. In the work of Brand et al. [29], scientists confirmed enhanced penetration of selected pesticides (e.g., 2,4-D (2,4-Dichlorophenoxyacetic acid), DEET (N,N-diethyl-m-toluamide), paraquat, parathion and malathion) when hairless mouse skin was co-exposed to titanium and zinc oxides [29]. In the work of Marrot [30], the possible explanation for elevated atopy or eczema during periods of increased pollution exposure (heavy metals or polycyclic aromatic hydrocarbons (PAHs)) include oxidative stress, inflammation and metabolic impairments correlated to more frequent use of sunscreens. It would be important to investigate if BPs can potentially contribute to such sunscreen-related effects.

The antagonistic actions of these pollutants at environmentally relevant doses are also potentially very important. Our finding that three-component mixtures of UV filters (4MBC, EMC and OXYB) act in an antagonistic manner highlights their potential underestimation for biological consequences. Moreover, we found that such antagonist actions could be replicated in a human breast cell model, suggesting this may also be relevant for mammalian models, something that is worth further experimental consideration. Such antagonist effects can potentially skew conclusions of testing in different biological models. We also highlight the importance of using different experimental models to confirm such findings. Our study highlights the importance of dose dependency for both these types of pollutants, justifying the necessity to perform mixture studies using a wide range of concentrations.

Mixtures of UV filters and BPs have never been studied before with the help of marine organisms, and these results open up a new field for experimental design. Until now, only single pollutants belonging to UV filters have been studied with microorganisms belonging to different trophic levels. OXYB appears to be the most toxic substance among the UV filters, considering studies on microalgae (*I. galbana*), where OXYB was found to have an EC50 of 13.87 µg/L, followed by EMC (EC50 74.73 µg/L) and 4MBC (EC50 171.5 µg/L) [31]. In another model, employing Mediterranean mussel (*M. galloprovincialis*), the 4MBC solution (EC50 587 µg/L) was found to be the most toxic among the UV filters studied, while slightly lower toxicities were found for EMC (EC50 3110 µg/L) and OXYB (EC50 3470 µg/L). Studies using sea urchins (*P. lividus*) showed that this organism was more susceptible to EMC (EC50 284 µg/L) and 4MBC (EC50 854 µg/L), when compared with exposure to OXYB (EC50 3280 µg/L) [32]. Based on these results, it can be suggested that each marine organism responds to BPs and sunscreens components in a different manner. Individual predispositions and environmental conditions may also contribute, but further studies investigating the co-present pollutants are warranted to understand real-world consequences.

4MBC is known to cause detrimental effects in several test models. For example, 4MBC has been extensively investigated in terms of impact on reproductive systems. The Japanese rice fish, also known as the medaka (*Oryzias latipes*), is a fresh and brackish water fish, making it a good model for upper-tier ecotoxicological battery testing. Exposing adult male medaka to 4MBC resulted in disruptive spermatogenesis at doses of 5 and 50 µg/L. While 5 µg/L solutions greatly impacted estradiol and vitellogenin concentrations in the female plasma [33]. Furthermore, exposure of medaka eggs to a 4MBC solution resulted in prolonged hatching times. Japanese mussels (*Ruditapes philippinarum*) exposed to 4MBC at concentrations from 1–100 µg/L resulted in physiological stress due to elevated levels of antioxidant enzymes (GST) and proteins related to the inhibition of apoptosis (BCL2) and cellular stress (GADD). Sea snail larvae (*Sinum vittatum*) exposed to 2.57 mg/kg solutions of 4MBC had a decreased harvesting rate, while oxygen stress indicators were not impaired in these test organisms [34]. Moreover, the cell viability, oxidative stress and growth impairments as toxicological endpoints were studied with *Tetrahymena thermophile* protozoans [35] exposed separately to 4MBC and EMC. The EC_{50} of 4MBC, after 24 h exposure, reached 5.125 mg/L, while toxicological endpoints of growth inhibition and cells vitality (5 mg/L at 6 h) were correlated with an ability to break down cellular membranes at 15 mg/L after 4 h post exposure. 4MBC has also been studied in mammals, including humans and rats, where plasma concentration levels were measured after application of a sunscreen product (containing 4% of 4MBC) to the skin [36]; the maximum plasma values reached 200 pmol/mL. Only a small portion of 4MBC glucuronide metabolite was detected in human urea samples, confirming the low metabolic capability of humans in 4MBC transformation processes. Bearing in mind all of the above, it becomes evident how important it is to study CEC mixtures in various organisms and cell lines, which belong to different trophic levels, as pollutant cocktails greatly lower effective concentration values in the case of synergy confirmation.

The UV filters and BPs we tested are commonly found together in numerous water bodies [37]. Large studies, performed to determine over 100 CECs, confirmed the co-presence of sunscreen active components and BPs in effluents from wastewater treatment plants [38], although in lower concentrations than studied here. Oxybenzone and BPs were also instrumentally determined in samples collected from the Niger River delta [39], confirming a necessity for considering any environmental sample as a cocktail of numerous unknown ingredients and their metabolites/transformation products. Certainly, more work is needed to learn the mechanism of the enhanced toxicity of mixtures studied here, but our findings suggest that the action of a mixture of UV filters and BPs interfere with each other. Mixtures containing any of the two BPs studied and one sunscreen show a clear tendency towards underestimation and synergy, while the effect for cocktails containing any two UV filters studied and one BP show a slightly weaker effect. Overall, we find that underestimation events seem to be more frequent and there is a clear concentration dependence trend. Our results suggest that more studies looking at the direct interaction of these sunscreen and BP molecules with their potential cellular protein targets are warranted.

4. Materials and Methods

4.1. Chemicals and Reagents

Three different organic UV-filters were selected in this study, i.e., OXYB (oxybenzone, CAS no. 131-57-7), 4MBC (4-methylbenzylidene-camphor, CAS no. 36861-47-9) and EMC (2-ethylhexyl 4-methoxycinnamate, CAS no. 5466-77-3), and three bisphenol-analogue compounds, i.e., BPA (bisphenol A, CAS no. 80-05-7), BPF (bisphenol F, CAS no. 620-92-8) and BPS (bisphenol S, CAS no. 80-09-1), which were purchased from Sigma Aldrich (Darmstadt, Germany). HPLC grade MeOH (methanol, CAS no. 67-56-1) was purchased from Merc (KGaA, Darmstadt, Germany). Ultrapure water was produced by a Milli-Q Gradient A10 system equipped with an EDS-Pak cartridge to remove endocrine disrupting compounds (Merck–Millipore, Darmstadt, Germany).

Chemicals used for Microtox® were purchased from Modern Water Ltd. (New Castle, DE, USA). These were 2% NaCl solution, lyophilized *Aliivibrio fischeri* bacteria, Microtox Diluent, Osmotic Adjusting Solution (OAS) and Reconstitution Solution (RS). The instruments and equipment used during the studies were Microtox® 500 analyser of Modern Water Ltd. (Modern Water Ltd., USA) and electronic pipettes (Mettler Toledo, Columbus, OH, USA; Eppendorf, Hamburg, Germany).

The breast cell line MCF10A was donated by Professor Anna-Karin Olsson from the Dept. of Medical Biochemistry and Microbiology (IMBIM), Uppsala Biomedical Center (BMC), Sweden. MCF10A cells were maintained as a monolayer in T-75 cell culture plastic flasks (Corning Life Science, Amsterdam, the Netherlands) containing 12 mL of growth medium, trypsinized (Trypsin, no phenol red, Gibco™, Catalog number: 15090046) and split 1:4 every 3 days. Complete growth medium consisted of Dulbecco's Modified Eagle Medium with F-12 (DMEM/F-12, Gibco™, Catalog number: 31331028) supplemented with 5% horse serum (Horse Serum, heat inactivated, Gibco™, Catalog number: 26050088), 0.5 mg/mL hydrocortisone (Sigma-Aldrich, H0888, (Stockholm, Sweden)), 100 ng/mL CT (Cholera Toxin, Sigma-Aldrich, C8052), 10 mg/mL insulin (Sigma-Aldrich, I6634), EGF (20 ng/mL, Human Epidermal Growth Factor (EGF) Recombinant Protein, Gibco™, Catalog number: PHG0311) and 5 mL P/S (Penicillin-Streptomycin, Gibco™, Catalog number: 15140148 ()). Cell cultures were maintained at 37 °C and 5% CO_2 in a humidified incubator.

4.2. Preparation of Model Solutions

Before starting the tests, working solutions (C1 = 5 µM, C2 = 10 µM, C3 = 20 µM) were prepared by diluting the stock solutions with MeOH and ultrapure Milli-Q water, previously made by dissolving accurately weighted amounts of analytical standards in HPLC grade MeOH (4 mg/mL). All individual stock solutions were stored in the dark at −20 °C. The working solutions were freshly prepared before each set of analysis. The amount of MeOH was set at a maximum of 2% in these solutions for all toxicity tests and was used as the solvent control medium. In Table 2, the basic information on the studied sunscreens and BPA analogues are given [21,40].

Table 2. Basic information on the UV filters and BPs studied.

Compound/Analyte	Acronym	CAS no.	Formula	Chemical Structure	Molecular Weight	logP	pKa	logK$_{ow}$	Water Solubility [mg/L]
Sunscreen components:									
Oxybenzone	OXYB	131-57-7	$C_{14}H_{12}O_3$		228.24	3.79	7.56	3.79	3.7 (at 25 °C)
4-Methylbenzylidene-camphor	4MBC	36861-47-9	$C_{18}H_{22}O$		254.37	5.14	>8	4.95	1.3 (at 20 °C)
2-Ethylhexyl 4-methoxycinnamate	EMC	5466-77-3	$C_{18}H_{26}O_3$		290.40	5.66	-	5.80	<0.15 (at 25 °C)
Bisphenols:									
2,2-Bis(4-hydroxyphenyl)propane	BPA	80-05-7	$C_{15}H_{16}O_2$		228.29	4.04	9.78–10.39	3.43	120 (at 20 °C)
4,4'-Methylenediphenol	BPF	620-92-8	$C_{13}H_{12}O_2$		200.23	2.9	9.84–10.45	2.76	200 (at 25 °C)
4,4'-Sulfonyldiphenol	BPS	80-09-1	$C_{12}H_{10}O_4S$		250.27	2.32	7.42–8.03	2.139	350 (at 25 °C)

4.3. Preparation of Ternary Mixtures

The mixtures were prepared via mixing three different compounds diluted to requested concentrations. The general scheme used to prepare the mixtures is shown below. The same approach followed suitably for C2 and C3 of S1 (C stands for concentration and S for substance/analyte in mixture). In this way, all mixtures were prepared and studied in duplicated experiments as follows:

Mixture 1: S1C1 + S2C1 + S3C1
Mixture 2: S1C1 + S2C1 + S3C2
Mixture 3: S1C1 + S2C1 + S3C3
Mixture 4: S1C1 + S2C2 + S3C1
Mixture 5: S1C1 + S2C2 + S3C2
Mixture 6: S1C1 + S2C2 + S3C3
Mixture 7: S1C1 + S2C3 + S3C1
Mixture 8: S1C1 + S2C3 + S3C2
Mixture 9: S1C1 + S2C3 + S3C3
Mixture 10: S1C2 + S2C1 + S3C1
Mixture 11: S1C2 + S2C1 + S3C2
Mixture 12: S1C2 + S2C1 + S3C3
Mixture 13: S1C2 + S2C2 + S3C1
Mixture 14: S1C2 + S2C2 + S3C2
Mixture 15: S1C2 + S2C2 + S3C3
Mixture 16: S1C2 + S2C3 + S3C1
Mixture 17: S1C2 + S2C3 + S3C2
Mixture 18: S1C2 + S2C3 + S3C3
Mixture 19: S1C3 + S2C1 + S3C1
Mixture 20: S1C3 + S2C1 + S3C2
Mixture 21: S1C3 + S2C1 + S3C3
Mixture 22: S1C3 + S2C2 + S3C1
Mixture 23: S1C3 + S2C2 + S3C2
Mixture 24: S1C3 + S2C2 + S3C3
Mixture 25: S1C3 + S2C3 + S3C1
Mixture 26: S1C3 + S2C3 + S3C2
Mixture 27: S1C3 + S2C3 + S3C3

In parallel with mixtures, studies using single compounds in respective, environmentally relevant concentrations were performed for every experiment run (experiments were performed on a daily routine for the same standards, solutions and bacterial reagents batch). The raw dose-response data for the studied compounds are presented in electronic Supplementary Materials as Figure S1.

4.4. Acute Toxicity Determination of the Mixtures

In order to determine the toxicity of the analytes of interest and their ternary mixtures, a standard Microtox® acute assay has been performed. To achieve a ready-to-use bacterial suspension, the lyophilized *Aliivibrio fischeri* bacteria were rehydrated with 1 mL of RS. The cell suspension was immediately transferred into a glass cuvette placed in the reagent well of the analyzer maintained at 5.5 ± 1.0 °C. Subsequently, 100 µL of bacterial solution and 900 µL of working solutions were added into the test vials. Before starting the test, an osmotic adjustment was performed, using a saline solution to make the sample salinity optimal (above 2%) for *Aliivibrio fischeri*. The cuvettes were incubated at 15 °C. The toxicity was determined based on the inhibition of the light naturally emitted (at 490 nm wavelength) by the bacteria after its exposure to a standard solution/mixture sample. The bioluminescence level was detected by Microtox® Model 500 Analyzer. Measurements of the luminescent output of the bacteria were recorded and compared with the light output of the control sample after 30 min. Each assay was performed in a duplicate. For quality control—according to delivery certificates—the phenol and copper (II) sulphate were used as positive controls.

Change of bioluminescence after t time was calculated according to Equation (1):

$$\% \text{ bioluminescence change}_t = [G_t/(1+ G_t)] \cdot 100 \tag{1}$$

where G (gamma correction factor) was calculated with Equation (2):

$$G_t = [(R_t \cdot I_0)/I_t] - 1 \tag{2}$$

where I_t is the bioluminescence of bacteria in real sample at time of t, I_0 is the initial bioluminescence of bacteria in real samples and R_t is calculated with Equation (3):

$$R_t = Ic_t/Ic_0 \tag{3}$$

where I_{ct} is the bioluminescence of bacteria in control sample after t time of incubation and I_{c0} is the initial bioluminescence of bacteria in control.

4.5. Modeling and MDRs Calculations

The modeling calculations and MDR have been done according to standardized procedures described in detail in [41]. Mathematical modeling has been performed with concentration addition (CA) and independent action (IA) approaches, followed by the data interpretation with the Model Deviation Ratio (MDR) evaluation as presented in details in [42]. Here, the CA was assessed by using Equation (4):

$$ECx_{mix} = \left(\sum_{i=0}^{n} \frac{p_i}{ECx_i}\right)^{-1} \tag{4}$$

where ECx_{mix} is the total concentration of the mixture that causes x effect, p_i indicates the proportion of component i in the mixture, n indicates the number of components in the mixture and ECx_i indicates the concentration of component i that would cause x effect.

The independent action model is used to test toxicants in a mixture for a dissimilar mode of action, assuming that they act independently; the IA model is a statistical approach to predict the chance that one of multiple events will occur. The total mixture effect is calculated using Equation (5):

$$E(c_{mix}) = 1 - \prod_{i=0}^{n}(1 - E(c_i)) \tag{5}$$

where $E(c_{mix})$ is the total concentration of the mixture and $E(c_i)$ is the concentration expected from component i.

The CA model does not account for a possible interaction between different chemicals in the mixture and the deviations of tested mixture toxicity from the predicted one could be evidence for synergistic or antagonistic interaction between chemicals. To outline significant deviations (interactions between chemicals) the model deviation ratio (MDR) approach is applied. MDR (unitless) is defined with Equation (6):

$$MDR = \frac{Expected\ toxicity}{Observed\ toxicity} \tag{6}$$

where *Expected toxicity* is the effective concentration toxicity for the mixture predicted by the CA/IA model and *Observed toxicity* is the effective concentration toxicity for the mixture obtained from toxicity testing.

MDR values below 0.5 indicate the presence of synergism, values in the range 0.5–0.71 indicate underestimation of the model, values in the range 1.4–2.0 indicate overestimation of the model and values above 2.0 indicate antagonism. All calculations were performed with Microsoft Excel 2016 standard set.

4.6. Methodology of Human Breast Cell Line Studies

Human breast epithelial cells (MCF10A) were seeded in 250 µL complete growth medium within the inner wells of a 96-well plate (Corning Life Science, Amsterdam, the Netherlands) at a density of approximately 2×10^4 cells/well. The outer wells were excluded from the experiment and filled in with PBS. To permit cell adhesion and adaption to the novel environment, the cells were placed in a humidified incubator at 37 °C and 5% CO_2 for 24 h. After 24 h, the cells were exposed to different combination of pollutant mixtures (C1 = 1 µM, C2 = 5 µM and C3 = 10 µM), which were dissolved in DMSO (dimethyl sulfoxide, Invitrogen™, Catalog number: D12345). Then, after 24 and 72 h (in different trials), PrestoBlue™ HS Cell Viability Reagent (Invitrogen™, Catalog number: P50200. Pub. No. MAN0018371) was employed to test the cell viability, which reflects cell proliferation. The data were further analyzed in Microsoft Excel and GraphPad Prism (version 9.0.0). Each treatment was conducted with at least three biological replicates.

5. Conclusions

The increasing emissions of CECs into the environment prompts scientists to intensify their research related to the determination of interactions between chemicals co-present in complex matrices. Mixtures of CECs are found in practically all industrial and public service branches; from pharmaceutics, health, wastewater treatment, catalyst applications, ecosystem monitoring, life cycle assessment and many others. Thus, it is increasingly important not only to develop new tools to instrumentally determine the content of given CEC mixtures, but also to validate and adjust already known approaches. Further development of advanced mathematical tools to confirm possible interactions that occur between pollutants are warranted, especially for mixtures of higher orders.

This study is one of the most comprehensive on the interactions of compounds in complex mixtures, providing evidence for the notion that concentrations of a given CEC plays a crucial, and sometimes unpredictable, role in exerting cellular or physiological impacts on a living organism. Our mathematical approach and experimental setup enable the collection of rational and reliable data, which prompt conclusions on how to assess the potential overadditive or canceling effects of cocktail components. The findings presented here are an important next step for a better understanding of how to perform complex toxicological studies in a systematic manner, and how to evaluate a model's validity relating to dissimilar groups of pollutants and cells/organisms belonging to different trophic levels.

In our opinion, this work adds new insights into the field of mixtures impact on biota and confirms the necessity to increase the number of studies on complex mixtures of chemicals affecting biota and, gradually, start introducing requirements on admissible concentrations of particular chemicals, bearing in mind their toxicological impact on biota when present in mixtures. In such cases, law and policy makers need reliable sources of information to present and suggest realistic and achievable goals in directives aimed at minimizing the impact of versatile pollutant cocktails on humans.

Supplementary Materials: The following supporting information can be downloaded at: https://www.mdpi.com/article/10.3390/molecules27103260/s1, Figure S1. The raw dose-response data for (a) BPA, (b) BPS, (c) BPF, (d) OXYB, (e) EMC and (f) 4-MBC that were used to pre-select concentration levels of mixtures components in this research (bars represent SD values for n = 2); Figure S2. MDR values of bioluminescent bacteria results for: (A) CA modeling of the BPS, OXYB and BPF mixture, (B) CA modeling of the BPS, OXYB and 4MBC mixture, (C) CA modeling of the BPS, OXYB and EMC mixture (D) CA modeling of the BPS, 4MBC and EMC mixture, (E) CA modeling of the BPS, EMC and BPF mixture (n = 2). Red color indicates confirmed synergy, blue indicates antagonism, while yellow and green refer to under- and overestimation, respectively; Figure S3. MDR values of bioluminescent bacteria results for: (A) CA modeling of the BPF, 4MBC and BPS mixture, (B) CA modeling of the BPF, 4MBC and EMC mixture, (C) CA modeling of the BPF, 4MBC and OXYB mixture (D) CA modeling of the BPF, OXYB and EMC mixture ($n = 2$). Red color indicates confirmed synergy, blue indicates antagonism, while yellow and green refer to under- and overestimation, respectively; Supplementary Table S1. MDR results of studies on the impact of three-component mixtures on BPA, BPS, BPF, 4MBC, EMC and OXYB toxicity (MDR values >2.0 exhibit antagonism, MDRs < 0.5 show synergism and MDR values of 0.50–0.71 and 1.40–2.00 mean, respectively, under- and overestimation of the presented models; for values of particular concentrations C1, C2 and C3 of all analytes, please refer to Section 2.2. in the main text) ($n = 2$); Supplementary Table S2. MDR results of studies on the impact of three-component mixtures on BPA, BPS, BPF, 4MBC and EMC toxicity (MDR values > 2.0 exhibit antagonism, MDRs < 0.5 show synergism and MDR values of 0.50–0.71 and 1.40–2.00 mean, respectively, under- and overestimation of the presented models; for values of particular concentrations C1, C2 and C3 of all analytes, please refer to Section 2.2. in the main text) ($n = 2$); Supplementary Table S3. MDR results of studies on the impact of three-component mixtures on BPA, BPS, BPF, 4MBC and EMC toxicity (MDR values > 2.0 exhibit antagonism, MDRs < 0.5 show synergism and MDR values of 0.50–0.71 and 1.40–2.00 mean, respectively, under- and overestimation of the presented models; for values of particular concentrations C1, C2 and C3 of all analytes, please refer to Section 4.2. in the main text) ($n = 2$); Supplementary Table S4. MDR results of studies on the impact of three-component mixtures on BPA, BPS, BPF, 4MBC and EMC toxicity (MDR values > 2.0 exhibit antagonism, MDRs < 0.5 show synergism and MDR values of 0.50–0.71 and 1.40–2.00 mean, respectively, under- and overestimation of the presented models; for values of particular concentrations C1, C2 and C3 of all analytes, please refer to Section 4.2. in the main text) ($n = 2$); Supplementary Table S5. MDR results of studies on the impact of three-component mixtures on BPA, BPS, BPF, 4MBC and EMC toxicity (MDR values > 2.0 exhibit antagonism, MDRs < 0.5 show synergism and MDR values of 0.50–0.71 and 1.40–2.00 mean, respectively, under- and overestimation of the presented models; for values of particular concentrations C1, C2 and C3 of all analytes, please refer to Section 4.2. in the main text) ($n = 2$); Supplementary Table S6. MDR results of studies on the impact of three-component mixtures on BPS, BPF and EMC toxicity (MDR values > 2.0 exhibit antagonism, MDRs < 0.5 show synergism and MDR values of 0.50–0.71 and 1.40–2.00 mean, respectively, under- and overestimation of the presented models; for values of particular concentrations C1, C2 and C3 of all analytes, please refer to Section 4.2. in the main text) ($n = 2$); Supplementary Table S7. MDR results of studies on the impact of three-component mixtures on BPF, 4MBC, BPS, OXB and EMC toxicity (MDR values > 2.0 exhibit antagonism, MDRs < 0.5 show synergism and MDR values of 0.50–0.71 and 1.40–2.00 mean, respectively, under- and overestimation of the presented models; for values of particular concentrations C1, C2 and C3 of all analytes, please refer to Section 4.2. in the main text) ($n = 2$); Supplementary Table S8. MDR results of studies on the impact of three-component mixtures on BPA, BPS, BPF, 4MBC and EMC toxicity (MDR values > 2.0 exhibit antagonism, MDRs < 0.5 show synergism and MDR values of 0.50–0.71 and 1.40–2.00 mean, respectively, under- and overestimation of the presented models; for values of particular concentrations C1, C2 and C3 of all analytes, please refer to Section 4.2. in the main text) ($n = 2$); Supplementary Table S9. MDR results of studies on the impact of three-component mixtures on OXYB, 4MBC and EMC toxicity (MDR values > 2.0 exhibit antagonism, MDRs < 0.5 show synergism and MDR values of 0.50–0.71 and 1.40–2.00 mean, respectively, under- and overestimation of the presented models; for values of particular concentrations C1, C2 and C3 of all analytes, please refer to Section 4.2. in the main text) ($n = 2$).

Author Contributions: Conceptualization: B.K., N.J., M.J.W., D.B. and H.B.S.; Data curation: B.K., N.J., W.L. and H.B.S.; Formal analysis: B.K. and H.B.S.; Funding acquisition: B.K. and H.B.S.; Investigation: B.K., N.J. and W.L.; Methodology: B.K., N.J., W.L. and H.B.S.; Project administration: B.K. and H.B.S.; Resources: B.K., W.L. and H.B.S.; Supervision: B.K., D.B. and H.B.S.; Software: N.J. and W.L.; Validation: B.K., N.J., W.L. and D.B.; Visualization: B.K., N.J., W.L. and H.B.S.; writing—original draft: B.K., N.J., W.L., M.J.W., D.B. and H.B.S.; Writing–review and editing: B.K., N.J., W.L., M.J.W., D.B. and H.B.S. All authors have read and agreed to the published version of the manuscript.

Funding: This research was funded by the Gdańsk University of Technology, grant number DEC-1/2020/IDUB/I.3.2 (Polish researchers) and Formas (HBS). The APC was funded by Uppsala University and the Gdańsk University of Technology.

Institutional Review Board Statement: Not applicable.

Informed Consent Statement: Not applicable.

Data Availability Statement: Data can be shared upon personal request to the authors.

Acknowledgments: Błażej Kudłak acknowledges the IDUB 'Excellence Initiative-Research University' program, DEC-1/2020/IDUB/I.3.2, for financial support. The technical support of Wiktoria Gełdon is also acknowledged. Helgi B. Schiöth is supported by the Swedish Research Council, Formas, the Swedish Cancer Foundation and the Novo Nordisk Foundation.

Conflicts of Interest: The authors declare no conflict of interest. The funders had no role in the design of the study; in the collection, analyses or interpretation of data; in the writing of the manuscript, or in the decision to publish the results.

Sample Availability: Samples of the compounds studied are available from the commercial producers.

References

1. Nilsen, E.; Smalling, K.L.; Ahrens, L.; Gros, M.; Miglioranza, K.S.B.; Picó, Y.; Schoenfuss, H.L. Grand challenges in assessing the adverse effects of contaminants of emerging concern on aquatic food webs. *Environ. Toxicol. Chem.* **2019**, *38*, 46–60. [CrossRef] [PubMed]
2. Pico, Y.; Belenguer, V.; Corcellas, C.; Diaz-Cruz, M.S.; Eljarrat, E.; Farré, M.; Gago-Ferrero, P.; Huerta, B.; Navarro-Ortega, A.; Petrovic, M.; et al. Contaminants of emerging concern in freshwater fish from four Spanish Rivers. *Sci. Total Environ.* **2019**, *659*, 1186–1198. [CrossRef] [PubMed]
3. Sánchez-Quiles, D.; Tovar-Sánchez, A. Are sunscreens a new environmental risk associated with coastal tourism? *Environ. Int.* **2015**, *83*, 158–170. [CrossRef] [PubMed]
4. Chen, F.; Huber, C.; Schröder, P. Fate of the sunscreen compound oxybenzone in Cyperus alternifolius based hydroponic culture: Uptake, biotransformation and phytotoxicity. *Chemosphere* **2017**, *182*, 638–646. [CrossRef]
5. Zhong, X.; Downs, C.A.; Che, X.; Zhang, Z.; Li, Y.; Liu, B.; Li, Q.; Li, Y.; Gao, H. The toxicological effects of oxybenzone, an active ingredient in suncream personal care products, on prokaryotic alga *Arthrospira* sp. and eukaryotic alga *Chlorella* sp. *Aquat. Toxicol.* **2019**, *216*, 105295. [CrossRef]
6. Coronado, M.; De Haro, H.; Deng, X.; Rempel, M.A.; Lavado, R.; Schlenk, D. Estrogenic activity and reproductive effects of the UV-filter oxybenzone (2-hydroxy-4-methoxyphenyl-methanone) in fish. *Aquat. Toxicol.* **2008**, *90*, 182–187. [CrossRef]
7. Negreira, N.; Rodríguez, I.; Rubí, E.; Cela, R. Determination of selected UV filters in indoor dust by matrix solid-phase dispersion and gas chromatography-tandem mass spectrometry. *J. Chromatogr. A* **2009**, *1216*, 5895–5902. [CrossRef]
8. Da Silva, A.P.; Trindade, M.A.G.; Ferreira, V.S. Polarographic determination of sunscreen agents in cosmetic products in micellar media. *Talanta* **2006**, *68*, 679–685. [CrossRef]
9. Ramos, S.; Homem, V.; Santos, L. Development and optimization of a QuEChERS-GC–MS/MS methodology to analyse ultraviolet-filters and synthetic musks in sewage sludge. *Sci. Total Environ.* **2019**, *651*, 2606–2614. [CrossRef]
10. Plagellat, C.; Kupper, T.; Furrer, R.; De Alencastro, L.F.; Grandjean, D.; Tarradellas, J. Concentrations and specific loads of UV filters in sewage sludge originating from a monitoring network in Switzerland. *Chemosphere* **2006**, *62*, 915–925. [CrossRef]
11. Chen, D.; Kannan, K.; Tan, H.; Zheng, Z.; Feng, Y.L.; Wu, Y.; Widelka, M. Bisphenol Analogues Other Than BPA: Environmental Occurrence, Human Exposure, and Toxicity—A Review. *Environ. Sci. Technol.* **2016**, *50*, 5438–5453. [CrossRef] [PubMed]
12. González, N.; Cunha, S.C.; Monteiro, C.; Fernandes, J.O.; Marquès, M.; Domingo, J.L.; Nadal, M. Quantification of eight bisphenol analogues in blood and urine samples of workers in a hazardous waste incinerator. *Environ. Res.* **2019**, *176*, 108576. [CrossRef] [PubMed]
13. Li, A.; Zhuang, T.; Shi, W.; Liang, Y.; Liao, C.; Song, M.; Jiang, G. Serum concentration of bisphenol analogues in pregnant women in China. *Sci. Total Environ.* **2020**, *707*, 136100. [CrossRef] [PubMed]
14. Wu, L.H.; Zhang, X.M.; Wang, F.; Gao, C.J.; Chen, D.; Palumbo, J.R.; Guo, Y.; Zeng, E.Y. Occurrence of bisphenol S in the environment and implications for human exposure: A short review. *Sci. Total Environ.* **2018**, *615*, 87–98. [CrossRef]
15. Williams, M.J.; Wang, Y.; Klockars, A.; Monica Lind, P.; Fredriksson, R.; Schiöth, H.B. Exposure to bisphenol a affects lipid metabolism in Drosophila melanogaster. *Basic Clin. Pharmacol. Toxicol.* **2014**, *114*, 414–420. [CrossRef]
16. Park, C.B.; Jang, J.; Kim, S.; Kim, Y.J. Single- and mixture toxicity of three organic UV-filters, ethylhexyl methoxycinnamate, octocrylene, and avobenzone on Daphnia magna. *Ecotoxicol. Environ. Saf.* **2017**, *137*, 57–63. [CrossRef]
17. Downs, C.A.; Kramarsky-Winter, E.; Segal, R.; Fauth, J.; Knutson, S.; Bronstein, O.; Ciner, F.R.; Jeger, R.; Lichtenfeld, Y.; Woodley, C.M.; et al. Toxicopathological Effects of the Sunscreen UV Filter, Oxybenzone (Benzophenone-3), on Coral Planulae and Cultured Primary Cells and Its Environmental Contamination in Hawaii and the U.S. Virgin Islands. *Arch. Environ. Contam. Toxicol.* **2016**, *70*, 265–288. [CrossRef]
18. Liu, H.; Sun, P.; Liu, H.; Yang, S.; Wang, L.; Wang, Z. Acute toxicity of benzophenone-type UV filters for Photobacterium phosphoreum and Daphnia magna: QSAR analysis, interspecies relationship and integrated assessment. *Chemosphere* **2015**, *135*, 182–188. [CrossRef]
19. Langford, K.H.; Reid, M.J.; Fjeld, E.; Øxnevad, S.; Thomas, K.V. Environmental occurrence and risk of organic UV filters and stabilizers in multiple matrices in Norway. *Environ. Int.* **2015**, *80*, 1–7. [CrossRef]
20. Astel, A.; Stec, M.; Rykowska, I. Occurrence and distribution of uv filters in beach sediments of the southern baltic sea coast. *Water* **2020**, *12*, 3024. [CrossRef]

21. Cadena-Aizaga, M.I.; Montesdeoca-Esponda, S.; Torres-Padrón, M.E.; Sosa-Ferrera, Z.; Santana-Rodríguez, J.J. Organic UV filters in marine environments: An update of analytical methodologies, occurrence and distribution. *Trends Environ. Anal. Chem.* **2020**, *25*, e00079. [CrossRef]
22. Hayden, C.G.J.; Cross, S.E.; Anderson, C.; Saunders, N.A.; Roberts, M.S. Sunscreen Penetration of Human Skin and Related Keratinocyte Toxicity after Topical Application. *Skin Pharmacol. Physiol.* **2005**, *18*, 170–174. [CrossRef] [PubMed]
23. Benson, H.A.E. Influence of anatomical site and topical formulation on skin penetration of sunscreens. *Ther. Clin. Risk Manag.* **2005**, *3*, 209–218.
24. Jiang, R.; Roberts, M.S.; Collins, D.M.; Benson, H.A.E. Absorption of sunscreens across human skin: An evaluation of commercial products for children and adults. *Br. J. Clin. Pharmacol.* **1999**, *48*, 635–637. [CrossRef] [PubMed]
25. Matta, M.K.; Florian, J.; Zusterzeel, R.; Pilli, N.R.; Patel, V.; Volpe, D.A.; Yang, Y.; Oh, L.; Bashaw, E.; Zineh, I.; et al. Effect of Sunscreen Application on Plasma Concentration of Sunscreen Active Ingredients A Randomized Clinical Trial. *JAMA* **2020**, *323*, 256–267. [CrossRef]
26. Mota, M.D.; da Boa Morte, A.N.; Silva, L.C.R.C.; Chinalia, F.A. Sunscreen protection factor enhancement through supplementation with Rambutan (*Nephelium lappaceum* L.) ethanolic extract. *J. Photochem. Photobiol. B Biol.* **2020**, *205*, 111837. [CrossRef]
27. Sarveiya, V.; Risk, S.; Benson, H.A.E. Liquid chromatographic assay for common sunscreen agents: Application to in vivo assessment of skin penetration and systemic absorption in human volunteers. *J. Chromatogr. B Anal. Technol. Biomed. Life Sci.* **2004**, *803*, 225–231. [CrossRef]
28. Egambaram, O.P.; Pillai, S.K.; Ray, S.S. Invited Review Materials Science Challenges in Skin UV Protection: A Review. *Photochem. Photobiol.* **2020**, *96*, 779–797. [CrossRef]
29. Brand, R.M.; Pike, J.; Wilson, R.M.; Charron, A.R. Sunscreens containing physical UV blockers can increase transdermal absorption of pesticides absorption of pesticides. *Toxicol. Ind. Health* **2003**, *19*, 9–16. [CrossRef]
30. Marrot, L. Pollution and Sun Exposure: A Deleterious Synergy. Mechanisms and Opportunities for Skin Protection Pollution and Sun Exposure: A Deleterious Synergy. Mechanisms and Opportunities for Skin Protection. *Curr. Med. Chem.* **2018**, *25*, 5469–5486. [CrossRef]
31. Kaiser, D.; Sieratowicz, A.; Zielke, H.; Oetken, M.; Hollert, H.; Oehlmann, J. Ecotoxicological effect characterisation of widely used organic UV filters. *Environ. Pollut.* **2012**, *163*, 84–90. [CrossRef] [PubMed]
32. Corinaldesi, C.; Damiani, E.; Marcellini, F.; Falugi, C.; Tiano, L.; Brugè, F.; Danovaro, R. Sunscreen products impair the early developmental stages of the sea urchin *Paracentrotus lividus*. *Sci. Rep.* **2017**, *7*, 7815. [CrossRef] [PubMed]
33. Liang, M.; Yan, S.; Chen, R.; Hong, X.; Zha, J. 3-(4-Methylbenzylidene) camphor induced reproduction toxicity and antiandrogenicity in Japanese medaka (*Oryzias latipes*). *Chemosphere* **2020**, *249*, 126224. [CrossRef] [PubMed]
34. Campos, D.; Gravato, C.; Quintaneiro, C.; Golovko, O.; Žlábek, V.; Soares, A.M.V.M.; Pestana, J.L.T. Toxicity of organic UV-filters to the aquatic midge *Chironomus riparius*. *Ecotoxicol. Environ. Saf.* **2017**, *143*, 210–216. [CrossRef]
35. Gao, L.; Yuan, T.; Zhou, C.; Cheng, P.; Bai, Q.; Ao, J.; Wang, W.; Zhang, H. Effects of four commonly used UV filters on the growth, cell viability and oxidative stress responses of the *Tetrahymena thermophila*. *Chemosphere* **2013**, *93*, 2507–2513. [CrossRef]
36. Schauer, U.M.D.; Völkel, W.; Heusener, A.; Colnot, T.; Broschard, T.H.; von Landenberg, F.; Dekant, W. Kinetics of 3-(4-methylbenzylidene)camphor in rats and humans after dermal application. *Toxicol. Appl. Pharmacol.* **2006**, *216*, 339–346. [CrossRef]
37. Tran, N.H.; Reinhard, M.; Gin, K.Y.H. Occurrence and fate of emerging contaminants in municipal wastewater treatment plants from different geographical regions—A review. *Water Res.* **2018**, *133*, 182–207. [CrossRef]
38. Pintado-Herrera, M.G.; González-Mazo, E.; Lara-Martín, P.A. Atmospheric pressure gas chromatography-time-of-flight-mass spectrometry (APGC-ToF-MS) for the determination of regulated and emerging contaminants in aqueous samples after stir bar sorptive extraction (SBSE). *Anal. Chim. Acta* **2014**, *851*, 1–13. [CrossRef]
39. Inam, E.; Offiong, N.A.; Kang, S.; Yang, P.; Essien, J. Assessment of the Occurrence and Risks of Emerging Organic Pollutants (EOPs) in Ikpa River Basin Freshwater Ecosystem, Niger Delta-Nigeria. *Bull. Environ. Contam. Toxicol.* **2015**, *95*, 624–631. [CrossRef]
40. Wang, H.; Liu, Z.; Zhang, J.; Huang, R.; Yin, H.; Dang, Z.; Wu, P.; Liu, Y. Insights into removal mechanisms of bisphenol A and its analogues in municipal wastewater treatment plants. *Sci. Total Environ.* **2019**, *692*, 107–116. [CrossRef]
41. Wieczerzak, M.; Kudłak, B.; Yotova, G.; Nedyalkova, M.; Tsakovski, S.; Simeonov, V.; Namieśnik, J. Modeling of pharmaceuticals mixtures toxicity with deviation ratio and best-fit functions models. *Sci. Total Environ.* **2016**, *571*, 259–268. [CrossRef] [PubMed]
42. Backhaus, T.; Faust, M. Predictive environmental risk assessment of chemical mixtures: A conceptual framework. *Environ. Sci. Technol.* **2012**, *46*, 2564–2573. [CrossRef] [PubMed]

Article

Combined Analytical Study on Chemical Transformations and Detoxification of Model Phenolic Pollutants during Various Advanced Oxidation Treatment Processes

Aleksander Kravos [1], Andreja Žgajnar Gotvajn [1], Urška Lavrenčič Štangar [1], Borislav N. Malinović [2] and Helena Prosen [1,*]

[1] University of Ljubljana, Faculty of Chemistry and Chemical Technology, 1000 Ljubljana, Slovenia; aleksander.kravos@fkkt.uni-lj.si (A.K.); andreja.zgajnar@fkkt.uni-lj.si (A.Ž.G.); urska.lavrencic.stangar@fkkt.uni-lj.si (U.L.Š.)

[2] University of Banja Luka, Faculty of Technology, 78000 Banja Luka, Bosnia and Herzegovina; borislav.malinovic@tf.unibl.org

* Correspondence: helena.prosen@fkkt.uni-lj.si; Tel.: +386-1-479-8556

Citation: Kravos, A.; Žgajnar Gotvajn, A.; Lavrenčič Štangar, U.; Malinović, B.N.; Prosen, H. Combined Analytical Study on Chemical Transformations and Detoxification of Model Phenolic Pollutants during Various Advanced Oxidation Treatment Processes. *Molecules* **2022**, *27*, 1935. https://doi.org/10.3390/molecules27061935

Academic Editors: Stefan Leonidov and Stefan Leonidov Tsakovski

Received: 31 January 2022
Accepted: 13 March 2022
Published: 16 March 2022

Publisher's Note: MDPI stays neutral with regard to jurisdictional claims in published maps and institutional affiliations.

Copyright: © 2022 by the authors. Licensee MDPI, Basel, Switzerland. This article is an open access article distributed under the terms and conditions of the Creative Commons Attribution (CC BY) license (https://creativecommons.org/licenses/by/4.0/).

Abstract: Advanced oxidation processes (AOPs) have been introduced to deal with different types of water pollution. They cause effective chemical destruction of pollutants, yet leading to a mixture of transformation by-products, rather than complete mineralization. Therefore, the aim of our study was to understand complex degradation processes induced by different AOPs from chemical and ecotoxicological point of view. Phenol, 2,4-dichlorophenol, and pentachlorophenol were used as model pollutants since they are still common industrial chemicals and thus encountered in the aquatic environment. A comprehensive study of efficiency of several AOPs was undertaken by using instrumental analyses along with ecotoxicological assessment. Four approaches were compared: ozonation, photocatalytic oxidation with immobilized nitrogen-doped TiO_2 thin films, the sequence of both, as well as electrooxidation on boron-doped diamond (BDD) and mixed metal oxide (MMO) anodes. The monitored parameters were: removal of target phenols, dechlorination, transformation products, and ecotoxicological impact. Therefore, HPLC–DAD, GC–MS, UHPLC–MS/MS, ion chromatography, and 48 h inhibition tests on *Daphnia magna* were applied. In addition, pH and total organic carbon (TOC) were measured. Results show that ozonation provides by far the most suitable pattern of degradation accompanied by rapid detoxification. In contrast, photocatalysis was found to be slow and mild, marked by the accumulation of aromatic products. Preozonation reinforces the photocatalytic process. Regarding the electrooxidations, BDD is more effective than MMO, while the degradation pattern and transformation products formed depend on supporting electrolyte.

Keywords: chlorophenols; *Daphnia magna*; electrooxidation; ozonation; phenol; photocatalysis

1. Introduction

Today's highly chemicalized world is far from reaching toxic-free environment. For example, in 2021 European Environment Agency reported that alarming share of European freshwaters during 2013–2019 had excessive levels of pesticides [1]. In a broader sense, less than 38% of waters are claimed to have good status and 75–96% of European seas exhibit contamination issues [2]. The latter points to a fact that no balance between anthropogenic pressure and waters' self-cleaning capabilities is yet established, even though more than 90% of urban wastewater across the EU is thought to be collected and treated [3]. Therefore, there is a growing commitment to understand and manage pollution, especially with persistent organic micropollutants.

Phenol and chlorophenols (CPs) are representative examples of a wider group of phenolic pollutants. Their presence in the environment is due to intensive historical use, drinking water chlorination, biodegradation of organochlorinated chemicals, and their importance in the chemical industry [4–7]. Phenol and some CPs are hyper-volume production chemicals, according to OECD. What is more, pentachlorophenol is believed to be the most common chlorinated industrial chemical in the EU [8]. In general, high acute toxicity and genotoxicity are reported, especially for polychlorinated CPs and their degradation products. Nevertheless, total global production of commercial CPs is estimated to tens of kilotons per year and phenol's production is only slightly less [9]. Used in industry, phenols are prevalent components of industrial wastewaters. Thus, they are continuously transferred into ecosystems and they accumulate in the sediments, as well as biota, where they appear to be ubiquitous. CPs in surface waters reach 2–2000 ng/L [5]; phenols, on the other hand, yet higher concentrations.

Advanced oxidation processes (AOPs) become well-established technology for water treatment in the last decades. Phenol concentrations > 5 mg/L [10] or even considerably smaller concentrations of CPs that are found in wastewaters are, in practice, biologically non-degradable, but their removal has been proven to be quickly achieved by many physical methods [4], wet-oxidations, ozonation, and many homo-/heterogenic AOPs so far, which include additions of catalysts and/or electro-, photo- or sonochemical treatment [6,7,11–14]. Considerable research has been taking place also regarding removal of other phenolic pollutants, e.g., nitrophenols, by AOPs that use sustainable materials [6,15]. Nevertheless, despite being optimized and highly effective for the removal of target phenols, complete mineralization is usually not readily achieved by most of the AOPs. Therefore, their chemical pathways from removal of parent compound to the mineralization in connection with the assessment of biological effects remain only rarely studied. For example, the latter aspect has so far been in the case of CPs reviewed by Karci, focusing on Fenton oxidation and UV/H_2O_2 treatment [6].

Ozonation (OZ) is a 'quasi' AOP that has one of the longest histories of use and research, reaching far back in the 20th century. A considerable amount of work has been reported so far on the removal of phenol (PHN) [16–20], 2,4-dichlorophenol (DCP) [18,21,22] and pentachlorophenol (PCP) [22–25], some reviewed, for example, by Pera-Titus et al. [12] Studies in majority concluded that OZ exhibits high effectivity in removal of target phenol and CPs by progressive formation of multiple C–O and C=O bonds before or after destruction of aromatic ring, as well as cleavage of labile C–Cl, C–H, and C–C bonds. This is possible through molecular or radical mechanism [14]. Yet only the minority of studies on OZ assess ecotoxicity specifically on water flea *Daphnia magna* [26] or study transformation products (TPs) by wide variety of analytical techniques [18–21,23], especially by mass spectrometry. One important study on degradation pathways—with a wide repertoire of identified TPs—was reported and discussed by Oputu et al. [20] A variety of TPs have been identified so far; the most significant and abundant already in the previous two decades [18,22,27,28].

The synthesis of advanced materials is a driving force for the development of new, increasingly more effective AOPs. Photocatalysis (PC) stands out as a perspective technology. A wide variety of photocatalysts are being assessed, but TiO_2-based are by far the most prevailing. Use of immobilized (less researched) TiO_2 thin films on various supports, e.g., glass, metal oxides, and fibres, represents a new alternative to the use of conventional powder forms. Photocatalysts can be, moreover, easily doped with, e.g., Pt, Sb, N, C, thus reinforcing photoactivity [11,29]. Focusing on immobilized TiO_2, the majority of research is placed on kinetics and target removal of PHN with optimizations of process parameters [30,31], such as those reported by Nickheslat et al. [32], Dougna et al. [33] or Sampaio et al. [34] According to our knowledge, only a minority of studies focuses on DCP [35,36] or PCP removal [36,37]. Moreover, assessments of ecotoxicity or induced chemical transformations are absent; only basic TPs (e.g., hydroxyphenols, organic acids) were identified solely by HPLC–UV [11,31,33,37].

During OZ of phenols, highly oxidized and hydrophilic ring-opening products, such as simple carboxylic acids, are selectively formed but accumulated. Their degradation could be faster by subsequent PC. Such complementarity has been stimulating interest in the sequential method (SQ) [38,39], but research on it is scarce, according to our survey.

AOPs are continuously being developed to reach a 'low-cost, high-tech, chemical-free' ideal. One such opportunity has been seen in electrochemical oxidation (EO), by which degradation is achieved at mild conditions during electrolysis [7,13]. Lab-scale optimizations of parameters—using boron-doped diamond [40–42] or metal-oxide [41–44] anodes in different supporting electrolytes—to reach optimal phenol removal efficiency versus energy consumption are being widely reported. Meanwhile, monitoring the evolution and toxicity of formed TPs is highly disregarded. Significant research on this matter (PHN only) has been, for example, performed by Jiang et al. [40], Barışçı et al. [44], Amado-Piña et al. [17], and Xing et al. [41], again mainly using HPLC–UV.

There is a lack of knowledge of induced degradation processes and pathways, for which there are no in-depth studies and comparisons available. As so, we focused on profound analyses of degradation processes of PHN, DCP and PCP induced by: OZ; PC with N-doped TiO_2 thin films on glass plates (N-$TiO_2^{im.}$) and photooxidation (PO); SQ (OZ followed by PC); and EO either with boron-doped diamond (BDD) or mixed-metal oxide RuO_2-IrO_2 (MMO) anode in two different supporting electrolytes. A wide range of analytical techniques were applied to obtain information about parent phenol removal, degradation progressivity, mineralization, and dechlorination, along with evolution of TPs. Moreover, ecotoxicological insight was gained by assessing ecotoxicity of treated fractions on water flea *Daphnia magna* for the first time for some of the above AOPs.

2. Materials and Methods

2.1. Chemicals and AOP Treatment

Chemicals. All the chemicals in the present study were bought from Sigma-Aldrich (Steinheim, Germany), Fluka (Seelze, Switzerland), Merck (Darmstadt, Germany), Kemika (Zagreb, Croatia), Fisher Scientific (Loughborough, UK), etc. Purity was at least *p.a.* or was not less than 98%. Further details about standards, reagents, solvents, materials for AOPs, and additional chemicals for identification can be found in Supplementary Materials (Section S1).

General AOP treatment procedure. Prepared test mixture (Table 1) was transferred into a reactor and treated with the selected method. During treatment, sampling was performed at exact time intervals. Collected aliquots (2–20 mL) were stored in plastic vials in freezer at −20 °C. Detailed descriptions of procedures are accessible in the Supplementary Materials (Sections S2–S5). PCP solutions were always prepared in concentrations less or equal to approximately 10 mg/L due to poor solubility in water.

AOP materials & configurations. During ozonation, the gaseous mixture O_2/O_3 was continuously introduced in the reactor containing test solutions. Photocatalysis was achieved by using N-doped TiO_2 synthesized by the sol-gel method from a $TiCl_4$ precursor. It was immobilized on glass plates in the form of thin films using the dip-coating technique. A photocell with a continuous flow of O_2 placed in a UVA-illuminator was used. Electrooxidation was achieved in an electrochemical cell with a mesh-type anode (BDD, MMO) and cathode (stainless steel). For details see the Supplementary Materials (Sections S2–S5).

Table 1. List of test mixtures in ultrapure water (MQ) containing 10 or 50 mg/L phenol (PHN), 2,4-dichlorophenol (DCP) or pentachlorophenol (PCP) treated with several AOPs, namely, ozonation (OZ), photocatalysis (PC), photooxidation (PO) ozonation followed by photocatalysis (sequential method, SQ), electrooxidation (EO) either with BDD or MMO anode in supporting electrolyte.

No.	AOP	Phenols	Approx. conc. (mg/L)	Solvent; Initial pH	Max. TT (min)
1	OZ	PHN	10	MQ; 8	
		DCP	10		60
		PCP	10		

Table 1. Cont.

2	OZ	PHN	50	MQ; 8	8
		DCP	50		
3	OZ	PHN, DCP, PCP	10, 10, 10 (mix)	MQ; 8	3
4	PC & PO	PHN	50	MQ, 8	180/300
		DCP	50		
		PHN	10		
		DCP	10		
		PCP	10		
5	SQ	PHN	20	MQ; 8	0.2 (OZ);
		DCP	20		180 (PC)
		PCP	10		
6	EO/BDD	PHN	50	2 g/L NaCl; 6	60
	EO/MMO	PHN	50		120
7	EO/BDD	PHN	50	2 g/L Na$_2$SO$_4$; 6	160
	EO/MMO	PHN	50		160

2.2. Instrumental Analysis

High-performance liquid chromatography coupled to diode-array UV detection. HPLC–DAD (HPLC System 1100 Series, Agilent Technologies, Santa Clara, CA, USA) was used for determination of target phenols (PHN, DCP, PCP) and chosen TPs (hydroquinone, catechol, tetrachlorohydroquinone), semiquantitative estimation of *p*-benzoquinone, and a number of chromatographic peaks (with UV-absorptivity). The used column was Kinetex XB-C18, 150 mm × 4.6 mm, 5 µm, 100 Å (Phenomenex, Torrance, CA, USA) and guard column Gemini-C18, (Phenomenex, Torrance, CA, USA). Flow rate was 0.7 mL/min. Conditions I were: mobile phase A—10% acetic acid in MQ; B—acetonitrile; separation programme (min-%A/%B): 0–90/10, 3–90/10, 10–60/40, 17–20/80, 23–20/80, 24–90/10, 25–90/10, 2 min post time; UV-Vis detection at 254, 270, 285, 305 nm. Conditions II were: mobile phase A—10 mM H$_3$PO$_4$ in MQ; B—acetonitrile; separation programme (min-%A/%B): 0–100/0, 6–100/0, 9–40/60, 12–20/80, 15–20/80, 16–100/0, 17–100/0, 2 min post time; UV-Vis detection at 254 and 210 nm.

Ultra-high-pressure liquid chromatography coupled to mass spectrometry. Specific TPs were tracked by LC–MS/MS (Vanquish LC System, TSQ Quantis, Thermo Fisher Scientific, Waltham, MA, USA) with negative electrospray ionization. The column, guard column and flow rate were the same as in HPLC–DAD. Mobile phase composition was A—0.1% formic acid in MQ and B—acetonitrile; separation programme (min-%A/%B): 0–90/10, 3–90/10, 10–60/40, 17–20/80, 23–20/80, 24–90/10, 25–90/10, 2 min post time. For MS analysis, N$_2$ (Messer, Bad Soden, Germany, 99.999%) was used as sheath/aux/sweep gas (AU): 70/24/0.5. The ion source was at 385 °C and the nebulizer gas was at 520 °C. Capillary voltage was set to −200 V. Spectra were recorded in TIC full Q3 scan mode (*m/z* 61–355) with no source fragmentation nor collision-induced dissociations in collision cell.

Gas chromatography coupled to mass spectrometry. Volatile and semipolar products were selectively extracted by solid-phase microextraction (SPME) by immersion of the CAR-PDMS, PA or PEG fibre (Supelco, Bellefonte, PA, USA) for 30 min at 30 °C in a 5 mL sample with 0.2 mL of 0.1 M H$_2$SO$_4$ added. Liquid–liquid extraction (LLE) in ethyl acetate and *n*-hexane was also conducted. Either LLE or SPME extracts were analysed by GC–MS (FOCUS GC, ISQ, Thermo Fisher Scientific, Waltham, MA, USA). Conditions I were: DB-624 column (30 m × 0.25 mm, 1 µm, Agilent J&W, Folsom, CA, USA); He flow was set to 0.8 mL/min; splitless injection with surge mode; inlet temperature 260 °C; temperature programme (50 °C, 5 min; 110 °C, 10 °C/min; 210 °C, 15 °C/min, 3 min; 230 °C, 10 °C/min, 8 min; 240 °C, 10 °C/min, 10 min; 250 °C, 10 °C/min); ion source temperature 250 °C; MS transfer temperature 250 °C; MS was operated in TIC mode in the range *m/z* 42–350. LLE extracts were also analysed with GC–MS/MS (TRACE 1300 GC, TSQ 9000, Thermo Fisher

Scientific, Waltham, MA, USA) at the following conditions II: HP-5MS column (30 m × 0.25 mm/0.25 μm, Agilent J&W, Folsom, CA, USA); He flow set to 1.0 mL/min; splitless injection; inlet temperature 280 °C; temperature programme (50 °C, 3 min; 60 °C, 2 °C/min, 1 min; 140 °C, 5 °C/min, 1 min; 320 °C, 10 °C/min, 1 min); ion source temperature 280 °C; MS transfer temperature 280 °C; MS was operated in TIC mode in the range m/z 42–350.

Ion chromatography. Protic species were identified (succinate/malate) and quantified (formiate, oxalate/fumarate, maleate, and Cl^-) or semiquantified (glyoxylate/glycolate/acetate) by anion-exchange IC (Dionex ICS 5000, Thermo Scientific, Sunnyvale, CA, USA), consisting of a gradient pump, an electrochemical suppressor (Dionex AERS 500, 4 mm) and a conductivity detector. The column was AS11-HC (4 × 250 mm, Dionex, Thermo Scientific, Sunnyvale, CA, USA) and flow rate 1.0 mL/min. Mobile phase composition was A—MQ, B—10 mM NaOH in MQ, C—100 mM NaOH in MQ. Separation programme (time-A/B/C): 0–95/5/0, 30–85/15/0, 35–70/15/15, 55–67/15/18, 60–60/15/25, 70–95/5/0, 10 min post time). Suppressor was set to 50 mA.

Total organic carbon (TOC), pH measurements, and UV spectroscopy. TOC (TOC multi N/C 3100, Analytik Jena, Jena, Germany) according to ISO 8245, 1999, pH (SevenEasy, Mettler Toledo, Columbus, OH, USA) and UV spectra (Agilent Cary 60 UV-Vis, Agilent Technologies, Santa Clara, CA, USA) in the range 200–450 nm were additionally determined in some cases for non-specific estimation of mineralization, evolution of acids or changes in chromophores.

2.3. Inhibition on Daphnia magna

Ecotoxicity was assessed by 48 h ecotoxicological testing of acute inhibition of mobility of water flea *Daphnia magna* (Cladocera, Crustacea) according to standard protocol described in OECD Guidelines No. 202 [45]. More details are given in the Supplementary Materials (Section S6).

Each treated sample (i.e., ozonated, photocatalyzed, photooxidized, electrooxidized) was previously diluted with OECD test medium (test mixture) by appropriate factor to reach the referential 'test' concentration of target phenol (γ_x). Chosen γ_x for PHN (50 mg/L), PHN (10 mg/L), PHN for EO by BDD/NaCl, DCP, and PCP were 10.0, 5.0, 5.0, 2.5, 0.6 mg/L, respectively. For example, ozonated 10 or 50 mg/L DCP sample at a chosen treatment time was diluted 4 or 20 times, respectively. The tests were conducted in 3 separate determinations in microtiter plates, each containing 10 mL of chosen test mixture and 10 less than 24 h old *Daphnia* offspring were added. Incubation of organisms in samples was conducted for 48 h in the dark at room temperature. After the incubation, inhibition ($\%_{inh}$) was determined according to Equation (4) in Section 2.4.

2.4. Numerical Evaluation

Approximate treatment time needed for reaching $x\%$ removal of target phenol ($TT_{x\%}$). Approximation of the estimated pseudo 1st kinetic order constant of degradation (k_r)

$$k_r = \frac{ln(2)}{estimated\ half\ time\ (t_{1/2})} \text{ OR estimated from graphs } ln[\text{parent phenol}] = f(TT) \quad (1)$$

Approximate level of mineralization at a certain time ($\%_{min}$)

$$\%_{min} = 100 - 100 \times \frac{TOC\ (\text{treated sample})}{TOC\ (\text{untreated sample})} \quad (2)$$

Dechlorination extent at a certain treatment time ($\%_{dec}$)

$$\%_{dec} = \frac{moles\ (Cl^-\ \text{in treated sample}) - moles\ (Cl^-\ \text{in untreated sample})}{y \times moles\ (\text{DCP and/or PCP in untreated sample})} \times 100 \quad (3)$$
$$y = 2 \text{ for DCP; } y = 5 \text{ for PCP}$$

Normalized relative amount of chosen transformation product (< product >)

$$<product> = \frac{peak\ area\ in\ chromatogram\ (\text{chosen product})}{the\ biggest\ peak\ area\ (\text{chosen product in the same set of samples})} \quad (4)$$

Relative descriptor of a chosen transformation product (RD)

$$RD_{prod.} = \frac{\gamma\ (\text{product in treated sample}) \times molar\ mass\ (\text{parent phenol})}{E_{ff} \times a \times \gamma\ (\text{parent phenol in untreated sample}) \times molar\ mass\ (\text{product})} \quad (5)$$
$$(E_{ff} = \frac{\gamma\ (\text{parent phenol in treated sample})}{\gamma\ (\text{parent phenol in untreated sample})})$$

Complete theoretical conversion '$C_6H_{5-x}OCl_x \to a$ chosen product' (a is 3 and 6 for oxalic and formic acid, respectively) is assumed, regardless of other chemical transformations and changes in the volume of test mixture due to sampling during AOP treatment.

Acute 48-h mobility inhibition with water flea *Daphnia magna* ($\%_{inh}$)

$$\%_{inh} = \frac{No.\ of\ immobilized\ Daphnia\ after\ incubation}{No.\ of\ Daphnia\ in\ the\ test\ mixture\ at\ the\ beginning} \times 100 \quad (6)$$

3. Results and Discussion

The motivation was to fully understand complex degradation processes of phenol (PHN), 2,4-dichlorophenol (DCP), and pentachlorophenol (PCP) in different matrices from chemical and ecotoxicological point of view. A focus was placed on profound analyses of treated fractions. This allowed a comparison of four approaches for chemical degradation: ozonation (OZ), photocatalytic oxidation with immobilized N-doped TiO$_2$ thin films on glass supports (PC), their sequence (SQ), and anodic electrooxidations (EO) by BDD and MMO anode, thus covering a wide range of three advanced technologies.

In order to collect data on target degradation progress, dechlorination, mineralization, changes in pH, chemical transformations, and evolution of selected by-products, numerous analytical methods and procedures were applied. Such as: HPLC–DAD, pH measurement, TOC determinations, UV spectroscopy, solid-phase microextraction (SPME) or liquid–liquid extraction (LLE) followed by GC–MS or GC–MS/MS, UHPLC–MS/MS, ion chromatography (IC), and ecotoxicological mobility inhibition tests on *Daphnia magna* water flea. All of these are further described in the following section.

3.1. Removal of Target Phenols, Mineralization, and Progressivity

To describe the efficiency of the removal, we used several descriptors that are explained in Section 2.4, namely: treatment time (TT) needed for reaching > 95% target phenol removal (TT$_{>95\%}$), estimated pseudo first-order kinetics constant (k_r; Equation (1)), and level of mineralization ($\%_{min}$; Equation (2)).

Ozonation (OZ). Results indicate that at the initial pH of 8, a favourable degradation process of phenols is possible, which includes rapid target degradation, depending on their chemical structure (correlation with nucleophilicity, mechanism, and intermediates), medium complexity (competition for O$_3$ consumption), and initial concentrations (substrate loads). All of that is reflected in Figure 1a and is in agreement with the literature [12,18]. In all cases, pseudo first-order kinetics could be approximated, e.g., estimated pseudo first-order constants (k_r) for PHN and DCP in separate solutions reached 0.6 and 2.1 min^{-1} (50 mg/L), respectively, whereas for the mixture they were 1.9, 2.8, and 4.9 min^{-1} for PHN, DCP, and PCP, respectively. The approximate TT$_{>95\%}$ were 4, 2, and ~0.1 min for PHN, DCP (50 mg/L), and PCP (10 mg/L), respectively (Figure 1a). The higher the initial concentration or the greater the amount of co-substrates (see OZ of mixture), the longer TT$_{>95\%}$. Described progressivity was primarily reflected in: (i) a sudden drop of pH from 8 to 3–4, with the formation of organic acids; (ii) compounds detected by HPLC–DAD (e.g., absorption decline < 255 nm, as well as evolution of extra polar products with minimum retention); and (iii) an increase in the number of IC peaks (showing a quick evolution of protic species), presented in the Supplementary Materials (Section S2.1). Since OZ proceeds in acidic medium, reactions of substrates with O_3 were assumed to be taking part on the gas–liquid interface, especially in the early stage [24,27]. Reaching strongly acidic pH was thought to greatly influence degradation process. Mineralization was, on the other hand, not readily achieved. For example, only 50, 40, and $40\%_{min}$ for PHN, DCP, and PCP (10 mg/L), respectively, was measured after 10 min.

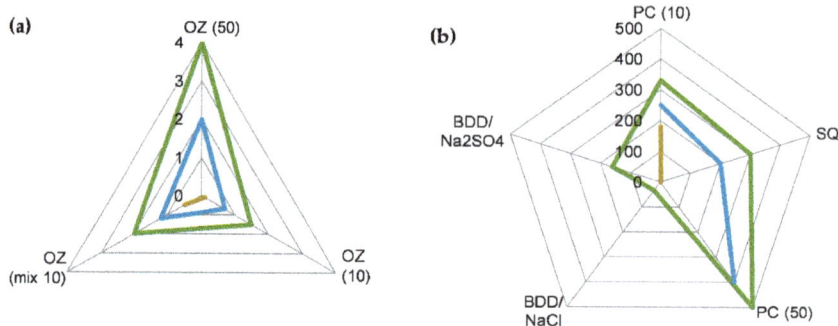

Figure 1. Treatment time for reaching > 95% removal (see Section 2.4) of phenol (PHN), 2,4-dichlorophenol (DCP) and pentachlorophenol (PCP) ($TT_{>95\%}$ in min) after treatment with (a) ozonation (OZ), and (b) photocatalysis (PC), sequential method (SQ), and electrooxidation with BDD anode in NaCl (BDD/NaCl) and Na_2SO_4 (BDD/Na_2SO_4) (note: numbers in parentheses are initial concentrations in mg/L).

Photocatalysis (PC) and photooxidation (PO). In general, selective, robust, and rapid degradation is considered for OZ, while mild and slow degradation is known for PC. Data concerning PC (i.e., $UV/O_2/TiO_2^{im.}$ system) indicate up to a 100-times longer treatment time (reaching few hours) than those characteristic for OZ (reaching few minutes). TT also depends on specific parameters, such as number of C–Cl bonds [35,36] and initial concentration [35]. This is in accordance with [12,27] which reported that PC is significantly less efficient than OZ. For example, estimated k_r was 0.01, 0.02, and 0.03 min^{-1} for PHN, DCP, and PCP (10 mg/L), respectively, but 0.005 and 0.008 min^{-1} for PHN and DCP (50 mg/L), respectively. Therefore, $TT_{>95\%}$ are above the 3-h time range. By increasing the number of C–Cl bonds and initial concentration, removal was faster (Figure 1b), which shows phenols' reactivity. PCP has the greatest number of labile C–Cl bonds, the smallest pK_a, and the highest UVA absorptivity (at 365 nm); therefore, its degradation was the fastest. PO (i.e., UV/O_2 system; absence of $N-TiO_2^{im.}$) was found to be the least effective and the slowest, since removal efficiency was up to 500 times lower than for PC. This is especially true in the case of PHN where only 2% removal was reached after 5 h treatment. In addition, estimated k_r were $<10^{-5}$, $<10^{-3}$, and 0.006 min^{-1} for PHN, DCP (50 mg/L), and PCP (10 mg/L), respectively (Figure 2). Effect of UV light was, therefore, the highest in the case of PCP (90% removal reached after 5 h) due to the progressive breakage of C–Cl bonds on the aromatic ring. Mineralization after >180 min of PC treatment was estimated to be 60, 70, and 60% for PHN, DCP, and PCP (10 mg/L), respectively, whereas no mineralization was induced by PO. UV spectroscopical data on PC of PHN and DCP (50 mg/L) further support the facts described above, as there was a slow decrease in characteristic peak absorptivity (270 nm for PHN, 285 nm for DCP) during treatment. Finally, assessment of HPLC and IC chromatogram peaks may give overall conclusion that PC and PO processes are more selective, and less dynamic in contrast to OZ (see Supplementary Materials Section S3.1).

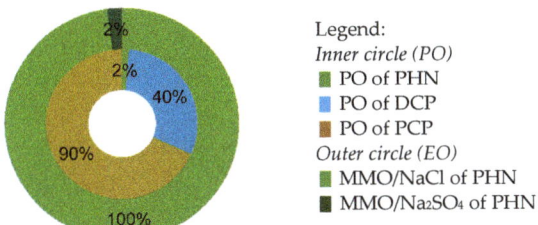

Figure 2. Maximum removal % of phenol (PHN), 2,4-dichlorophenol (DCP), and pentachlorophenol (PCP) achieved by 300 min of photooxidation (PO), 120 min of electrooxidation with MMO in NaCl, and 160 min with MMO in Na$_2$SO$_4$.

Sequential method (SQ). In the case of SQ, a direct comparison to PC or OZ alone is not possible due to different initial concentrations used in experiments. Nevertheless, experiments show a contrast between OZ and PC, where OZ was considerably faster (Figure 1b), as already discussed. For example, a 0.2 min 'flash' of OZ removed almost all PCP and approximately 75% of DCP and 25% of PHN. So, SQ may provide a faster option as opposed to slow PC. Examples of chromatograms are given in the Supplementary Materials (Section S4.1). On the other hand, it should be noted that during OZ, the pH suddenly dropped to <6 while, interestingly, during PC the pH did not change significantly.

Electrooxidation (EO). Only PHN degradation (50 mg/L) was investigated by EO. Removal was the fastest in NaCl as the supporting electrolyte. In the case of BDD/NaCl, TT$_{>95\%}$ was only below 35 min (Figure 1b) but chlorinated aromatics were formed which were almost completely removed after 60 min. Whereas in the case of MMO, TT$_{>95\%}$ was up to 120 min; also due to mild polychlorination (Figure 2). The process was the most progressive with BDD, which can be estimated also from HPLC chromatograms. On the other hand, EO in Na$_2$SO$_4$ was slower, selective, and less dynamic. Less HPLC–DAD-detectable peaks were generated in comparison to EO/NaCl (see the Supplementary Materials, Section S5.1). If BDD and MMO are compared, treatment with BDD was effective on a long run, as 96% removal after 160 min was achieved (Figure 1b). In contrast, an MMO anode provided only less than 10% removal (Figure 2). Data are comparable to the literature [17,40,44].

3.2. Dechlorination of Chlorophenols

Monitoring of Cl$^-$ concentration allowed us to track the breakage of C–Cl bonds, referred to as dechlorination. From it, the overall amount of remaining chlorinated compounds can be estimated. The applied descriptor was the dechlorination extent (%$_{dec}$), explained in Section 2.4 (Equation (3)).

Ozonation. During OZ, fast (<3 min) and complete dechlorination could be observed. Curves for dechlorination extent (Figure 3) are in all cases of the same shape, marked by a fast increase in Cl$^-$ concentration, but more steady changes in the later stage of OZ. This points to the probability that more labile C–Cl bonds on C atoms that are part of aromatic structures (Ar–Cl) are quickly broken. The same is valid for OZ of mixture, where 100% combined dechlorination of PCP and DCP was reached after only 3–4 min (data not shown). All in all, non-chlorinated TPs were in the majority expected during and after OZ, which is favourable.

Photocatalysis and photooxidation. During PC, the dechlorination process was much slower than in OZ, expanding to 2–5 h (Figure 3), which is proportional to slower removal efficiency (see Section 3.1). For example, in the first hour of PC there was still >50% of chlorinated organic compounds (including non-degraded DCP and PCP), and then from the third hour onwards, most of the chlorine was already in the form of Cl$^-$, which is favourable (Figure 3). Nevertheless, long-term dechlorination efficiency could be predicted since more than 80%$_{dec}$ was achieved after a 5-h treatment. In the PO experiments, cleavages of C–Cl also occurred, especially in the case of PCP, rather than DCP (Figure 3b), suggesting that UVA irradiation plays an important role not only in PCP removal but also in its dechlorination (<60%$_{dec}$ reached after 5 h). In comparison, Gunlazuardi and Lindu [37] reported on the much slower PCP's release of Cl$^-$ ions.

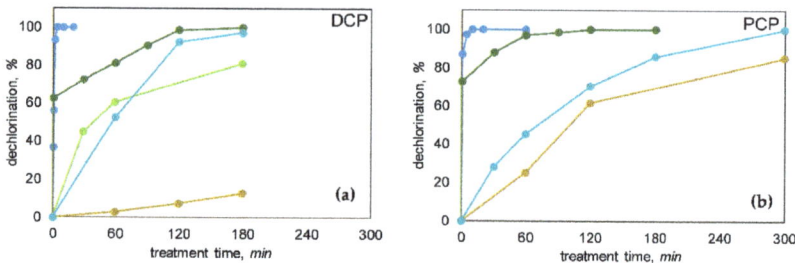

Legend: ■ OZ, ■ SQ, ■ PC (10 mg/L), ■ PC (50 mg/L), ■ PO.
Note: In the case of sequential method (SQ), 1st point represents Cl$^-$ concentration after ozonation (OZ).

Figure 3. Dechlorination (i.e., conversion of organic chlorine into chloride; see Equation (3) in Section 2.4) of (**a**) 2,4-dichlorophenol (DCP) and (**b**) pentachlorophenol (PCP) during ozonation (OZ), photocatalysis (PC), photooxidation (PO), and sequential method (SQ).

Sequential method. Similar to removal efficiency (Figure 1), OZ with subsequent PC also reinforced dechlorination process (Figure 3). With 0.2 min of OZ, it was possible to declare >60%$_{dec}$ and fast dechlorination. However, with subsequent PC, the process steadily slowed down.

3.3. Transformation Products and Possible Degradation Pathways

3.3.1. General Identification

Several analytical techniques were applied to separate and identify the transformation products (TPs) forming during treatment: HPLC–DAD, UHPLC–MS/MS, liquid–liquid extraction (LLE) or solid-phase microextraction (SPME) followed by GC–MS, LLE followed by GC–MS/MS, and IC (see Section 2.2). A list of some identified TPs is given in Table 2, where they are numbered from A1 to D16. A part of TPs were identified by several analytical techniques, while some of them only by one of the applied techniques, thus lowering the identification certainty.

Ozonation. OZ is considered very effective in destruction of the phenols' stable aromatic structure in contrast to other AOPs due to direct reaction with bipolar O$_3$ molecule [14,17,20]. In general, degradation is characterized by many stages, as suggested by Figure 4. The first indicator of any chemical transformation was the change in colour of solutions. Firstly (Table 2), rapid and simultaneous formation of 1,2- and 1,4-aromates, e.g., hydroxyphenols, benzoquinoids (B12,14,15, C1–6), and others (A10–16, B2–6,7) characteristically occurred, reported also in [20,21,23,24], due to substitutions on *orto* and *para* sites with further oxidations [24]. In smaller extent, partially dechlorinated CPs were formed (B10), also mentioned in [18,25]. Opening of the oxidized aromatics' benzene ring may have led to the formation of multifunctional carbonyl compounds, and from there on to condensation and cyclization reactions leading to furans, cyclopentanones etc. (C9–15), similarly as in [16,17,20,27]. A degradation process is highly progressive, but it was terminated by the formation of C1–4 simple organic acids (e.g., oxalic, formic, acetic; D9,12–16) that are known to accumulate because of their stability towards O$_3$ [17,24,28]. Further reactions between acids, oxidations [20], cyclizations, cycloadditions, peroxidations, redox transformations [17], addition reactions with bond breakages [28], and further ozonolysis might proceed as well, as suggested from literature [20]. In addition, the initial stages of OZ and higher initial concentrations of phenols brought about oxidative coupling reactions, resulting in formation of chlorinated and/or polyhydroxylated coupling products in traces (most probably phenoxyphenols, biphenyls, dibenzodioxins and others; A1–5,7). Their m/z with isotopic fingerprints were specifically detected by UHPLC–MS/MS (Table 2) due to good ionization by negative ESI and are therefore only suggested as possible, but those TPs were highlighted also by Oputu et al. [20] and Hirvonen et al. [22]. The respective products were more numerous in the case of OZ of CPs. For example, number of coupling products was 1 and 8 for PHN and DCP, respectively. These reactions were also non-specifically confirmed by a sudden drop in pH and by number and peak types detected by HPLC–DAD and IC. The above highlighted and many other transformations are described also in literature [6,16,18–21,24,28], especially recently by Oputu et al. in 2020 [20]. For example, decrease in absorption < 255 nm indicated a rapid formation of simple acyclic compounds, such as organic acids.

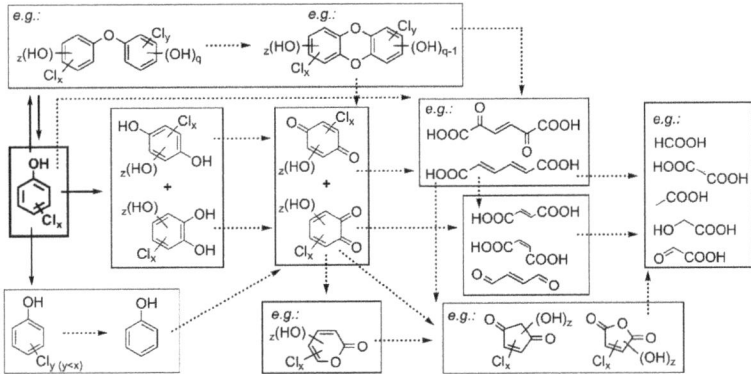

Description: aromatic coupling products (C_{6+}) ⇆ parent phenol (C_6) → hydroxyphenols, benzoquinoid species and condensates (C_6) → C_{4-6} organic acids or condensation products → C_{2-3} unsaturated organic acids → simple C_{1-3} saturated organic acids.

Figure 4. General scheme of the suggested degradation process during ozonation (partially adopted from [6,16–21,24,28]).

Photocatalysis and photooxidation. In the literature, target identifications of TPs in PC are prevailing, using only HPLC–UV. A look into identified TPs (Table 2) suggests that PC was incapable to efficiently open the aromatic ring. Thus, there was a prevalence of reactions on the aromatic ring [27]. For example, degradation was marked by formation of benzoquinones, hydroxyphenols (B10,12,14,15), less chlorinated chlorophenols (B10) [36], as well as dimers, adducts, biphenyls or phenoxyphenols which are thought to be product of (oxidative) coupling, i.e., formations of C–O and C–C bonds (A1–5,7). These predominated in the first 2 h of PC, and mostly in the case of PO, so their formation was accelerated by UVA irradiation. The formation of all mentioned aromatics was a result of radical hydroxylations, photoinduced oxidations, reductive/hydroxylative dechlorinations, and other substitution reactions. Consequently, CPs, hydroxyphenols, and benzoquinones are by far the most frequently identified in the literature, such as hydroquinone (C4), catechol (C6) and *p*-benzoquinone (C2) [11,31,33,37]. UVA irradiation plays a significant role in the cleavage of C–Cl bonds and coupling reactions, but only the presence of photocatalyst allows for increased destruction of aromatic structures. Radical reactions induce also other transformations, such as photooxidations, cyclizations, coupling reactions, and condensations resulting in evolution of simpler compounds, e.g., hydroxycyclopentanediones (C13,14), reported also in [35]. Cyclized organic oxygen derivatives (C11,13,14, D1), and organic acids (D9) also appeared during PC, but in very low concentrations, and are reported also in [11,12]. As dechlorination was slow, more chlorinated aromatics were detected. In the case of PC of PHN, UV spectroscopy showed a gradual increase in the secondary absorption peak ranging 280–310 nm, which could have possibly indicated the formation of various quinoid or condensed aromatic species. During treatment, the overall absorptivity gradually decreased, and the absorption peaks were no longer clearly defined. Moreover, there were less HPLC–DAD peaks, which would indicate formation of simple non-aromatic compounds.

Sequential method. TPs in SQ treated samples were similar to those in OZ which were then further degraded by PC (Table 2). Interestingly, there were fewer coupling products detected although they are otherwise typical of PC; possibly because of preozonation.

Electrooxidation. As already mentioned, EO in electrolyte NaCl is the most effective for the removal of PHN, yet the least successful since there was unfavourable formation of chlorinated aromatics due to in situ electrogeneration of chlorinating agents. In the case of BDD/NaCl, chlorinated aromatics (mostly chlorophenols, B10, reported also by Chatzisymeo et al. [43]) were preferentially formed (Table 2). However, after 60 min they were broken down into chlorinated carbonyl compounds and polychlorohydrocarbons (C8–10,14,16, D2,3,6–8,10,11), e.g., chloroform, tetra/pentachloropropenes, tetrachloroetene, tetrachlorocyclopropanes, etc. Thus, ring-opening reactions effectively occurred. MMO/NaCl treatment was characterized by an even greater generation of chlorinated aromatics which were accumulated (A1–5,7, B10,12, C3). For example, even after 120 min of EO they still prevailed; contrarily, in the case of BDD they were quickly degraded. Despite extensive and rapid target degradation of PHN in NaCl, we cannot speak of a successful process.

On the other hand, slower EO in Na_2SO_4 provided more acceptable chemical transformations. Non-chlorinated, hydroxylated and/or highly oxidized aromatics were preferentially formed (Table 2), such as hydroquinone, catechol, *p*-benzoquinone, and organic acids [40,44], as well as some coupling products. On the long run, aromatic ring might have been opened. MMO was an exception since progressivity was slower. Thus, *p*-benzoquinone accumulated in a relatively big proportion. Moreover, there were fewer TPs identified than during EO/NaCl. In addition, probably relatively greater amount of *p*-benzoquinone was generated by MMO than by BDD, as estimated from peak areas in HPLC–DAD. Hydroquinone, catechol, *p*-benzoquinone, and organic acids (e.g., fumaric, oxalic, maleic) formation were in majority identified by target analysis with HPLC–UV also in the literature [17,40,42,44].

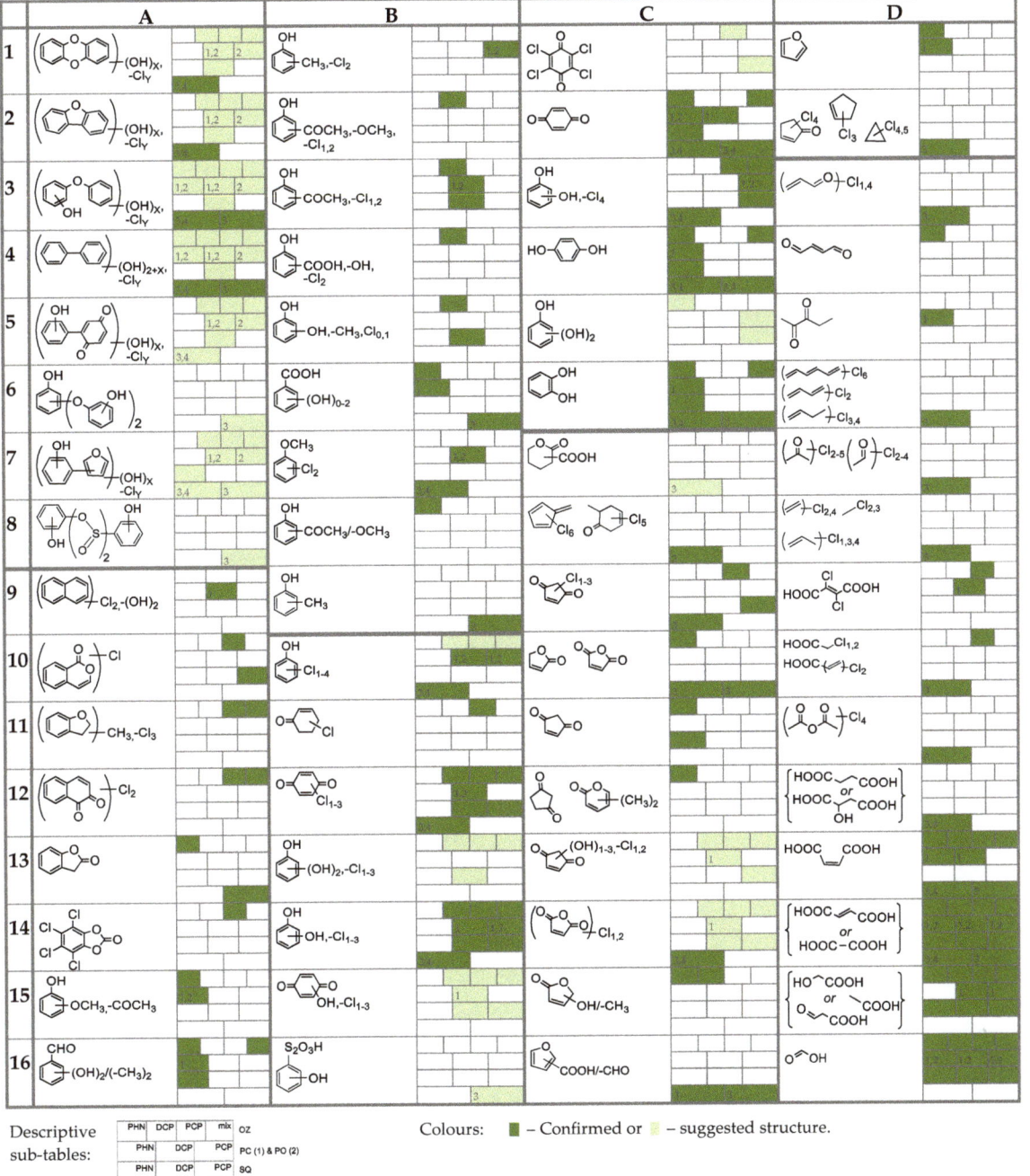

Table 2. Transformation by-products (TPs) of phenol (PHN), 2,4-dichlorophenol (DCP), pentachlorophenol (PCP) and their mixture during ozonation (OZ), photocatalysis (PC; #1), photooxidation (PO; #2), sequential method (SQ), and electrooxidation (EO) with BDD (#3) and MMO (#4) anode in NaCl and Na_2SO_4 (for decription of table's structure and colours, see footnotes).

3.3.2. Monitoring of the Selected Products

Hydroquinone (HQ), catechol (CT), tetrachlorohydroquinone (TH), and organic acids (oxalic, OX; formic acid, FO) were quantified from their respective calibration curves obtained by HPLC–DAD and IC (Figures 5 and 6). By IC, acetic/glyoxylic/glycolic, maleic, succinic/malic, propionic, lactic, fumaric acid (FM; note: coelution with oxalic acid but differentiated by HPLC–DAD) could also be detected. Relative abundance of p-benzoquinone (BQ), as well as other identified TPs was monitored according to Equation (4) (Section 2.4).

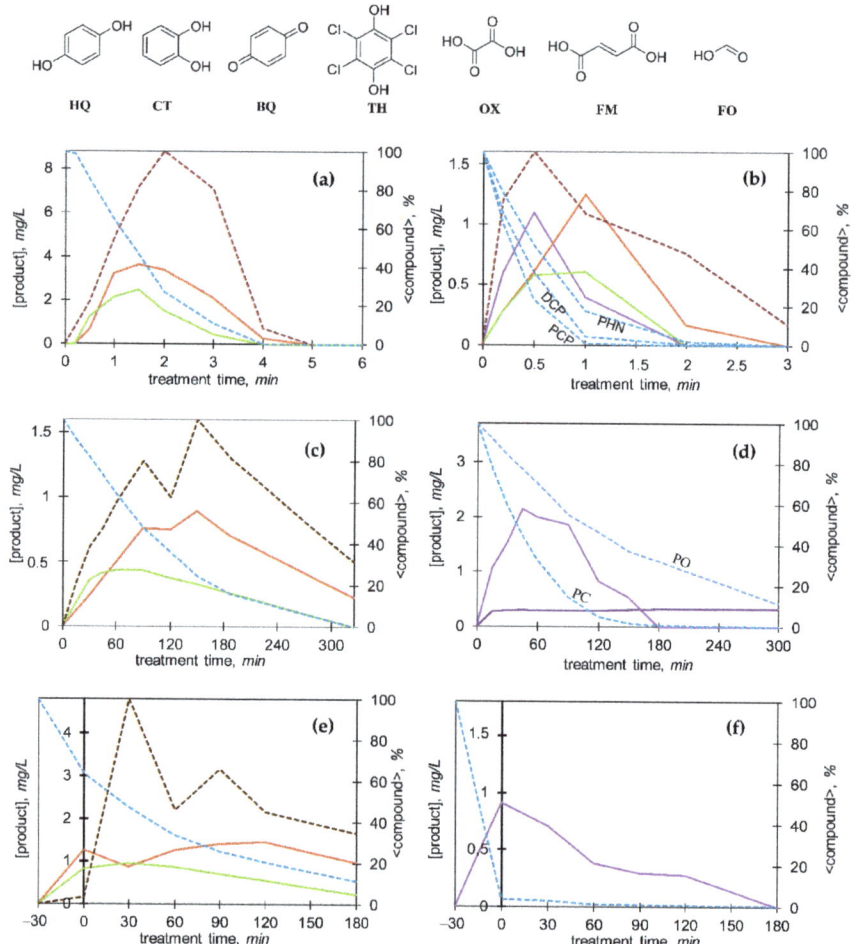

Legend: — [HQ], — [CT], — [TH], --- < BQ >, --- < PHN >, --- < DCP >, --- < PCP > (notation [product] represents concentration in mg/L, whereas notation <product> represents relative amount in %).
Note: '−30 min' in graphs (e) and (f) actually represents 0.2 min OZ.

Figure 5. Concentrations (left y-axis) of hydroquinone (HQ), catechol (CT), and tetrachlorohydroquinone (TH), and normalized relative amount (right y-axis; see Equation (4) in Section 2.4) of parent phenols (PHN, DCP, PCP) and p-benzoquinone (BQ) during (**a**) ozonation of phenol (PHN) (50 mg/L); (**b**) ozonation of mixture (10 mg/L each); (**c**) photocatalysis of PHN (10 mg/L); (**d**) photocatalysis and photooxidation of pentachlorophenol (PCP) (10 mg/L); (**e**) sequential method of PHN (approximately 20 mg/L); (**f**) sequential method of PCP (approximately 10 mg/L).

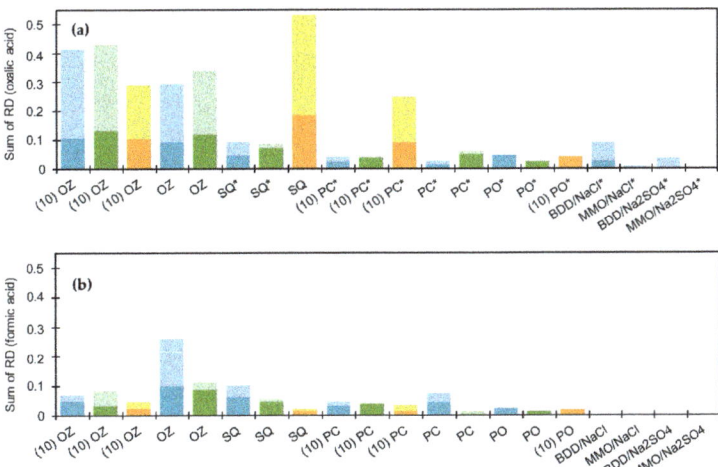

Legend: ■ PHN, ■ DCP, ■ PCP; darker colour (early degradation stage; i.e., relative descriptor, RD, at treatment time when <80% removal was reached or after 'flash' ozonation in the case of SQ); lighter colour (later degradation stage; i.e., relative descriptor, RD, at treatment time when >80% was reached or after photocatalysis in the case of SQ); * Possibly a mixture of oxalic and fumaric acid; values in parenthesis (x-axis) are initial concentrations of target phenols.

Figure 6. Presence of (**a**) oxalic and (**b**) formic acid in early and later degradation stages (quantified by relative descriptor, RD, according to Equation (5) in Section 2.4) of phenol (PHN), 2,4-dichlorophenol (DCP) and pentachlorophenol (PCP) reached by ozonation (OZ), photocatalysis (PC), photooxidation (PO), and electrooxidation (EO) with BDD and MMO anode in NaCl and Na_2SO_4.

Ozonation. The results generally show that organic acids were rapidly formed during OZ (mostly oxalic, but also formic, maleic, and acetic/glyoxylic/glycolic acid); those were slowly degraded afterwards (Figure 6). Monitoring of aromatic representatives (HQ, CT, TH, BQ; Figure 5a,b) indicate that their formation was favourable only in the initial OZ stages, which means that the aromatic ring was later opened due to ozonolysis, resulting in acyclic compounds and/or their condensates. Interestingly, HQ evolution was more favourable than CT's (Figure 5a,b).

Photocatalysis and Photooxidation. Unlike with OZ, the amounts of HQ, CT, BQ, and TH were considerable, and they were long-lasting. For example, during PC of PHN, concentrations of HQ and CT were reaching up to approximately 7 and 3 mg/L, respectively, while during PO of PHN concentrations were only 0.3 and <0.2 mg/L (data not shown), respectively. Similar examples are given in Figure 5c,d. Thus, the formation of organic acids was much slower (similar to slower degradation process), reaching lower concentration ranges (Figure 6). CT formation was, again, slower than HQ's and it reached lower concentrations in both PC and PO processes, reported also in [34]. In the case of PCP degradation, formation of quite persistent TH stands out, indicating substitution reaction on the *para* site. UVA irradiation in the absence of photocatalyst still triggered the formation of HQ, and CT, but to a much lower extent. This indicates that photolysis of C–H and/or C–Cl bonds on the aromatic ring and possibly the incorporation of oxygen may have also led to oxidations to some extent. Additionally, several findings can be obtained from Figure 7. In the first phase, chlorohydroxyphenol was formed directly from DCP (oxidative dechlorination), and in the later phase it might have been also formed by hydroxylation of monochlorophenol itself, which had been previously generated by reductive dechlorination of DCP. The following pathways could be proposed: DCP ⇉ dichlorohydroxyphenols + PHN + chlorohydroxyphenol + [monochlorophenol ⇉ chlorohydroxyphenol].

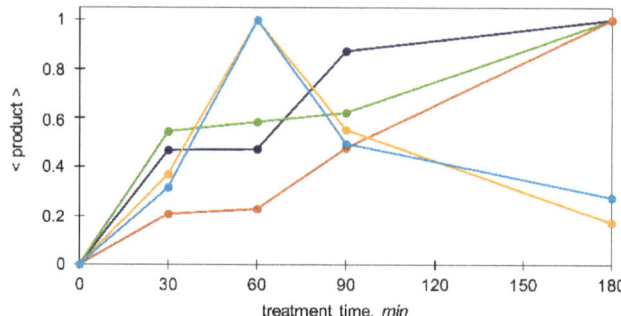

Legend: ● <PHN>, ● <monochlorophenol>, ● <chlorohydroxyphenol>, ● and ● <dichlorohydroxyphenol, isomer 1 and 2>.

Figure 7. Normalized relative amount of chosen products (see Equation (4) in Section 2.4) in ozonated 2,4-dichlorophenol (DCP) (50 mg/L) determined by SPME/GC–MS.

Sequential method. In comparison, the decomposition pattern of PHN and DCP by SQ might have been similar to PC's, except the concentrations of aromatic TPs were lower, as these are rapidly formed and destructed during OZ. Production of acids was also lower due to long PC (Figure 6), but SQ of PCP is an exception since most of it had been already effectively degraded by preozonation, and thus, acids' production was higher. In the case of PHN (Figure 5e), HQ and CT were formed immediately with OZ, and during PC their concentrations remained mostly unchanged. Moreover, in the case of PCP (Figure 5f), with previous OZ, it was possible to lower the TH evolution extent effectively, which would have otherwise been formed by PC.

Electrooxidation. In the case of BDD/NaCl, 2,4-dichlorophenol was detected up to concentration 29.0 mg/L; in addition, approximately six times higher concentrations were determined after 160 min of MMO/NaCl than after 35 min of EO with BDD/NaCl. A similar trend can be concluded for other chlorophenols (Figure 8). During EO/Na$_2$SO$_4$, HQ was detected in all cases in concentrations < 0.5 mg/L, whereas lower concentrations of CT were found only in the BDD/160 min sample. BQ evolved in all samples, regardless of the anode and electrolyte used. In addition, degraded PHN was almost entirely converted into BQ after 160 min of MMO/Na$_2$SO$_4$ treatment. Acid formation was relatively low, which is proportional to a slow degradation process (Figure 6).

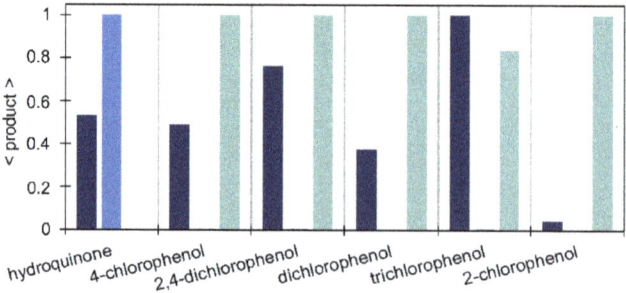

Samples: ■ PHN treated with BDD in NaCl for 35 min, ■ PHN treated with BDD in NaCl for 60 min, ■ PHN treated with MMO in NaCl for 120 min.

Figure 8. Relative amount of aromatic products (see Equation (4) in Section 2.4) in electrooxidized samples PHN determined by SPME/GC–MS (normalized for each product to the highest peak area of the three analyzed samples).

3.4. Detoxification Estimated from Acute Ecotoxicity Tests with Daphnia magna

Detoxification of pollutants is the ultimate goal of all degradation technologies, and not just the removal of parent phenols. Therefore, tests on aquatic invertebrates, such as water fleas, are extremely important, as these organisms frequently come into contact with pollutants, which can affect the whole freshwater ecosystem. The advantage of the applied tests is the ability to non-specifically evaluate the inhibitory and biological effects of mixtures of all known and also unknown TPs. Results are shown in Table 3, showing determined 48-h acute inhibition of the mobility of *Daphnia magna* (see Equation (6) in Section 2.4) after incubation of test organisms in diluted samples (AOP-treated phenols).

Table 3. Comparison of the inhibition ($\%_{inh}$) on *D. magna* organisms (i.e., detoxification extent; see Equation (6) in Section 2.4) of treated phenol (PHN), 2,4-dichlorophenol (DCP), pentachlorophenol (PCP), and mixture solutions with ozonation (OZ), photocatalysis (PC), photooxidation (PO), sequential method (SQ) and electrooxidation with BDD and MMO in NaCl and N_2SO_4 at different treatment times (TT).

PHN	$TT_{0\%}$	$TT_{0-25\%}$	$TT_{25-50\%}$	$TT_{50-100\%}$	$>TT_{100\%}$
OZ (50 mg/L)					
OZ (10 mg/L)					
PC (50 mg/L)					
PC (10 mg/L)					
PO					
SQ		OZ	PC	PC	PC
BDD/NaCl					
MMO/NaCl					
BDD/Na_2SO_4					
MMO/Na_2SO_4					
DCP	$TT_{0\%}$	$TT_{0-25\%}$	$TT_{25-50\%}$	$TT_{50-100\%}$	$>TT_{>100\%}$
OZ (50 mg/L)					
OZ (10 mg/L)					
PC (50 mg/L)					
PC (10 mg/L)					
PO					
SQ		OZ	OZ	OZ & PC	PC
PCP	$TT_{0\%}$	$TT_{0-25\%}$	$TT_{25-50\%}$	$TT_{50-100\%}$	$>TT_{>100\%}$
OZ					
PC					
PO					
SQ		OZ	OZ	OZ & PC	PC
Mixture	$TT_{0\%}$	$TT_{0-25\%}$	$TT_{25-50\%}$	$TT_{50-100\%}$	$>TT_{>100\%}$
OZ *					

Legend: ■ $< 20\%_{inh}$, ■ $20-39\%_{inh}$, ■ $40-59\%_{inh}$, ■ $60-79\%_{inh}$, ■ $> 80\%_{inh}$; – not determined. Note: Notations 'OZ' and 'PC' were added in SQ. * $TT_{x\%}$ for PHN degradation.

Ozonation. Rapid and complete dechlorination (see [21,24]), high preference to break down phenols' aromatic ring and progressive formation of accumulative organic acids, seen as a sudden drop in pH [16–18,27], are the reasons for rapid detoxification (Table 3). OZ allowed for quick detoxification which was achieved after less than 1 min treatment, which is similar to findings in [26]. This is true for OZ of 50 and also 10 mg/L of chosen phenols (Table 3). Moreover, inhibition on *D. magna* did not increase during treatment, which means that less toxic TPs—relative to DCP and PCP—were formed (Table 3). However, this is not the case for PHN, where the first increase in inhibition might have been the result of the formation of oxidized aromatics (e.g., 48hEC$_{50}$ ratio for HQ and 4,4′-biphenydiol relative to PHN is 0.004 and 0.1 for *D. magna*, respectively [46]), and the second increase perhaps due to oxalic acid formation [17].

Photocatalysis and Photooxidation. Described facts on PC in sections above are the reason for slow decline in inhibition of the treated samples, and even slower detoxification reached by PO (Table 3). Uniquely for PHN, an increase in inhibition was characteristic for both concentrations used. This was due to the formation of highly oxidized aromatics, such as (chloro)hydroxyphenols, CPs, and (polychlorinated) phenoxyphenols/biphenyls, which tend to be more toxic and bioavailable to *D. magna*. For example, 48hEC$_{50}$ ratio for DCP, PCP, HQ, and triclosan relative to PHN is 0.1, 0.007, 0.004, and 0.02 for *D. magna*, respectively [46]. If transversely compared, inhibitions were higher and longer lasting (despite the same dilution factor) when higher initial concentrations were used (Table 3).

Sequential method. SQ provided immediate detoxification by 'flash' OZ, which otherwise could not be readily achieved by PC alone. Only in the case of PHN, again, inhibition increased and persisted (Table 3); most likely due to coupling products and photolysis reactions of TPs themselves.

Electrooxidation. During EO/NaCl, inhibition was significantly increased, e.g., MMO inhibition was 100% even after 120 min treatment (Table 3). This was most likely due to the presence of chlorophenols, which are much more toxic than PHN (e.g., 48hLC$_{50}$ for 2-chlorophenol, 2,4-dichlorophenol, and 2,4,6-trichlorophenol relative to PHN is 0.35, 0.21, and 0.17, respectively, for *D. magna* [47]). Such a remarkable increase in toxicity represents an unfavourable decomposition process of PHN. On the other hand, despite the initial 94% inhibition, complete detoxification followed in the case of BDD after 35–60 min electrolysis (Table 3). This was due to the gradual decomposition of chlorinated aromatics into much less toxic and volatile chlorinated alkenes and carbonyl compounds, according to 48hLC$_{50}$ data for *D. magna* in [47]. In the case of EO/Na$_2$SO$_4$, after 160 min of treatment with BDD, a final 0%$_{inh}$ could be achieved, but with MMO as much as 100%$_{inh}$ was measured (Table 3). Nevertheless, there was an intermediate increase in inhibition, which was possibly a result of the increased amount of BQ and other analogous aromatics. For example, in MMO, BQ was the predominant TP that accumulated. In conclusion, results discussed in the previous sections greatly reflect in ecotoxicological data, that can be also compared to tests on other organisms, e.g., *L. sativa* [17] and *V. fischeri* [44].

4. Conclusions

The study is focused on a representative family of ubiquitous and genotoxic pollutants: phenolic compounds, namely, phenol, 2,4-dichlorophenol and pentachlorophenol. Their chemical fate and potential impact on water organisms during four types of AOP treatments (ozonation, photocatalytic oxidation with immobilized N-TiO$_2$ thin films, their sequence, and anodic electrooxidation) were investigated, using a variety of complementary techniques for instrumental analysis along with ecotoxicological assessment.

Results indicate that ozonation causes a favourable decomposition process from all viewpoints, which includes rapid target degradation (depending on structure, presence of co-substrates and initial concentration), fast and complete dechlorination, the rupture of aromatic structure and the progressive formation of organic acids. Incomplete mineralization but rapid detoxification is characteristic. Photocatalysis is a much slower degradation processes compared to ozonation. It shows the inability to open the aromatic ring efficiently and quickly, which reflects in the persistence of reactions on the aromatic ring and coupling of aromatics. The latter is manifested in increased inhibition of treated samples on *D. magna*, which only slowly decreases. An alternative approach for the destruction of pollutants might be seen in ozonation followed by photocatalysis, i.e., the sequential method, which is demonstrated to shorten the required time of degradation processes and to reinforce the dechlorination along with detoxification. As for electrochemical degradation, our results justify the use of more efficient BDD anode but also a need for proper electrolyte selection since the latter affect chemical transformations. Phenol removal is by far the fastest and most efficient in electrolyte NaCl, but it is accompanied by unfavourable formation of chlorinated transformation products. On the other hand, less effective degradation yet more favourable reactions and detoxification trends are provided by electrooxidation in Na_2SO_4.

Such a multidisciplinary approach to research the chemical degradation of pollutants induced by various AOPs is important for environmental protection. The use of many aspects of sample analysis with the combined expertise of analytical chemistry, environmental chemistry, environmental engineering, materials science, and ecotoxicology is rarely reported in the literature but is essential for evaluation of any AOP. In the future, in-depth research shall be performed in the field of (i) combination of complementary AOPs, such as ozonation followed by photocatalysis, in order to develop optimal sequential AOPs and to, therefore, minimize their limitations. Furthermore, (ii) there is also a necessity to assess phenols' fate during treatment with AOPs in (semi)real matrices, suchlike model and real waste/surface waters. Finally, (iii) their natural chemical transformations and stability shall be investigated so as to model their real persistency.

Supplementary Materials: The following supporting information can be downloaded at: https://www.mdpi.com/article/10.3390/molecules27061935/s1, Table S1: Process parameters used for ozonation; Figure S1: HPLC–DAD chromatograms of (a) ozonation of DCP (initial conc. ~50 mg/L) at TT 0 min, (b) at TT 1.5 min, and (c) at TT 4 min, as well as (d) IC chromatograms at TT 0.5 min, and (e) at TT 4 min; Table S2: Process parameters used for photocatalysis and photooxidation (reference [48]); Figure S2: HPLC–DAD chromatograms of photocatalytic treatment of (a) PHN (initial conc. ~50 mg/L) at TT 180 min, of (b) DCP (initial conc. ~50 mg/L) at TT 90 min, and of (c) PCP (initial conc. ~10 mg/L) at TT 180 min, as well as (d) IC chromatogram of photocatalytic treatment of DCP (initial conc. ~50 mg/L) at TT 180 min; Figure S3: HPLC–DAD chromatograms of photooxidation of (a) PHN (initial conc. ~50 mg/L) at TT 180 min, (b) DCP (initial conc. ~50 mg/L) at 180 min, and of (c) PCP (initial conc. ~10 mg/L) at TT 180 min, as well as (d) IC chromatogram of photooxidation of DCP (initial conc. ~50 mg/L) at TT 180 min; Figure S4: HPLC–DAD chromatograms of the sequential method of: (a) PHN after flash ozonation and (b) after 120 min of photocatalysis; (c) DCP (initial conc. ~20 mg/L) after flash ozonation and (d) after 120 min of photocatalysis; Table S3: Process parameters used for anodic electrooxidations; Figure S5: HPLC–DAD chromatograms of electrooxidation of PHN (initial conc. ~50 mg/L) (a) by BDD in Na_2SO_4 at TT 160 min, (b) by MMO in Na_2SO_4 at TT 160 min, (c) by BDD in NaCl at TT 35 min, (d) by BDD in NaCl at TT 60 min, and (e) by MMO in NaCl at TT 120 min; Figure S6: General procedure of ecotoxicity tests following OECD Guidelines No. 202.

Author Contributions: Conceptualization, A.K. and H.P.; methodology, A.K. and H.P.; software, A.K.; validation, A.K., A.Ž.G., U.L.Š., B.N.M. and H.P.; formal analysis, A.K.; investigation, A.K.; resources, A.Ž.G., U.L.Š., B.N.M. and H.P.; data curation, A.K.; writing—original draft preparation, A.K.; writing—review and editing, A.Ž.G., U.L.Š., B.N.M. and H.P.; visualization, A.K.; supervision, A.Ž.G., U.L.Š. and H.P.; project administration, H.P.; funding acquisition, A.Ž.G., U.L.Š., B.N.M. and H.P. All authors have read and agreed to the published version of the manuscript.

Funding: This research was funded by Slovenian Research Agency through research programmes P1-0153 (Research and development of analytical methods and procedures), P1-0134 (Chemistry for sustainable development), P2-0191 (Chemical engineering), and bilateral research project ARRS-BI-BA/19-20-040.

Institutional Review Board Statement: Not applicable.

Data Availability Statement: The data presented in this study are available on request from the corresponding author.

Acknowledgments: Gratitude is expressed to Boštjan Žener for the synthesis and supply of the TiO_2-based catalyst used in experiments.

Conflicts of Interest: The authors declare no conflict of interest. The funders had no role in the design of the study; in the collection, analyses, or interpretation of data; in the writing of the manuscript, or in the decision to publish the results.

Sample Availability: Not applicable.

Abbreviations

AOP (advanced oxidation process), BDD (boron-doped diamond), BQ (*p*-benzoquinone), CP (chlorophenol), CT (catechol), DCP (2,4-dichlorophenol), EO (electrooxidation), FM (fumaric acid), FO (formic acid), HQ (hydroquinone), MMO (mixed-metal oxide), MQ (ultrapure water), OX (oxalic acid), OZ (ozonation), PCP (pentachlorophenol), PHN (phenol), PO (photooxidation), SQ (sequential method, i.e., ozonation followed by photocatalysis), RD (relative descriptor calculated by Equation (5) in Section 2.4), TH (tetrachlorohydroquinone), TOC (total organic carbon), TP (transformation by-product), TT and $TT_{x\%}$ (treatment time and treatment time for reaching $x\%$ phenol removal).

References

1. New Indicator on Pesticides in European Waters (European Environment Agency, EEA News). Available online: https://www.eea.europa.eu/ims/pesticides-in-rivers-lakes-and (accessed on 1 January 2022).
2. European Environment Agency. *Contaminants in Europe's Seas: Moving towards a Clean, Non-Toxic Marine Environment*, 1st ed.; (Report No 25/2018); Reker, J., Ed.; Publications Office of the European Union: Copenhagen, Denmark, 2018; pp. 8, 49–55. [CrossRef]
3. Waste Water Treatment Improves in Europe but Large Differences Remain (European Environment Agency, EEA News). Available online: https://www.eea.europa.eu/highlights/waste-water-treatment-improves-in (accessed on 1 January 2022).
4. Garba, Z.N.; Zhou, W.; Lawan, I.; Xiao, W.; Zhang, M.; Wang, L.; Chen, L.; Yuan, Z. An overview of chlorophenols as contaminants and their removal from wastewater by adsorption: A review. *J. Environ. Manag.* **2019**, *241*, 59–75. [CrossRef] [PubMed]
5. Czaplicka, M. Sources and transformations of chlorophenols in the natural environment. *Sci. Total Environ.* **2004**, *322*, 21–39. [CrossRef] [PubMed]
6. Karci, A. Degradation of chlorophenols and alkylphenol ethoxylates, two representative textile chemicals, in water by advanced oxidation processes: The state of the art on transformation products and toxicity. *Chemosphere* **2014**, *99*, 1–18. [CrossRef]
7. Villegas, L.G.C.; Mashhadi, N.; Chen, M.; Mukherjee, D.; Taylor, K.E.; Biswas, N. A Short Review of Techniques for Phenol Removal from Wastewater. *Curr. Pollut. Rep.* **2016**, *2*, 157–167. [CrossRef]
8. European Environment Agency. *Industrial Waste Water Treatment—Pressures on Europes's Environment*, 1st ed.; No 23/2018; Granger, M., Ed.; Publications Office of the European Union: Copenhagen, Denmark, 2019; p. 39. [CrossRef]
9. Igbinosa, E.O.; Odjadjare, E.E.; Chigor, V.N.; Igbinosa, I.H.; Emoghene, A.O.; Ekhaise, F.O.; Igiehon, N.O.; Idemudia, O.G. Toxicological profile of chlorophenols and their derivatives in the environment: The public health perspective. *Sci. World J.* **2013**, *2013*, 460215. [CrossRef]
10. Espinoza-Montero, P.J.; Vasquez-Medrano, R.; Ibanez, J.G.; Frontana-Uribe, B.A. Efficient Anodic Degradation of Phenol Paired to Improved Cathodic Production of H_2O_2 at BDD Electrodes. *J. Electrochem. Soc.* **2013**, *160*, G3171–G3177. [CrossRef]
11. Ahmed, S.; Rasul, M.G.; Martens, W.N.; Brown, R.; Hashib, M.A. Heterogeneous photocatalytic degradation of phenols in wastewater: A review on current status and developments. *Desalination* **2010**, *261*, 3–18. [CrossRef]
12. Pera-Titus, M.; García-Molina, V.; Baños, M.A.; Giménez, J.; Esplugas, S. Degradation of chlorophenols by means of advanced oxidation processes: A general review. *Appl. Catal. B Environ.* **2004**, *47*, 219–256. [CrossRef]
13. Teodosiu, C.; Gilca, A.F.; Barjoveanu, G.; Fiore, S. Emerging pollutants removal through advanced drinking water treatment: A review on processes and environmental performances assessment. *J. Clean. Prod.* **2018**, *197*, 1210–1221. [CrossRef]
14. Wang, J.; Chen, H. Catalytic ozonation for water and wastewater treatment: Recent advances and perspective. *Sci. Total Environ.* **2020**, *704*, 135249. [CrossRef]
15. Mangindaan, D.; Lin, G.Y.; Kuo, C.J.; Chien, H.W. Biosynthesis of silver nanoparticles as catalyst by spent coffee ground/recycled poly(ethylene terephthalate) composites. *Food Bioprod. Process.* **2020**, *121*, 193–201. [CrossRef]
16. Suzuki, H.; Araki, S.; Yamamoto, H. Evaluation of advanced oxidation processes (AOP) using O_3, UV, and TiO_2 for the degradation of phenol in water. *J. Water Process Eng.* **2015**, *7*, 54–60. [CrossRef]
17. Amado-Piña, D.; Roa-Morales, G.; Barrera-Díaz, C.; Balderas-Hernandez, P.; Romero, R.; Martín del Campo, E.; Natividad, R. Synergic effect of ozonation and electrochemical methods on oxidation and toxicity reduction: Phenol degradation. *Fuel* **2017**, *198*, 82–90. [CrossRef]
18. Poznyak, T.; Vivero, J. Degradation of aqueous phenol and chlorinated phenols by ozone. *Ozone Sci. Eng.* **2005**, *27*, 447–458. [CrossRef]
19. Jaramillo-Sierra, B.; Mercado-Cabrera, A.; Peña-Eguiluz, R.; Hernández-Arias, A.N.; López-Callejas, R.; RodríGuez-Méndez, B.G.; Valencia-Alvarado, R. Assessing some advanced oxidation processes in the abatement of phenol aqueous solutions. *Environ. Prot. Eng.* **2019**, *45*, 23–38. [CrossRef]
20. Oputu, O.U.; Fatoki, O.S.; Opeolu, B.O.; Akharame, M.O. Degradation Pathway of Ozone Oxidation of Phenols and Chlorophenols as Followed by LC-MS-TOF. *Ozone Sci. Eng.* **2020**, *42*, 294–318. [CrossRef]
21. Van Aken, P.; Van den Broeck, R.; Degrève, J.; Dewil, R. The effect of ozonation on the toxicity and biodegradability of 2,4-dichlorophenol-containing wastewater. *Chem. Eng. J.* **2015**, *280*, 728–736. [CrossRef]

22. Hirvonen, A.; Trapido, M.; Hentunen, J.; Tarhanen, J. Formation of hydroxylated and dimeric intermediates during oxidation of chlorinated phenols in aqueous solution. *Chemosphere* **2000**, *41*, 1211–1218. [CrossRef]
23. Quispe, C.; Valdés, C.; Delgadillo, A.; Villaseñor, J.; Cheel, J.; Pecchi, G. Toxicity studies during the degradation of pentachlorophenol by ozonation in the presence of MnO_2/TiO_2. *J. Chil. Chem. Soc.* **2018**, *63*, 4090–4097. [CrossRef]
24. Hong, P.K.A.; Zeng, Y. Degradation of pentachlorophenol by ozonation and biodegradability of intermediates. *Water Res.* **2002**, *36*, 4243–4254. [CrossRef]
25. Anotai, J.; Wuttipong, R.; Visvanathan, C. Oxidation and detoxification of pentachlorophenol in aqueous phase by ozonation. *J. Environ. Manag.* **2007**, *85*, 345–349. [CrossRef] [PubMed]
26. Trapido, M.; Hirvonen, A.; Veressinina, Y.; Hentunen, J.; Munter, R. Ozonation, ozone/UV and UV/H_2O_2 degradation of chlorophenols. *Ozone Sci. Eng.* **1997**, *19*, 75–96. [CrossRef]
27. Beltrán, F.J.; Rivas, F.J.; Gimeno, O. Comparison between photocatalytic ozonation and other oxidation processes for the removal of phenols from water. *J. Chem. Technol. Biotechnol.* **2005**, *80*, 973–984. [CrossRef]
28. Turhan, K.; Uzman, S. Removal of phenol from water using ozone. *Desalination* **2008**, *229*, 257–263. [CrossRef]
29. Zangeneh, H.; Zinatizadeh, A.A.L.; Habibi, M.; Akia, M.; Hasnain Isa, M. Photocatalytic oxidation of organic dyes and pollutants in wastewater using different modified titanium dioxides: A comparative review. *J. Ind. Eng. Chem.* **2015**, *26*, 1–36. [CrossRef]
30. Scotti, R.; D'Arienzo, M.; Morazzoni, F.; Bellobono, I.R. Immobilization of hydrothermally produced TiO_2 with different phase composition for photocatalytic degradation of phenol. *Appl. Catal. B Environ.* **2009**, *88*, 323–330. [CrossRef]
31. Arana, J.; Dona-Rodriguez, J.M.; Portillo-Carrizo, D.; Fernández-Rodríguez, C.; Pérez-Pena, J.; Gonzalez Diaz, O.; Navio, J.A.; Macias, M. Applied Catalysis B: Environmental Photocatalytic degradation of phenolic compounds with new TiO_2 catalysts. *Appl. Catal. B Environ.* **2010**, *100*, 346–354. [CrossRef]
32. Nickheslat, A.; Amin, M.M.; Izanloo, H.; Fatehizadeh, A.; Mousavi, S.M. Phenol photocatalytic degradation by advanced oxidation process under ultraviolet radiation using titanium dioxide. *J. Environ. Public Health* **2013**, *2013*, 815310. [CrossRef]
33. Dougna, A.A.; Gombert, B.; Kodom, T.; Djaneye-Boundjou, G.; Boukari, S.O.B.; Leitner, N.K.V.; Bawa, L.M. Photocatalytic removal of phenol using titanium dioxide deposited on different substrates: Effect of inorganic oxidants. *J. Photochem. Photobiol. A Chem.* **2015**, *305*, 67–77. [CrossRef]
34. Sampaio, M.J.; Silva, C.G.; Silva, A.M.T.; Vilar, V.J.P.; Boaventura, R.A.R.; Faria, J.L. Photocatalytic activity of TiO_2-coated glass raschig rings on the degradation of phenolic derivatives under simulated solar light irradiation. *Chem. Eng. J.* **2013**, *224*, 32–38. [CrossRef]
35. Gar Alalm, M.; Samy, M.; Ookawara, S.; Ohno, T. Immobilization of $S-TiO_2$ on reusable aluminum plates by polysiloxane for photocatalytic degradation of 2,4-dichlorophenol in water. *J. Water Process Eng.* **2018**, *26*, 329–335. [CrossRef]
36. Yin, L.; Shen, Z.; Niu, J.; Chen, J.; Duan, Y. Degradation of Pentachlorophenol and 2,4-Dichlorophenol by Sequential Visible-Light Driven Photocatalysis and Laccase Catalysis. *Environ. Sci. Technol.* **2010**, *44*, 9117–9122. [CrossRef] [PubMed]
37. Gunlazuardi, J.; Lindu, W.A. Photocatalytic degradation of pentachlorophenol in aqueous solution employing immobilized TiO_2 supported on titanium metal. *J. Photochem. Photobiol. A Chem.* **2005**, *173*, 51–55. [CrossRef]
38. Ratnawati, R.; Enjarlis, E.; Husnil, Y.A.; Christwardana, M.; Slamet, S. Degradation of Phenol in Pharmaceutical Wastewater using TiO_2/Pumice and O_3/Active Carbon. *Bull. Chem. Reac. Eng. Cat.* **2020**, *15*, 146–154. [CrossRef]
39. Xie, Y.; Chen, Y.; Yang, J.; Liu, C.; Zhao, H.; Cao, H. Distinct synergetic effects in the ozone enhanced photocatalytic degradation of phenol and oxalic acid with Fe^{3+}/TiO_2 catalyst. *Chin. J. Chem. Eng.* **2018**, *26*, 1528–1535. [CrossRef]
40. Jiang, H.; Dang, C.; Liu, W.; Wang, T. Radical attack and mineralization mechanisms on electrochemical oxidation of p-substituted phenols at boron-doped diamond anodes. *Chemosphere* **2020**, *248*, 126033. [CrossRef]
41. Xing, X.; Ni, J.; Zhu, X.; Jiang, Y.; Xia, J. Maximization of current efficiency for organic pollutants oxidation at BDD, Ti/SnO_2-Sb/PbO_2, and Ti/SnO_2-Sb anodes. *Chemosphere* **2018**, *205*, 361–368. [CrossRef]
42. Weiss, E.; Groenen-Serrano, K.; Savall, A. A comparison of electrochemical degradation of phenol on boron doped diamond and lead dioxide anodes. *J. Appl. Electrochem.* **2008**, *38*, 329–337. [CrossRef]
43. Chatzisymeon, E.; Fierro, S.; Karafyllis, I.; Mantzavinos, D.; Kalogerakis, N.; Katsaounis, A. Anodic oxidation of phenol on Ti/IrO_2 electrode: Experimental studies. *Catal. Today* **2010**, *151*, 185–189. [CrossRef]
44. Barışçı, S.; Turkay, O.; Öztürk, H.; Şeker, M.G. Anodic Oxidation of Phenol by Mixed-Metal Oxide Electrodes: Identification of Transformation By-Products and Toxicity Assessment. *J. Electrochem. Soc.* **2017**, *164*, E129–E137. [CrossRef]
45. OECD. Test No. 202: Daphnia sp. Acute Immobilisation Test. In *OECD Guidelines for the Testing of Chemicals*; Section 2; OECD Publishing: Paris, France, 2004; pp. 1–12. [CrossRef]
46. Results of Eco-Toxicity Tests of Chemicals Conducted by Ministry of the Environment in Japan. Available online: https://www.env.go.jp/chemi/sesaku/02e.pdf (accessed on 13 October 2020).
47. Martins, J.; Teles, L.O.; Vasconcelos, V. Assays with *Daphnia magna* and *Danio rerio* as alert systems in aquatic toxicology. *Environ. Int.* **2007**, *33*, 414–425. [CrossRef] [PubMed]
48. Žener, B.; Matoh, L.; Carraro, G.; Miljević, B.; Korošec, R.C. Sulfur-, nitrogen- and platinum-doped titania thin films with high catalytic efficiency under visible-light illumination. *Beilstein J. Nanotechnol.* **2018**, *9*, 1629–1640; (in Supplementary Materials). [CrossRef] [PubMed]

Article

Green Downscaling of Solvent Extractive Determination Employing Coconut Oil as Natural Solvent with Smartphone Colorimetric Detection: Demonstrating the Concept via Cu(II) Assay Using 1,5-Diphenylcarbazide

Kullapon Kesonkan [1], Chonnipa Yeerum [1], Kanokwan Kiwfo [2], Kate Grudpan [2,3,*] and Monnapat Vongboot [1,*]

1 Department of Chemistry, Faculty of Science, King Mongkut's University of Technology Thonburi, Bangkok 10140, Thailand
2 Center of Excellence for Innovation in Analytical Science and Technology for Biodiversity-Based Economic and Society (I-ANALY-S-T_B.BES-CMU), Chiang Mai University, Chiang Mai 50200, Thailand
3 Department of Chemistry, Faculty of Sciences, Chiang Mai University, Chiang Mai 50200, Thailand
* Correspondence: kgrudpan@gmail.com (K.G.); sumalee.tan@kmutt.ac.th (M.V.)

Abstract: Coconut oil as a natural solvent is proposed for green downscaling solvent extractive determination. Determination of Cu(II) using 1,5-Diphenylcarbazide (DPC) was selected as a model for the investigation. Cu(II)-DPC complexes in aqueous solution were transferred into coconut oil phase. The change of the color due to Cu(II)-DPC complexes in coconut oil was followed by using a smartphone and image processing. A single standard concept was used for a series of Cu(II) standard solutions. A downscaling procedure using a 2 mL vial provided a calibration: color intensity = −142 [Cu(II)] + 222, (R^2 = 0.98), 10% RSD. Using a well plate, a calibration was: color intensity = 61 [Cu(II)] + 68 (R^2 = 0.91), 15% RSD. Both were for the range of 0–1 ppm Cu(II). Application of the developed procedure to water samples was demonstrated. The developed procedures provided a new approach of green chemical analysis.

Keywords: coconut oil; natural solvent; downscaling extraction; colorimetric solvent extraction determination; copper (II); diphenylcarbazide; diphenylcarbazone; green chemical analysis

1. Introduction

Solvent extraction spectrophotometric determination has long been deployed for various analytes. Spectrophotometric determination of Cd(II) using dithizone was performed by extracting the complex into chloroform [1]. Using 1,5-diphenylcabazide (DPC), Cu(II) could be determined spectrophotometrically after extracting into benzene [2]. Fe(II) and Fe(III) could be extracted into 1-pentanol using anthranilic acid for spectrophotometric determination [3]. Complexing with 4-(2-pyridylazo)-resocinol, Pb(II) was spectrophotometrically determined in isobutyl methyl ketone [4]. Extraction of Zn(II) with ethylthioacetoactate into ethyl acetate allowed Zn(II) determination [5]. For Sb(III) and Sb(V), the determination was made by using mandelic acid and malachite green with extraction into chlorobenzene solvent [6]. Spectrophotometric solvent extraction determination of Ni(II) with di-2-pyridyl ketone benzoylhydrazone reagent and chloroform was reported [7].

Due to concerns about toxicity of the organic solvents used for the extraction, replacement of once popular solvents was gradually implemented. For example, benzene was at one point replaced by toluene, then by carbon tetrachloride or chloroform. And gradually use of such solvents for these purposes has been declining [8,9]. One of the green analytical chemistry approaches engages with research for greener alternatives for toxic organic solvents to minimize unfavorable impacts [10–13].

Recently, vegetable oils have been viewed favorably, in applications for food [14–16], health [17–19], and cosmetic [20–22] purposes, due to their properties in biodegradability, and excellent environmental aspects such as low ecotoxicity and low toxicity toward humans, apart from their global availability [23]. Some vegetable oils, namely, rapeseed oil, peanut oil, and coconut oil, were reported for sample preparation of high performance liquid chromatography (HPLC) determination of polycyclic aromatic hydrocarbons (PAHs) by using the oils to extract PAHs from polluted quagmires [24]. Coconut oil offers some characteristics suitable to an organic solvent for extracting some species from the aqueous to the organic phase, compared to the conventional solvents, such as its dielectricity (dielectric constants of 2.2 for coconut oil, compared with 2.3 and 2.4 for benzene and toluene, respectively, see Table S1). Coconut oil contains higher unsaturated fatty acids among vegetable oils, especially lauric acid (C12:0) [23], apart from other fatty acids.

Recently, use of smartphones with image processing has been incorporated in cost-effective colorimetric determination as a part of modern green chemical analysis [25,26].

In this work, considering the high local abundance of coconut oil not only in Southeast Asia, but also in various tropical areas, and although it has been applied for solvent extraction in a sample preparation step for HPLC analysis of PAHs as above mentioned [24], our attempts were to introduce coconut oil as a natural alternative solvent to enable a new greener approach for solvent extraction colorimetric determination. Determination of Cu(II) using DPC was selected as a model for demonstrating this approach. Investigation of downscaling procedures employing a smartphone colorimetric determination was made for solvent extractive determination employing natural coconut oil solvent. This approach would lead to greener chemical analysis that provides benefits to the environment.

2. Results and Discussion

2.1. Reinvestigation for Cu(II) Extraction toward the Reaction between Cu(II) and DPC

Solvent extraction determination of Cu(II) employing DPC color reagent and benzene as solvent was reported [2]. Cu(II) reacts with DPC to form complexes, possibly with the oxidized form of DPC, diphenylcarbazone, in an alkaline aqueous medium [27,28]. The complexes would be then extracted into an organic phase. The method was selected as a model for this work. Following the conditions reported in [2] but with modifications, an investigation was made. Figure 1 illustrates the spectrum of the benzene extract. The absorption maximum exhibited at 549 nm, which agreed with the previous findings [2]. Other organic solvents, namely, toluene, DCM, and DCE were also investigated for extraction of Cu(II)-DPC complexes under the same concentrations and conditions. Some properties of the solvents are represented in Table S1. The absorption maxima of the extracts using the investigated solvents exhibited practically the same wavelength (545 ± 4 nm).

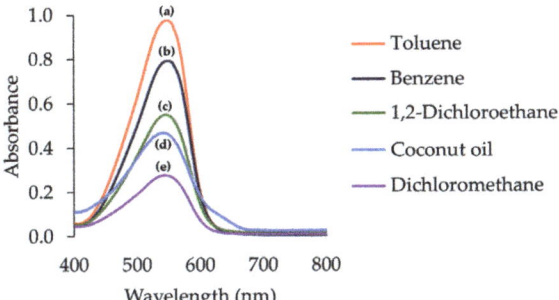

Figure 1. Visible spectra of Cu(II)-DPC complexes in solvents. Spectra: (a) toluene, (b) benzene, (c) 1,2-dichloroethane, (d) coconut oil, and (e) dichloromethane; for coconut oil, was normalized using OriginPro 2023 program (adjacent-averaging, points of window: 30).

Coconut oil was investigated as an alternative solvent for this purpose. Using the same conditions as previously, the spectrum (in Figure 1) exhibited an absorption maximum at practically the same wavelength as with the above conventional organic solvents. With gas chromatographic information, it was found that all the chromatograms were identical (see Figure S1).

The absorbance (0.518) of Cu(II)-DPC extract in coconut oil was observed to be lower than that in toluene (0.977) and that in benzene (0.794) but it was in the same order as with in DCE (0.551) which was higher than with DCM (0.279).

Considering the toxicity of the solvents, coconut oil is considered to be nontoxic (hazard rating = 0), while the others are in the rating of level 2 apart from DCE being level 3. In the development of the solvent extraction determination techniques, benzene, toluene, and carbon tetrachloride were popularly used in the early stages. Coconut oil behaves differently from the other solvents in term of viscosity (see the values in Table S1). With this viscosity behavior, after the extraction, coconut oil may entail difficulty in transferring its aliquot for absorbance measurement by a conventional spectrophotometer. To overcome the difficulty, solvent extraction colorimetric determination of Cu(II) with DPC using coconut oil as solvent could make use of a smartphone camera, which is nowadays employed for colorimetry. Such a procedure with downscaling would also lead to green chemical analysis.

2.2. Downscaling Using a Small Vial

The design of downscaling the extraction was aimed to microliter scale operation. A small vial (2 mL capacity) that is usually used for chromatographic purposes, was used as an extraction container. The volumes of both aqueous and organic phases were designed for downscaling and were equally of 600 μL each. For solvent extraction, DCM has been used more favorably than toluene, benzene, and DCE; however, from the results in Table S1, DCE provided similar results to coconut oil. For this study, DCE and coconut oil were further investigated, and a single standard approach was incorporated. The coloring reagent (DPC) was added into the aqueous solution containing Cu(II) in buffer, instead of having DPC in organic phase as earlier. By doing this, Cu(II) should react with excess DPC to form Cu(II)-DPC complexes, which would be more efficiently extracted into the organic layer. The procedure was modified from earlier experiments. A standard Cu(II) solution (2.4 ppm) was used as a single standard solution for various Cu(II) standard concentrations in aqueous phase. For the 1 ppm Cu(II) final concentration of aqueous phase (600 μL), 250 μL of the standard Cu(II) solution (2.4 ppm) was pipetted into a small vial, followed by 150 μL DI water, then 100 μL Na_2HPO_4 buffer (1 M) and 100 μL DPC (10 mM), resulting in the final concentrations of 0.2 M and 1.67 mM for buffer and DPC, respectively. The mixture was mixed well before equilibrating with 600 μL DCE for 1 min by hand. A series of 0.2, 0.4, 0.6, 0.8, and 1 ppm Cu(II) final concentrations were prepared similarly by varying the volumes (50, 100, 150, 200, and 250 μL) of the single standard Cu(II) solution (2.4 ppm) and DI water volumes of 350, 300, 250, 200, and 150 μL, respectively. For blank, only 400 μL DI water without the standard Cu(II) solution was pipetted. After equilibrating, each vial was left to stand for phase separation. It was observed the clear phase septation could be observed nearly immediately after standing. Agitation was not observed in the experiments performed at the temperature of 35 ± 5 °C. The vials were photographed under the conditions described in Section 3.3 as illustrated in Figure S2a. After taking the photograph, each organic layer was transferred for absorbance measurement. The photograph was further processed for RGB mode [25], by ImageJ Software (version 13.0.6, National Institutes of Health, Bethesda, MD, USA) with RGB profile function. G intensity is observed to be the highest among Red (R), Green (G), and Blue (B) values. The calibration graphs can be obtained by a plot of G intensity versus Cu(II) concentrations while for conventional spectrophotometry, a calibration graph was from a plot of absorbance versus Cu(II) concentrations.

Similarly, coconut oil was used for the downscaling operation using a small vial for extraction under the same conditions as the procedure using DCE. Using coconut oil provided better sensitivity (considering from the slopes of calibrations), when using a smartphone detection, as shown in Table 1. Although coconut oil provided the better observed sensitivity, due to its viscosity, it is not practical to transfer the organic phase from the extraction vial into a cuvette for absorbance measurement using a spectrophotometer. The cumbersomeness was in addition to the fact that phase separation of the aqueous-coconut oil phases needed a longer period than that of aqueous-DCE phases. In Table 1, the results obtained by taking photographs are presented.

Table 1. Calibration graphs obtained from extraction of Cu(II) using DPC in a small vial.

Solvent	Calibration Equation	
	Conventional Spectrophotometry	Image Processing
1,2-Dichloroethane	Absorbance = 0.58 [Cu(II)] + 0.08, (R^2 = 0.98)	G intensity = −134 [Cu(II)] + 234, (R^2 = 0.97)
Coconut oil	—	G intensity = −142 [Cu(II)] + 222, (R^2 = 0.98)

As detecting using a smartphone and image processing offers a conventional means for investigating, factors affecting the downscaling extraction determination of Cu(II) using DPC were further investigated.

The effect of DPC concentration was investigated by using the previous conditions for extraction of 1 ppm Cu(II) (final concentration of 600 μL), with varying final DPC concentrations in 600 μL aqueous solutions of 0.17, 0.83, 1.67, and 2.50 mM, providing mole ratios of DPC:Cu(II) of 11, 53, 106, and 160. It was observed that the G intensity of the extract became constant after the mole ratio (DPC:Cu(II)) of 53. The DPC concentration of 1.7 mM was used for further experiments. It was indicated that the ratio of DPC:Cu(II) was already in excess to extract Cu(II) into the DCE phase with maximum extraction efficiency. This was indicated by the constant G intensity.

Under the above conditions (with 1.7 mM DPC in 600 μL aqueous 1 ppm Cu(II) solution), apart from 1 min, shaking times of 0.5, 2, and 3 min were studied. It was observed that shaking times of 1 min or more resulted in the same G intensity. So, 1 min hand shaking time should be suitable for the extraction.

The volumes of DCE solvent were studied for 300 and 200 μL in addition to 600 μL for the extraction using the previous conditions. Using DCE of 600, 300, and 200 μL provided calibration equations (plot of G intensity versus Cu(II) concentration): G intensity = −134 [Cu(II)] + 234 (R^2 = 0.97), G intensity = −236 [Cu(II)] + 219 (R^2 = 0.99), and G intensity = −322 [Cu(II)] + 219 (R^2 = 0.99), respectively. As expected, the lower volumes of DCE used would result in higher sensitivity (slopes). Although the sensitivity may be gained, however, there may be a problem in obtaining clear phase separation, as well as problems in transferring the organic phase for absorbance measurement using a conventional spectrophotometer.

It should be noted that using a smartphone camera for detection, 12 extraction vials can be handled for one run. It could be designed for 4 extraction vials for standards together with 4 samples in duplicate to be operated simultaneously for the extraction process. The 12 extraction vials after phase separation could be one-shot-photographed under the light-control-set up (see description in the Materials and Methods section). Figure S2 represents a one-shot-photograph of a set of 12 extraction vials. With this handling, a sample throughput would be gained compared to the conventional spectrophotometric procedure.

2.3. Downscaling Using a Well Plate

The downscaling extraction was further designed using a well plate aiming for lower volume (microliter scale operation) with higher sample throughput. In a 96-well plate, each well (capacity of 300 μL) would serve as an extraction container. An extraction was performed by having an equal volume of 120 μL each of aqueous and organic phases. Some preliminary investigations on light control for photography were made (see Section S3.2.1 in Supplementary Materials). For extraction, intensity (ΔG) of the extract could be then evaluated by ΔG intensity = G intensity$_{empty\ well}$ − G intensity$_{extract\ well}$.

In such a well, 120 μL aqueous solution containing Cu(II) standard (0.6 ppm) in Na_2HPO_4 buffer (0.2 M) and DPC (1.7 mM) was prepared by using an auto-pipet with a single standard solution approach similar to the above manipulation but with 5 folds lesser amounts. With using the auto-pipet, the solution would be mixed well. The aqueous solution was then equilibrated with 120 µL organic solvent. Using a multichannel auto-pipet, the equilibrating could be done for 8 extractions (8 wells) simultaneously. This would lead to enhanced sample throughput.

DCE as well as coconut oil were studied for extraction. As DCE has higher density than water, it would be expected that after phase separation, the DCE phase would be at the bottom of the well plate. Photographing of the experiment was made at the bottom. For coconut oil, it would be expected that the organic phase would be at the upper layer, due to its lower density. For this, the photograph was taken at the top of the well plate. Figure 2 represents the intensity signal (G intensity) profiles obtained using the above-described conditions. The profiles were at 1 and 5 min after the equilibrating. The profiles of both solvent extractions, in each well, exhibited two parts, namely at the middle and at the edge of the well, although it could be observed more on DCE than that of coconut oil. For coconut oil, a plateau shaped signal profile was pronounced more than the observation on the DCE's. For coconut oil, a more reproducible profile pattern was observed. The profiles at 1 and 5 min were the same. Less reproducible profiles among the wells were seen for DCE and the signal profiles at 1 min showed a difference from 5 min. During the experiments, for DCE, it was observed that the DCE extract became gradually more viscous as a function of time. It could be that a component of the well plate (polystyrene) may dissolve (see Figure S4). The DCE was not used further.

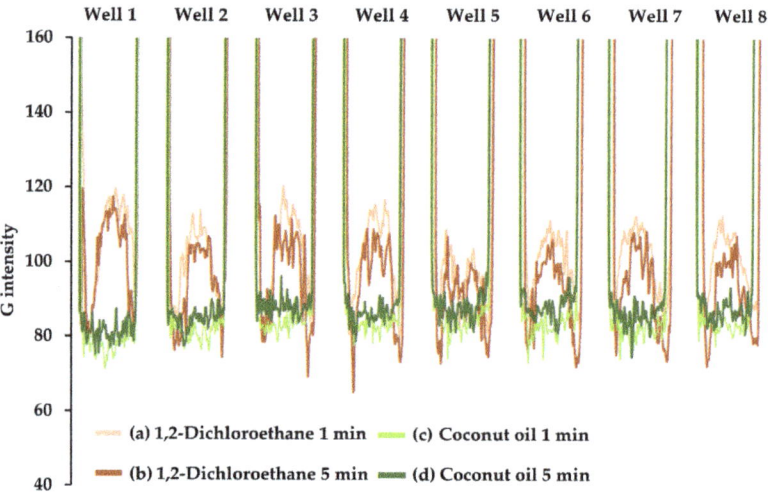

Figure 2. The intensity signal (G intensity) profiles of organic extract (from aqueous Cu(II)-DPC) using DCE at 1 min (a) and at 5 min (b) and coconut oil at 1 min (c) and at 5 min (d).

Further study was made for standing time after the extraction. Under the above conditions, a standing period between 1–5 min was investigated. It was found that the same color intensity value of the organic (coconut oil) phase was obtained after 1 min. A standing time of 1 min was selected for further experiments.

For equilibrating aqueous-organic phases, this was done via using an auto-pipet by aspirating and dispensing from and to the well. The number of aspirating and dispensing cycles was studied. It was found that 15 or more cycles resulted in constant color intensity value of the organic extract. So, an operation with 15 cycles was selected for equilibrating the aqueous-organic phases.

Fixing the aqueous solution (Cu(II) and DPC in the buffer as above condition) of 120 µL, a series of the aqueous solutions containing 0, 0.2, 0.4, 0.6, and 0.8 ppm Cu(II) was extracted with 120 µL coconut oil. Experiments were repeated with the coconut oil of volumes of 60 and 40 µL. Three calibration graphs were obtained for the extractions with 120, 60, and 40 µL coconut oil. It was found that the calibration graph due to 40 µL coconut oil had the highest slope followed by the calibration graph of 60 µL coconut oil. The calibration graph with 120 µL coconut oil showed the lowest slope value but higher precision. This could be due to the coconut oil in the extraction playing a role in the organic phase adhering to the wells' sidewalls and affecting the photography.

2.4. Analytical Performance

2.4.1. Analytical Characteristics of the Downscaling Extraction Using Vial

The downscaling for extraction determination of Cu(II) using DPC involve handling solutions of less than 1 mL. Extraction performance and determination of Cu(II) were comparable to the conventional milliliter level extraction. A linear calibration (0–1 ppm Cu(II)) was absorbance = 0.90 [Cu(II)] + 0.12 (R^2 = 0.99) when using a spectrophotometer, while the smartphone provided: G intensity = −133 [Cu(II)] + 173 (R^2 = 0.99). The relative standard deviations (RSD) of both detections were less than 10% (0.4 ppm Cu(II)). Limits of detection (LODs) (3σ [29]) were estimated to be 0.02 and 0.1 ppm, while limits of quantitation (LOQs) (10σ [29]) being 0.1 and 0.4 ppm for spectrophotometric and smartphone detection, respectively. A set of 12 extraction vials could be handled in one run for extraction and detection by smartphone, see Figure S2. This made the procedure with smartphone detection more convenient than measuring absorbance by a spectrophotometer as no organic extract has to be transferred from the extraction vial to a cuvette. The 12 extraction vials of the set may comprise four standards for calibration and eight vials for four samples in duplicate, or various possibilities of other arrangements could be made. For the set of 12 extraction vials, a one run experiment may take 20 min, including building calibration and evaluation of the results.

2.4.2. Analytical Characteristics of the Downscaling Extraction Using Well Plate

With a well plate, further downscaling could be operated with even smaller volumes (120 µL). The number of extraction units can be increased; 32 wells out of 96 wells would provide reasonably good results due to different light distribution occurring to some wells, see Figure S3. The 32 wells could be arranged for four standards in duplicate for duplicate calibration, in order to check the photographic conditions, and then 24 wells could hold 12 samples in duplicate (as illustrated in Figure S3b). Arrangement in other patterns may also be designed. After extraction, the plate was photographed in one shot. The operation of 32 extractions on one plate took 20 min, including image processing and evaluation. A calibration (0–1 ppm Cu(II)) was ΔG intensity = 61 [Cu(II)] + 68 (R^2 = 0.91) with RSD of less than 15% (0.4 ppm Cu(II)). LOD (3σ) and LOQ (10σ) were evaluated to be 0.1 and 0.2 ppm, respectively. The operation using the well plate provided higher sample throughput than the throughput by the procedure using vials.

2.5. Applications

The proposed procedures have been applied to six local tap and drinking water samples. It was found that Cu(II) were less than LODs. All the samples were spiked with 0.4 ppm Cu(II). The results are presented in Table 2. The extraction procedure using a spectrometer was treated as a reference method. The results by all the procedures agreed each other and with the reference method.

Table 2. Assay of Cu(II) in tap and drinking water samples spiked with 0.4 ppm Cu(II).

Samples	Vial Procedure				Well Plate Procedure	
	Visible Spectrophotometer		Smartphone Camera		Smartphone Camera	
	Cu Found (ppm)	%Recovery	Cu Found (ppm)	%Recovery	Cu Found (ppm)	%Recovery
1	0.41 ± 0.02	103 ± 3.70	0.41 ± 0.02	103 ± 4.88	0.34 ± 0.01	85 ± 1.49
2	0.34 ± 0.00	85 ± 0.00	0.44 ± 0.00	110 ± 0.00	0.40 ± 0.00	103 ± 1.05
3	0.40 ± 0.01	100 ± 1.27	0.45 ± 0.01	113 ± 1.12	0.40 ± 0.03	100 ± 7.41
4	0.37 ± 0.01	93 ± 1.37	0.40 ± 0.01	100 ± 2.50	0.45 ± 0.00	113 ± 0.97
5	0.37 ± 0.01	93 ± 1.37	0.43 ± 0.02	108 ± 3.53	0.40 ± 0.01	100 ± 3.23
6	0.41 ± 0.01	103 ± 2.44	0.44 ± 0.01	110 ± 1.15	0.34 ± 0.01	85 ± 3.18

3. Materials and Methods

3.1. Chemicals and Reagents

All chemicals and reagents were analytical grade. Deionized (DI) water was used throughout. All glassware was soaked overnight in 10% w/v nitric acid solution (QRëC, Auckland, New Zealand), followed by rinsing with DI water prior to use.

A pure virgin coconut oil was obtained from Theppadungporn Coconut Co., Ltd., Bangkok, Thailand, and used without further treatment. The other chemicals included benzene (99.7%, Merck, Darmstadt, Germany), toluene (99.5%, VWR Chemicals, Solon, OH, USA), dichloromethane (DCM) (99.8%, Fisher Scientific, Waltham, MA, USA) and 1,2-dichloroethane (DCE) (99%, Loba Chemie, Mumbai, India).

As coconut oil posts freezing temperature of 25 °C [30], all experiments were carried out at a temperature of 35 ± 5 °C so that a clear organic solvent would be established.

A stock solution of 1000 ppm Cu(II) was prepared from copper sulfate ($CuSO_4 \cdot 5H_2O$) (Kemaus, New South Wales, Australia). Working standard solutions of Cu(II) were prepared daily by appropriate dilutions of the stock solution with DI water. A solution of 10 mM DPC (Loba Chemie, Mumbai, India) was prepared by dissolving of 60 mg of the solid DPC in 12 mL of acetone (99.5%, Loba Chemie, Mumbai, India), followed by making up the volume with DI water in a 25 mL volumetric flask. A Na_2HPO_4 buffer solution (pH 9) was prepared by dissolving 8.90 g $Na_2HPO_4 \cdot 2H_2O$ (Kemaus, New South Wales, Australia) in hot DI water; after complete dissolution, the volume was made up with DI water in a 50 mL volumetric flask.

3.2. Reinvestigation of Cu(II)-DPC Extracted into Organic Solvents

In a 60 mL polypropylene bottle, a mixture (5 mL) containing Cu(II) standard (0.6 ppm) and Na_2HPO_4 buffer (0.4 M) was equilibrated with DPC in benzene (0.4 mM, 7 mL). After vortexing for 1 min, it was left to stand for 10 min. An aliquot of benzene layer was transferred into a cuvette for recording visible absorption spectra, having a reagent blank as reference. The other organic solvents, namely, toluene, DCM, and DCE were also investigated for extraction of Cu(II)-DPC complexes followed the same concentration and conditions.

3.3. Downscaling Solvent Extraction Determination of Cu(II) Using DPC

3.3.1. Using a Small Vial

Using an extraction vial (2 mL, clear color, KIMA, Bangkok, Thailand), 600 µL of aqueous solution containing Cu(II) with DPC in Na_2HPO_4 buffer was equilibrated with 600 µL of organic solvent. It was left to stand for phase separation before colorimetric measurement of the organic layer. The colorimetric measurements were made by using a smartphone camera (Galaxy Note 10 plus, Samsung, Seoul, Republic of Korea) and a spectrophotometer (Lambda35, PerkinElmer, Waltham, MA, USA).

For smartphone detection, the digital image of the extract vials was taken using smartphone camera with manual mode (ISO 200, Shutter Speed 3000, Zoom 1x and F1.5), under the light-controlled box see Figure 3a. The image was processed for the RGB profile using ImageJ Software (version 13.0.6, National Institutes of Health, Bethesda, MD, USA) with RGB profile function, by creating region of interest (ROI) with line across organic layer of every vial under investigated. Average G value was obtained from the ROI of each organic extract in the vial.

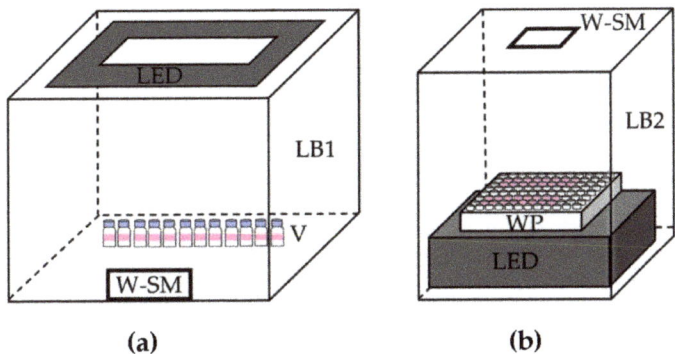

Figure 3. Illustration of the set ups for: (**a**) the extraction procedure using a small vial; (**b**) the extraction procedure using a well plate. LB1 = light control box (UDIOBOX UDIO BIZ, 40 × 40 × 40 cm, Bangkok, Thailand); V = small vial; LED = light emitting diode (LED) light source; W-SM = window for smartphone, LB2 = light control box (15 × 11 × 25 cm); LED = LED light source (LD-160, 14 × 5.7 × 9.5 cm, Lightdow, Shenzhen, China); WP = well plate; W-SM = window for smartphone. Note: the dimensions are not to scale.

For spectrophotometric detection, after extraction, the 500 µL of DCE extract was transferred into a cuvette for absorbance (543 nm) measurement by a spectrophotometer.

3.3.2. Using a Well Plate

Using a well plate (Nunc™ MicroWell™ 96-Well, Nunclon Delta-Treated, Flat-Bottom Microplate (167008), Thermo Scientific, Waltham, MA, USA), 120 µL of aqueous solution containing Cu(II) with DPC in Na_2HPO_4 buffer was equilibrated with 120 µL of organic solvent using multichannel auto-pipet (8 Channel pipettor (AP-8-200), Axygen, CA, USA). It was then photographed for colorimetric measurement of the organic layer. The colorimetric measurement was made by using a smartphone camera with manual mode as mentioned above, under the light-controlled box (see Figure 3b). The image was processed for the RGB profile using ImageJ Software (version 13.0.6, National Institutes of Health, Bethesda, MD, USA) with the function of RGB profile by having ROI with line across the well plate. Average G value could be obtained from the ROI of each organic extract in each well.

4. Conclusions

Coconut oil as a natural organic solvent is proposed as an alternative for solvent extraction colorimetric determination. Proof of the concept was demonstrated by the downscaling procedure for the determination of Cu(II) with DPC using a smartphone detector. Development of the downscaling procedures was based on using a small 2 mL vial and a 96 well microplate. Application to water samples was demonstrated. This new approach of green chemical analysis should be explored further for routine analysis in the real world. Various benefits would be obtained including cost-effective analysis, simple operation, on-site analysis, and environmentally friendly processes that are also less hazardous to the operator. Evaluating the developed procedures by the Green Analytical Procedure Index (GAPI) approach [31] (see Figure S5), extraction with coconut oil using well plates was the greenest, while the downscaling procedure using vials was greener compared to the traditional procedure [2]. It is very much of interest to investigate using other vegetable oils for green solvent extraction colorimetric determination, as a new approach for cost-effective green chemical analysis. This approach also supports the United Nations Sustainable Development Goals (SDGs).

Supplementary Materials: The following supporting information can be downloaded at: https://www.mdpi.com/article/10.3390/molecules27238622/s1, Table S1: Some properties of the solvents and spectrophotometric information of Cu(II)-DPC complexes in the solvents; Figure S1: Chromatographic chromatograms of different coconut oils: (a) CHAOKOH brand lot1 (manufacture date February 2020), (b) CHAOKOH brand lot2 (manufacture date August 2020), (c) COCOLOVE brand, and (d) KING ISLAND brand; Figure S2: Extraction of Cu(II) via vial procedures using (a) DCE solvent and (b) coconut oil solvent; Figure S3: The well plate used for the extraction, (a) the empty well plate; (b) the well plate after extraction indicating with labels of standards and samples; Figure S4: DCE extract observed to be more viscous as function of time (extraction of aqueous Cu(II)-DPC into DCE in a polystyrene well plate; Figure S5: Evaluating the developed procedures by GAPI: (a) traditional [2]; (b) the developed procedure using vial with DCE; (c) the developed procedure using well plate with coconut oil [32–42].

Author Contributions: Conceptualization, K.G. and M.V.; methodology, K.K. (Kullapon Kesonkan), K.G. and M.V.; validation, K.K. (Kullapon Kesonkan); formal analysis, K.K. (Kullapon Kesonkan), C.Y., K.K. (Kanokwan Kiwfo), K.G. and M.V.; investigation, K.K. (Kullapon Kesonkan); resources, K.G. and M.V.; data curation, K.K. (Kullapon Kesonkan), C.Y., K.K. (Kanokwan Kiwfo), K.G. and M.V.; writing—original draft preparation, K.K. (Kullapon Kesonkan), C.Y., K.K. (Kanokwan Kiwfo), K.G. and M.V.; writing—review and editing, K.K. (Kullapon Kesonkan), C.Y., K.K. (Kanokwan Kiwfo), K.G. and M.V.; visualization, K.K. (Kullapon Kesonkan), C.Y., K.K. (Kanokwan Kiwfo), K.G. and M.V.; supervision, K.G. and M.V.; project administration, K.K. (Kullapon Kesonkan), K.G. and M.V.; funding acquisition, K.G. and M.V. All authors have read and agreed to the published version of the manuscript.

Funding: This work was supported by the Science Achievement Scholarship of Thailand (SAST) (102/2560), the Thailand Research Fund (TRF) Distinguished Research Professor Award Grant (DPG6080002 to K.G.), and Chiang Mai University (through Center of Excellence for Innovation in Analytical Science and Technology for Biodiversity-based Economic and Society (I-ANALY-S-T_B.BES-CMU)).

Institutional Review Board Statement: Not applicable.

Informed Consent Statement: Not applicable.

Data Availability Statement: All the data are reported in this manuscript and supplementary materials.

Acknowledgments: Department of Chemistry, Faculty of Science, King Mongkut's University of Technology Thonburi, Science Achievement Scholarship of Thailand (SAST) K.K. (Kullapon Kesonkan), TRF Distinguished Research Professor Award Grant (DPG6080002 to K.G., and Chiang Mai University are acknowledged. We are grateful to Piyanat Issarangkura Na Ayutthaya for useful discussions.

Conflicts of Interest: The authors declare no conflict of interest.

References

1. Saltzman, B.E. Colorimetric microdetermination of cadmium with dithizone. *Anal. Chem.* **1953**, *25*, 493–496. [CrossRef]
2. Stoner, R.E.; Dasler, W. Spectrophotometric determination of copper following extraction with 1,5-diphenylcarbohydrazide in benzene. *Anal. Chem.* **1960**, *32*, 1207–1208. [CrossRef]
3. Dinsel, D.L.; Sweet, T.R. Separation and determination of iron(II) and iron(III) with anthranilic acid using solvent extraction and spectrophotometry. *Anal. Chem.* **1963**, *35*, 2077–2081. [CrossRef]
4. Dagnall, R.M.; West, T.S.; Young, P. Determination of lead with 4-(2-pyridylazo)-resorcinol—I: Spectrophotometry and solvent extraction. *Talanta* **1965**, *12*, 583–588. [CrossRef]
5. Chennuri, S.L.; Haldar, B.C. Solvent extraction of zinc(II) with ethylthioacetoacetate into ethyl acetate. *J. Radioanal. Nucl. Chem.* **1984**, *84*, 197–200. [CrossRef]
6. Sato, S. Differential determination of antimony(III) and antimony(V) by solvent extraction-spectrophotometry with mandelic acid and malachite green, based on the difference in reaction rates. *Talanta* **1985**, *32*, 341–344. [CrossRef]
7. Terra, L.H.S.Á.; da Cunha Areias, M.C.; Gaubeur, I.; Encarnación, M.; Suárez-iha, V. Solvent extraction-spectrophotometric determination of nickel(II) in natural waters using di-2-pyridyl ketone benzoylhydrazone. *Spectrosc. Lett.* **1999**, *32*, 257–271. [CrossRef]

8. Byrne, F.P.; Jin, S.; Paggiola, G.; Petchey, T.H.M.; Clark, J.H.; Farmer, T.J.; Hunt, A.J.; McElroy, C.R.; Sherwood, J. Tools and techniques for solvent selection: Green solvent selection guides. *Sustain. Chem. Process.* 2016, *4*, 1–24. [CrossRef]
9. Joshi, D.R.; Adhikari, N. An overview on common organic solvents and their toxicity. *J. Pharm. Res. Int.* 2019, *28*, 1–18. [CrossRef]
10. Yavir, K.; Konieczna, K.; Marcinkowski, Ł.; Kloskowski, A. Ionic liquids in the microextraction techniques: The influence of ILs structure and properties. *Trends Anal. Chem.* 2020, *130*, 115994. [CrossRef]
11. Płotka-Wasylka, J.; Mohamed, H.M.; Kurowska-Susdorf, A.; Dewani, R.; Fares, M.Y.; Andruch, V. Green analytical chemistry as an integral part of sustainable education development. *Curr. Opin. Green Sustain. Chem.* 2021, *31*, 100508. [CrossRef]
12. Pacheco-Fernández, I.; Pino, V. Green solvents in analytical chemistry. *Curr. Opin. Green Sustain. Chem.* 2019, *18*, 42–50. [CrossRef]
13. Armenta, S.; Garrigues, S.; de la Guardia, M. The role of green techniques in green analytical chemistry. *Trends Anal. Chem.* 2015, *71*, 2–8. [CrossRef]
14. Abdul Halim, H.S.a.; Selamat, J.; Mirhosseini, S.H.; Hussain, N. Sensory preference and bloom stability of chocolate containing cocoa butter substitute from coconut oil. *J. Saudi Soc. Agric. Sci.* 2019, *18*, 443–448. [CrossRef]
15. Rizzo, G.; Masic, U.; Harrold, J.A.; Norton, J.E.; Halford, J.C.G. Coconut and sunflower oil ratios in ice cream influence subsequent food selection and intake. *Physiol. Behav.* 2016, *164*, 40–46. [CrossRef] [PubMed]
16. Li, Y.; Fabiano-Tixier, A.S.; Ginies, C.; Chemat, F. Direct green extraction of volatile aroma compounds using vegetable oils as solvents: Theoretical and experimental solubility study. *LWT-Food Sci. Technol.* 2014, *59*, 724–731. [CrossRef]
17. Ramesh, S.V.; Krishnan, V.; Praveen, S.; Hebbar, K.B. Dietary prospects of coconut oil for the prevention and treatment of Alzheimer's disease (ad): A review of recent evidences. *Trends Food Sci. Technol.* 2021, *112*, 201–211. [CrossRef]
18. Woolley, J.; Gibbons, T.; Patel, K.; Sacco, R. The effect of oil pulling with coconut oil to improve dental hygiene and oral health: A systematic review. *Heliyon* 2020, *6*, e04789. [CrossRef]
19. Sandupama, P.; Munasinghe, D.; Jayasinghe, M. Coconut oil as a therapeutic treatment for Alzheimer's disease: A review. *J. Future Foods* 2022, *2*, 41–52. [CrossRef]
20. Mahbub, M.; Octaviani, I.D.; Astuti, I.Y.; Sisunandar, S.; Dhiani, B.A. Oil from *kopyor* coconut (*Cocos nucifera* var. *Kopyor*) for cosmetic application. *Ind. Crops Prod.* 2022, *186*, 115221. [CrossRef]
21. Pham, T.L.-B.; Thi, T.T.; Nguyen, H.T.-T.; Lao, T.D.; Binh, N.T.; Nguyen, Q.D. Anti-aging effects of a serum based on coconut oil combined with deer antler stem cell extract on a mouse model of skin aging. *Cells* 2022, *11*, 597. [CrossRef] [PubMed]
22. Kamairudin, N.; Gani, S.S.A.; Masoumi, H.R.F.; Hashim, P. Optimization of natural lipstick formulation based on pitaya (*Hylocereus polyrhizus*) seed oil using D-optimal mixture experimental design. *Molecules* 2014, *19*, 16672–16683. [CrossRef] [PubMed]
23. Yara-Varón, E.; Li, Y.; Balcells, M.; Canela-Garayoa, R.; Fabiano-Tixier, A.-S.; Chemat, F. Vegetable oils as alternative solvents for green oleo-extraction, purification and formulation of food and natural products. *Molecules* 2017, *22*, 1474. [CrossRef]
24. Nadjet, B.; Bakoz, P.Y.; Adbellah, G.; Boudjema, H. Vegetable oils extraction of polycyclic aromatic hydrocarbons from an Algerian quagmire. *Am. J. Appl. Chem.* 2014, *2*, 6–9. [CrossRef]
25. Fan, Y.; Li, J.; Guo, Y.; Xie, L.; Zhang, G. Digital image colorimetry on smartphone for chemical analysis: A review. *Measurement* 2021, *171*, 108829. [CrossRef]
26. Khoshmaram, L.; Mohammadi, M. Combination of a smart phone based low-cost portable colorimeter with air-assisted liquid-liquid microextraction for speciation and determination of chromium (III) and (VI). *Microchem. J.* 2021, *164*, 105991. [CrossRef]
27. Shishehbore, M.R.; Nasirizadeh, N.; Haji, S.A.M.; Tabatabaee, M. Spectrophotometric determination of trace copper after preconcentration with 1,5- diphenylcarbazone on microcrystalline naphthalene. *Can. J. Anal. Sci. Spectrosc.* 2005, *50*, 130–134.
28. Crespo, G.A.; Andrade, F.J.; Iñón, F.A.; Tudino, M.B. Kinetic method for the determination of trace amounts of copper(II) in water matrices by its catalytic effect on the oxidation of 1,5-diphenylcarbazide. *Anal. Chim. Acta* 2005, *539*, 317–325. [CrossRef]
29. Miller, J.N.; Miller, J.C. *Statistics and Chemometrics for Analytical Chemistry*, 6th ed.; Pearson Education Limited: Essex, UK, 2010; pp. 110–126.
30. Eller, Z.; Varga, Z.; Hancsók, J. Advanced production process of jet fuel components from technical grade coconut oil with special hydrocracking. *Fuel* 2016, *182*, 713–720. [CrossRef]
31. Płotka-Wasylka, J. A new tool for the evaluation of the analytical procedure: Green analytical procedure index. *Talanta* 2018, *181*, 204–209. [CrossRef]
32. Marczenko, Z.; Balcerzak, M. *Separation, Preconcentration and Spectrophotometric Inorganic Analysis*, 1st ed.; Elsevier Science: Amsterdam, The Netherlands, 2000; p. 6.
33. Jarusuwannapoom, T.; Hongrojjanawiwat, W.; Jitjaicham, S.; Wannatong, L.; Nithitanakul, M.; Pattamaprom, C.; Koombhongse, P.; Rangkupan, R.; Supaphol, P. Effect of solvents on electro-spinnability of polystyrene solutions and morphological appearance of resulting electrospun polystyrene fibers. *Eur. Polym. J.* 2015, *41*, 409–421. [CrossRef]
34. Dong, T.; Knoshaug, E.P.; Pienkos, P.T.; Laurens, L.M.L. Lipid recovery from wet oleaginous microbial biomass for biofuel production: A critical review. *Appl. Energy* 2016, *177*, 879–895. [CrossRef]
35. Alamu, O.J.; Dehinbo, O.; Sulaiman, A.M. Production and testing of coconut oil biodiesel fuel and its blend. *Leonardo J. Sci.* 2010, *16*, 95–104.
36. Dahim, M.; Al-Mattarneh, H.; Ismail, R. Simple capacitor dielectric sensors for determination of water content in transformer oil. *Int. J. Eng. Res. Technol.* 2018, *7*, 157–160.
37. Safety Data Sheet of Benzene. Available online: https://www.airgas.com/msds/001062.pdf (accessed on 1 November 2022).

38. Safety Data Sheet of Toluene. Available online: https://www.airgas.com/msds/001063.pdf (accessed on 1 November 2022).
39. Safety Data Sheet of Dichloromethane. Available online: https://pdf4pro.com/fullscreen/dichloromethane-material-safety-data-56924.html (accessed on 1 November 2022).
40. Safety Data Sheet of 1,2-Dichloroethane. Available online: https://www.airgas.com/msds/001068.pdf (accessed on 1 November 2022).
41. Safety Data Sheet of Coconut Oil. Available online: https://www.mccsd.net/cms/lib/NY02208580/Centricity/Shared/Material%20Safety%20Data%20Sheets%20_MSDS_/MSDS%20Sheets_Coconut_Oil_205_00.pdf (accessed on 1 November 2022).
42. Khan, A.I. *A GC-FID Method for the Comparison of Acid-and Base-Catalyzed Derivatization of Fatty Acids to FAMEs in Three Edible Oils*; Thermo Fisher Scientific: Runcorn, UK, 2013; p. 20733.

Article

Sulfur-Doped Binary Layered Metal Oxides Incorporated on Pomegranate Peel-Derived Activated Carbon for Removal of Heavy Metal Ions

Binta Hadi Jume [1], Niloofar Valizadeh Dana [2], Marjan Rastin [3], Ehsan Parandi [4,*], Negisa Darajeh [5] and Shahabaldin Rezania [6,*]

1. Department of Chemistry, College of Science, University of Hafr Al Batin, Al Jamiah District, P.O. Box 1803, Jeddah 39524, Saudi Arabia
2. Department of Applied Chemistry, Faculty of Pharmaceutical Chemistry, Tehran Medical Sciences, Islamic Azad University, Tehran 1913674711, Iran
3. Department of Metallurgy and Materials Engineering, Faculty of Engineering, University of Kashan, Kashan 8199696555, Iran
4. Department of Food Science & Technology, Faculty of Agricultural Engineering and Technology, University of Tehran, Karaj 6719418314, Iran
5. Department of Soil and Physical Sciences, Faculty of Agriculture and Life Sciences, Lincoln University, Lincoln, Christchurch 7647, New Zealand
6. Department of Environment and Energy, Sejong University, Seoul 05006, Republic of Korea
* Correspondence: ehsan_parandi@ut.ac.ir (E.P.); shahab.rezania@sejong.ac.kr (S.R.)

Abstract: In this study, a novel biomass adsorbent based on activated carbon incorporated with sulfur-based binary metal oxides layered nanoparticles (SML-AC), including sulfur (S_2), manganese (Mn), and tin (Sn) oxide synthesized via the solvothermal method. The newly synthesized SML-AC was studied using FTIR, FESEM, EDX, and BET to determine its functional groups, surface morphology, and elemental composition. Hence, the BET was performed with an appropriate specific surface area for raw AC (356 $m^2 \cdot g^{-1}$) and modified AC-SML (195 $m^2 \cdot g^{-1}$). To prepare water samples for ICP-OES analysis, the suggested nanocomposite was used as an efficient adsorbent to remove lead (Pb^{2+}), cadmium (Cd^{2+}), chromium (Cr^{3+}), and vanadium (V^{5+}) from oil-rich regions. As the chemical structure of metal ions is influenced by solution pH, this parameter was considered experimentally, and pH 4, dosage 50 mg, and time 120 min were found to be the best with high capacity for all adsorbates. At different experimental conditions, the AC-SML provided a satisfactory adsorption capacity of 37.03–90.09 $mg \cdot g^{-1}$ for Cd^{2+}, Pb^{2+}, Cr^{3+}, and V^{5+} ions. The adsorption experiment was explored, and the method was fitted with the Langmuir model (R^2 = 0.99) as compared to the Freundlich model (R^2 = 0.91). The kinetic models and free energy (<0.45 KJ·mol^{-1}) parameters demonstrated that the adsorption rate is limited with pseudo-second order (R^2 = 0.99) under the physical adsorption mechanism, respectively. Finally, the study demonstrated that the AC-SML nanocomposite is recyclable at least five times in the continuous adsorption–desorption of metal ions.

Keywords: biomass; food waste; removal; heavy metal ions; nanocomposite; layered metal–sulfur

1. Introduction

Ions of heavy metals are frequently discovered in products generated by the oil industry. Vanadium (V^{3+}), lead (Pb^{2+}), chromium (Cr^{3+}), and cadmium (Cd^{2+}) are some examples of heavy metal ions that can be found [1]. Significant opportunities exist for these metal ions to enter groundwater, surface water, and drinking water because of a wide range of processes, including oil processing (extraction, shipping, and storage), mining, and other activities; as a result, the danger posed to the environment as well as to the health of humans is increased [2]. To prevent diseases and syndromes in humans, metabolic processes require trace amounts of heavy metals [3]. However, because heavy metals are used in a wide variety of industrial processes and are typically deposited in water, an excess amount of heavy metals can cause significant health issues in humans, such as degenerative processes in the muscles, body, and nervous system [3,4]. Therefore, monitoring and removing heavy metal ions from environmental water supplies is essential. The allowed levels for Cd^{2+}, Cr^{3+}, Pb^{2+}, and V^{5+} in drinking water have been established at 5 µg L^{-1}, 50 µg L^{-1}, 10 µg L^{-1}, and 100 mg L^{-1}, respectively [2].

Several methods including bioremediation, electrocoagulation, reverse osmosis, oxidation, filtration, nanofibrous membrane, biochar, and adsorption [5–8] have been employed to remove heavy metal ions from water samples. Numerous benefits and drawbacks exist for each of these methods. For example, chemical precipitation creates hazardous waste, photocatalysis and reverse osmosis create secondary pollution, and ion exchange and filtering require advanced technology. Oxidation is also non-regenerative. Electrocoagulation and membrane technology are both intricate processes that result in unfavorable sludge. Even though bioremediation is a technology that is recognized to be safe, it is costly and cannot be applied in extreme conditions [5,6]. Due to their many advantages, researchers continue to be intrigued by adsorption-based approaches as highly efficient methods for treating water, especially for removing heavy metal ions. These advantages include the fact that they do not introduce secondary pollutants into the environment, are inexpensive, can be regenerated, require little complex equipment, and have high adsorption capacity [1,4,5].

Adsorbents for removing heavy metal ions have been made from various micro/nanoscale materials, including biomass, montmorillonite, carbon-based material, polymers, metal–organic framework, and metal oxides. Most proposed materials have drawbacks such as high price, lack of stability, insufficient sorption capacity, ineffective removal effectiveness, poor selectivity, and the generation of secondary pollutants [4,7]. Activated carbon (AC) is a highly porous adsorbent used in the adsorption process to control air and water pollution. This is because activated carbon possesses cationic and anionic nature [9], a high surface area, is cost-effective, has rapid biodegradability [10], is structurally reliable and thermally stable [2,6]. Coal-derived commercial activated carbon is commonly used to remove dissolved solids and ionic contaminants from potable water and industrial effluent, including heavy metal ions and synthetic colors. However, its high price, difficulty in regeneration, and disposal problems restrict its usefulness. Producing activated carbon from inexpensive agricultural products or materials (biochar) such as pomegranate peels and other food waste is one option to cut expenses [6]. There is a significant amount of potential for biochar AC as an adsorbent to be utilized in the process of modifying or reactivating the surface. The addition of metal oxide nanoparticles to AC can result in an improvement in both the efficiency and stability of the adsorption process [2].

The cations present between the layers of negatively charged metal-layered solids can be exchanged effectively. They are rigid and unaffected by temperature fluctuations. Additionally, compared to compounds with a single layer, they exhibit improved properties as a result of the synergistic interaction between the several phases [11]. The layers only have a weak connection to one another, but the ions that are intercalated between them help to keep them stable. The fact that these layered heterostructures have multiple functions makes them an extremely attractive topic of study. When distinct cations are intercalated, they can generate a wide variety of diverse characteristics and structures. They are less hazardous and less leachable due to features including biocompatibility, pH-dependent solvability, and the special intercalating property [2]. Metal-layered ion exchangers based on sulfides have remarkable adsorption and selectivity capabilities towards metal ions, and their removal kinetics are extremely fast. Highly selective and excellent ion exchange properties are displayed by inorganic ligands composed of sulfide metals because (1) the incorporated ions diffuse freely and can access the interior of the metal–ligand layers with relative ease and (2) the basic ligand forms strong interactions with the inserted metal ions [12]. Since soft Lewis basic sulfides have a high affinity for soft Lewis acidic metal ions, the ions are strongly adsorbed by the sulfides [2].

Herein, highly mesoporous activated carbon made from pomegranate peels was chemically precipitated with sulfur-based metal-layered manganese and tin ions to boost its adsorption capacity and efficiency. According to the literature reviews, the diffusion and nonisotopic exchanger processes are likely to be responsible for the ability of metal cation-doped adsorbent to selectively extract cations from complex matrix [2]. Utilizing FTIR, SEM, EDX, and BET to characterize the SML-AC nanocomposite, newly synthesized materials were employed as an adsorbent to remove Pb^{2+}, Cd^{2+}, Cr^{3+}, and V^{5+} ions from oil-field water samples. Langmuir, Freundlich isotherm, and kinetic models were used to verify the experimental procedure and adsorption capacity.

2. Results and Discussion

2.1. Adsorbent Characterization

2.1.1. FTIR Spectroscopy

Surface functional groups of pristine AC and AC-SML were considered with FTIR, as shown in Figure 1. According to the AC spectrum, the band at 3400 cm^{-1} corresponds to the O-H groups anchored onto the carbon material. The IR bands at 2905 cm^{-1}, 1712 cm^{-1}, and 1450 cm^{-1} and attributed to C-H, C=O, and C-C/C=C stretching of AC functionals and the skeleton of the carbon framework. After incorporating sulfur-layered metal oxide into AC, the IR spectrum of AC-SML showed a new sharp peak at 1020 cm^{-1} and small peaks at around 500 cm^{-1}. It should be noted that the peaks that appeared in the wide range of 500–1050 cm^{-1} corresponded to the presence of metal–carbon (M-O-C) and sulfur–metal (M-S-M) linkage in the nanocomposite. This claim is confirmed by a previous study which demonstrated that the extensive bands 500–900 cm^{-1} are attributed to the metallic layered derived from the M-O stretching and O-M-O bonding [13,14]. Hence, the proposed characteristics of IR bands in both spectra of AC and AC-SML imply the successful incorporation of sulfur–metal layered into the activated carbon.

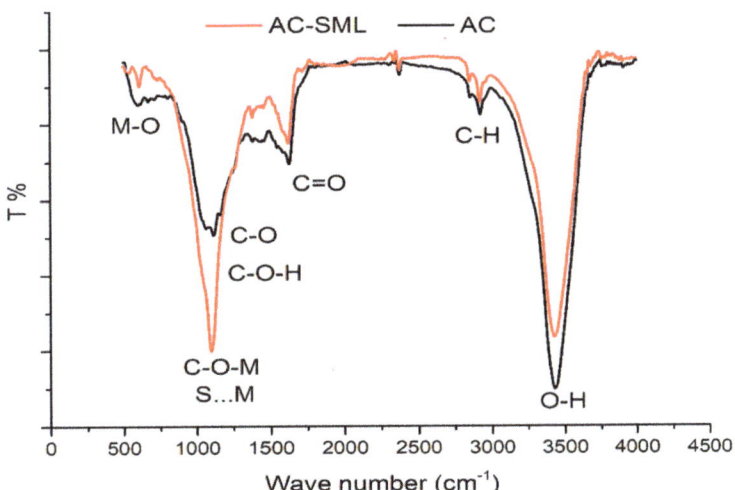

Figure 1. FTIR spectroscopy of pomegranate peel AC and AC-SML.

2.1.2. SEM Microscopy

The newly synthesized pristine AC and AC-SML surface morphology were investigated using FESEM. Figure 2a illustrates the micrograph of AC. It is the external surface of the activated carbon which is quite irregular and full of cavities due to alkaline activation. After the immobilization of sulfur-layered metal oxide onto AC (Figure 2b), the micrograph indicates the formation of circular and bulk substances over AC. This trend confirms the formation of AC-SML nanocomposite.

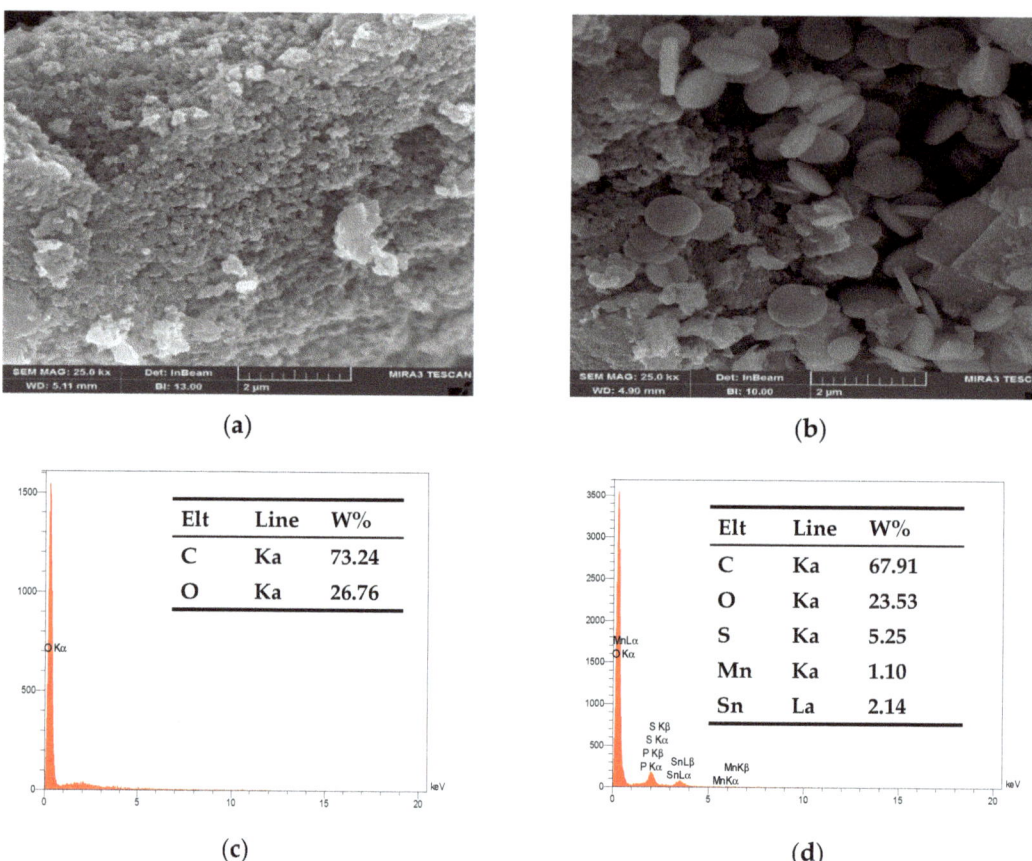

Figure 2. FESEM micrograph for raw nano AC (**a**) and modified AC-SML (**b**). EDX spectrum and elemental composition for raw AC (**c**) and AC-SML nanocomposite (**d**).

2.1.3. EDX Spectroscopy

To verify the presence of the preferred elements on plain AC and AC-SML nanocomposite, the EDX technique was utilized. Figure 2c,d represents the EDX signals and weight percentage of AC and AC-SML nanocomposite elements. The elemental analysis demonstrates the presence of two main elements over plain AC: carbon (73.24%) and oxygen (26.76%). Hence, after incorporating metal-S-oxide nanoparticles onto the AC surface, the expected elements were observed such as C, O, S, Mn, and Sn, with a weight percentage of 67.91%, 23.53%, 5.25%, 1.10%, and 2.14%, respectively. The EDX additionally performed the good dispersion of sulfur and layered metal (Sn-S-Mn) oxides on the activated carbon.

2.1.4. BET Surface Area

Specific surface area is an imperative factor in adsorption chemistry since it directly reflects the sorption capacity. BET technique based on the N_2 adsorption–desorption process utilized to record the surface area of plain AC and AC-SML nanocomposite. The N_2 adsorption–desorption curve is shown in Figure 3 for AC (a) and AC-SML nanocomposite (b). The specific surface area values were obtained at 356 m$^2 \cdot$g^{-1} and 195 m$^2 \cdot$g^{-1}, respectively. The specific surface area is decreased after incorporating sulfur-based metal oxide into the AC. The high specific surface area value for plain AC is probably due to the porous structure, which also provided a higher pore diameter (3.38 nm) than the modified nanocomposite (2.92 nm). The pore volume and diameter indicate that the prepared materials are mesoporous, which is appropriate for adsorption.

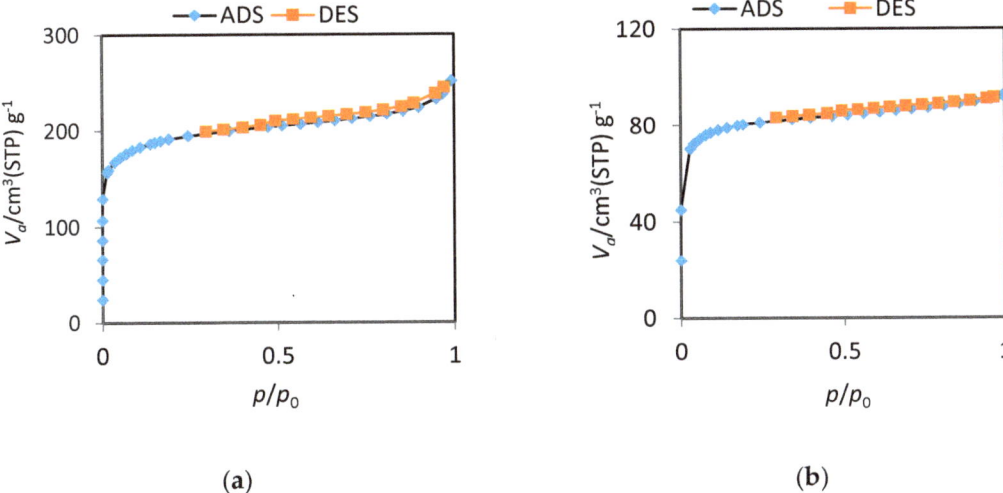

Figure 3. BET-N_2 adsorption-desorption process for plain AC (**a**) and AC-SML nanocomposite (**b**).

2.2. Adsorption Parameters

2.2.1. Types of Materials

The influence of the types of adsorbent material (raw Ac and AC-SML) on adsorbent efficiency was studied in similar conditions (dosage of 100 mg, pH 4, and time 60 min). According to Figure 4, the raw AC shows high efficiency for Cd^{2+} and Pb^{2+} ions and low efficiency for Cr^{3+} and V^{5+}. This probably is due to the occupation of AC active sites with Cd^{2+} and Pb^{2+} ions, since it is rich in oxygenated functional groups. Moreover, after modification of AC with doping SML (Sn-S-Mn) nanoparticles, an increment in the adsorption percentage was observed for selected metal ions (Cd^{2+}, Cr^{3+}, Pb^{2+}, and V^{5+}). This phenomenon can be associated with the enhancement of the electrostatic interaction between sulfur (S^{2-}) and a selected metal cation. Hence, the trend (Figure 4) reflects the synergetic effect of AC-SML nanocomposite for the uptake of Cd^{2+}, Cr^{3+}, Pb^{2+}, and V^{5+} cations.

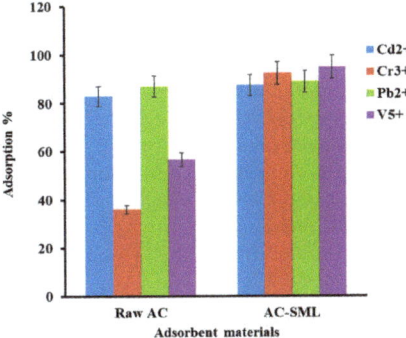

Figure 4. Influence types of materials on adoption efficiency.

2.2.2. Effect of Solution pH

The solution pH was investigated from 2 to 7 since metal ions are highly influenced by solution pH. In the survey literature, the selected metal ions are existing in different states, including cadmium: 2–8.5 (Cd^{2+}), 8.6–10.5 ($Cd(OH)^{-}$), 10.6–12 ($Cd(OH)_2)_{aq}$ [15]; chromium: 1–3 (Cr^{3+}), 3.1–6 ($Cr(OH)^{2+}$), 6.1–8 ($Cr(OH)^{+}_2$), 8–2 ($Cr(OH)_3$) [16]; lead: 1–6.1 (Pb^{2+}), 6.2–8.5 ($Pb(OH)^{-}$), 8.6–11 ($Pb(OH)_2$ aq) [17], and vanadium: pH 2–3 (VO_2), 3.1–4 ($VO(OH_3)^{0}$), 4.1–8.5 ($VO_2(OH)_2{}^{-}$), 8.6–10 ($VO_3(OH)^{2-}$). Removal efficiency is experimentally investigated for selected heavy metal ions and shown in Figure 5. As can be seen, Cd^{2+} and Pb^{2+} provided low efficiency at low pH, which can be defined through the undesired protonation of the adsorbent surface followed by electrostatic repulsion. It is also noteworthy that the proton ions (H^{+}) trade competition with metal cations to adsorb over nanocomposite. Due to electrostatic interaction and coordination, the favorable removal efficiency was obtained for Cd^{2+} and Pb^{2+} at pH 4 to 6. A slight decrease in efficiency at pH 6–7 is due to the formation of metal hydroxide ($Pb(OH)^{-}$). Chromium and vanadium showed different trends due to their neutralization in the presence of a hydrogen wealthy medium. The other reason that may additionally interact in such pH-dependent behavior is the specific constructions and geometries of the metal cations with pH, which increased the adsorption efficiency. However, pH 4 was the best experimental condition for all metal cations.

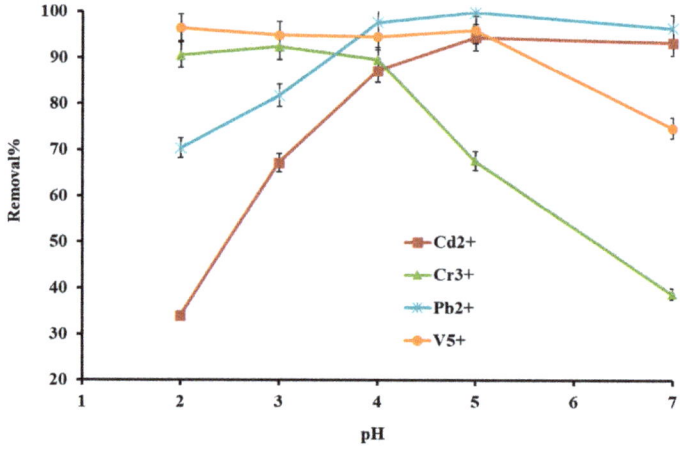

Figure 5. Effect of solution pH on metal ion removal efficiency.

2.2.3. Effect of Adsorbent Dosage

The amount of adsorbent is vital for the quantitative removal of adsorbate in the adsorption process. This is described via the adsorbent dosage in the range of 10–150 mg, as shown in Figure 6. The removal profile reflected that the adsorption efficiency is increasing for all selected metal cations. However, the quantitative adsorption efficiency increased from 52% to 95% when sorbent material varied from 10 to 80 mg. After that, the efficiency slightly changed up to 150 mg. Hence, the efficiency expansion can be attributed to more adsorption sites on the adsorbent since the solid phase surface area increases exponentially with the regular degree of adsorbent dosage. Thus, 80 mg was selected for all metal cations with maximum adsorption efficiency.

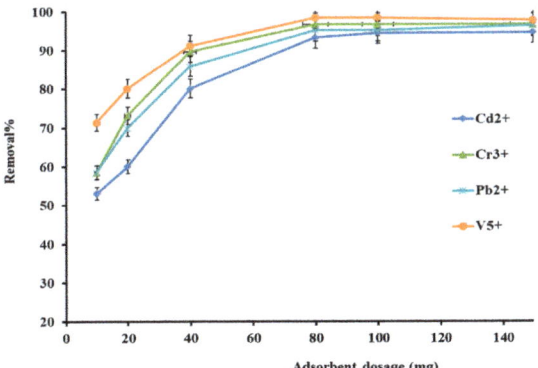

Figure 6. Effect of AC-SML dosage on metal ion removal efficiency.

2.2.4. Effect of Time

Adsorption's contact time is another crucial parameter for the effective uptake of metal ions from an aqueous solution. The impact of contact time for the adsorption of the selected heavy metal cation was investigated in the exclusive times in the range of 5–240 min. Figure 7 indicates that the removal increased gradually from 25% to 80% for all metal cations by increasing the time up to 120 min, except vanadium with 96%. With additional increase in the contact time up to 240 min, adsorption efficiency no longer extends significantly for vanadium, but the adsorption efficiency of lead, cadmium, and chromium ions increases to >92%. The first stage (5–120 min) is performing the availability of the active site to uptake metal cations, and after that, the adsorption sites are saturated, and the system reaches equilibrium [18]. Finally, the 120 min was selected for the further procedure due to the high percent removal.

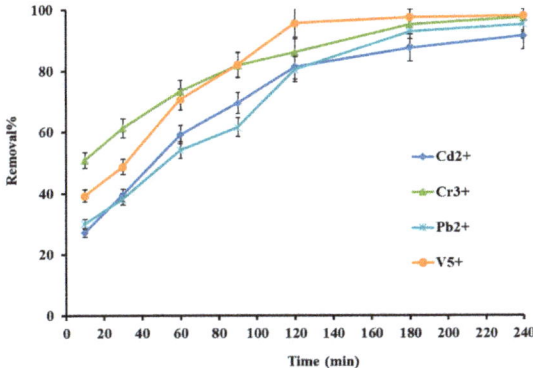

Figure 7. Effect of contact time on metal ion removal efficiency.

2.3. Adsorption Kinetics

The adsorption time limitation is performed with kinetic models of pseudo-first-order and pseudo-second-order models. The linear varieties of the proposed kinetic models can be estimated with Equations (1) and (2), respectively.

$$Ln(Q_e - Q_t) = LnQ_e - k_1 t, \tag{1}$$

$$t/Q_t = 1/k_2 Q_e^2 + t/Q_e \tag{2}$$

The parameters are the following: Q_e (mg/g) is equilibrium adsorption capacity, Q_t (mg/g) adsorption capability at any time t. The time constates of k_1 (1/min) and k_2 (g/mg/min) corresponds to the first-order and second-order models, respectively.

Figure 8a,b reflects the linear plots of kinetic models, which are plotted $Ln(Q_e-Q_t)$ versus time and t/Q_t versus time, and the values of parameters are listed in Table 1. Based on the determination coefficient (R^2), the contact time was fitted with pseudo-second order (>0.99) as compared to pseudo-first order (0.88–0.96). In addition, Q_e values (theory) of second-order models were in good agreement with experimental adsorption capacity. Therefore, the kinetic rate is limited with pseudo-second order, in which the adsorption process has been controlled with electron sharing and electrostatic interactions.

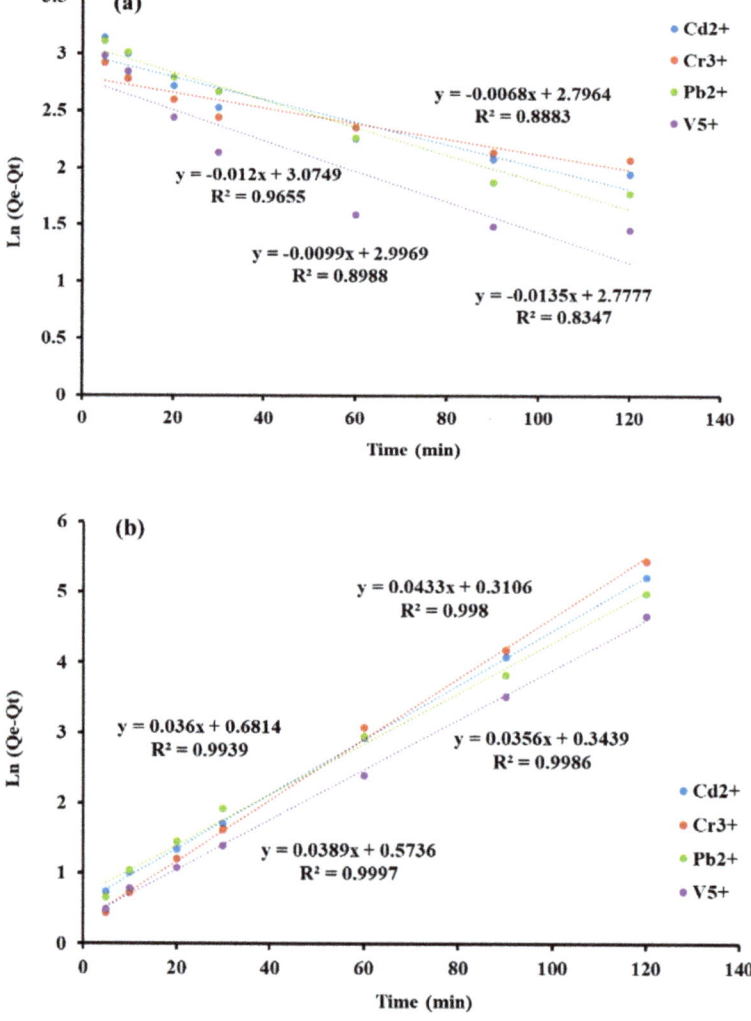

Figure 8. Kinetic models of pseudo-first order (a) and pseudo-second order (b).

Table 1. Kinetic models and parameters for selected heavy metal ions.

Kinetic Models	Parameters	Cd^{2+}	Cr^{3+}	Pb^{2+}	V^{5+}
pseudo-first order	R^2	0.898	0.888	0.965	0.834
	k_1 (1/min)	0.009	0.006	0.0121	0.0135
	q_e (mg·g^{-1})	19.16	16.74	22.07	16.31
pseudo-second order	R^2	0.999	0.998	0.993	0.998
	k_2 (g/mg/min)	0.0012	0.0018	0.0012	0.0015
	q_e (mg·g^{-1})	25.71	23.25	27.77	28.08

2.4. Adsorption Equilibrium and Isotherm Models

The impact of the initial concentration of selected heavy metal ions is experimented on to consider the equilibrium capability of AC-SML nanocomposite. For this, 10 mg of AC-SML was applied to adsorb the Cd^{2+}, Cr^{3+}, Pb^{2+}, and V^{5+} with a concentration range of 5–300 mg·L^{-1} in 120 min contact time. The equilibrium isotherm (Q_e vs. C_e) was recorded and plotted as shown in Figure 9. The isotherm graph shows the expected Qe increases from 3 mg·g^{-1} to 85 mg·g^{-1} via increasing the concentration of selected heavy metal ions until it reaches equilibrium. The isothermal evaluation revealed that the adsorption process follows the IUPAC regular pattern (Type II) [19]. This pattern described the adsorption process following a monolayer pattern.

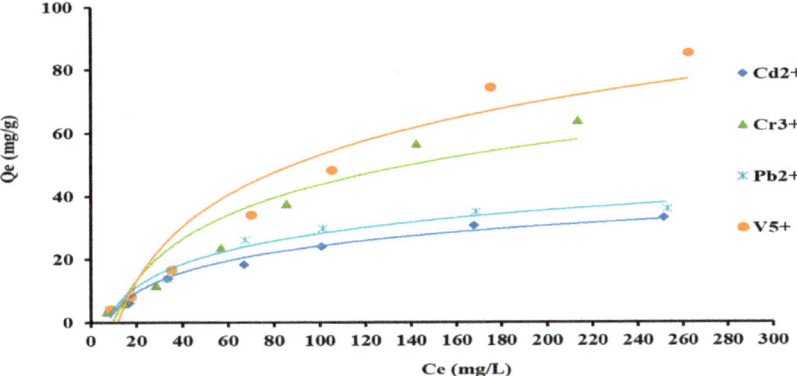

Figure 9. Equilibrium adsorption capacity versus the initial concentration.

Hence, isotherm models were utilized to explore the adsorption pattern and mechanism. Therefore, the well-known isotherm models, namely Langmuir, Freundlich, Dubinin–Radushkevich (DR), and free energy were used to describe the adsorption pattern and adsorption mechanism. Langmuir is attributing the monolayer adsorption onto the homogenous surface. Freundlich reflects the multilayer adsorption onto the heterogeneous surface. DR confirms the multilayer adsorption over either homogenous or heterogeneous surfaces. Free energy suggests the adsorption of machoism, physical or chemical. Hence, the proposed models' linear mechanism formed by the following Equations (3)–(5):

$$C_e/Q_e = C_e/Q_m + 1/K_L Q_m \qquad (3)$$

$$\ln Q_e = \ln K_F + (1/n) \ln C_e \qquad (4)$$

$$\ln Q_e = \ln Q_S - K_{ad}\left(\varepsilon^2\right) \qquad (5)$$

$$\varepsilon = RT \ln(1 + 1/C_e) \qquad (6)$$

$$E = (2K_{ad})^{-1/2} \qquad (7)$$

In the equations, Q_m (mg·g^{-1}) is maximum adsorption capacity, K_L (L·mg^{-1}), K_{ad}, and K_F [(mg/g)(L/mg)$^{1/n}$] correspond to Langmuir, DR, and Freundlich constants, respectively. C_e (mg·L^{-1}) equilibrium concentration of heavy metal ions and 1/n reflect the favorability of the adsorption process. Q_s (mg·g^{-1}) is the DR theoretical sorption capacity. Finally, E is sorption energy (Equation (7)), directly displaying the adsorption mechanism. The E values < 40 kJ/mol are the physisorption mechanism, and E > 80 kJ/mol is the chemisorption mechanism [20,21].

The corresponding linear models of every isotherm have been plotted (Figure 10a,b), and the parameters of every model were calculated in Table 2. Based on the values of R^2, it used to be assumed that both Langmuir ($R^2 > 0.92$) and Freundlich ($R^2 > 0.90$) models observe to describe the adsorption of selected heavy metal ions onto the AC-SML adsorbent nanocomposite. Hence, adsorption can be either monolayer or multilayer, with a maximum adsorption capacity of 37.03 mg·g^{-1} to 90.09 mg·g^{-1} for all selected metal ions. Moreover, the appropriate values of R^2 of the DR model confirm the favourability of the multilayer pattern. The values of energy approved the physisorption mechanism for both monolayer and multilayer adsorption patterns for selected heavy metal ions over AC-SML nanocomposite.

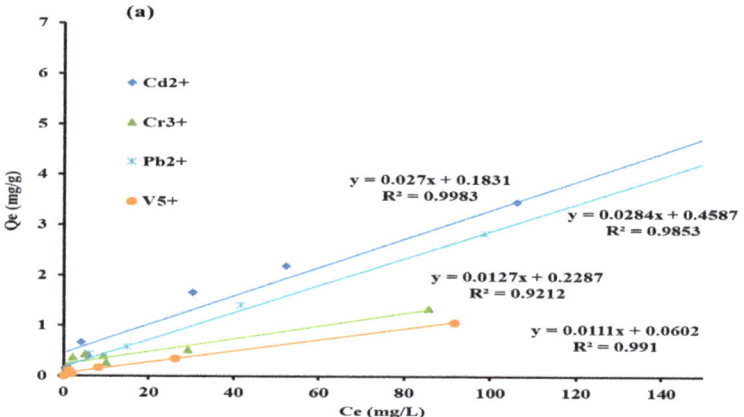

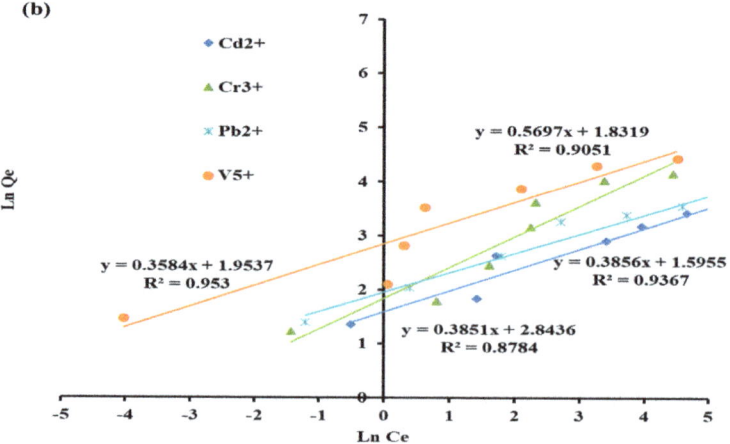

Figure 10. Adsorption isotherm models of Langmuir (**a**) and Freundlich (**b**).

Table 2. Langmuir, Freundlich, DR models and free Energy for adsorption of selected heavy metal ions over AC-SML nanocomposite.

Isotherms	Parameters	Cd^{2+}	Cr^{3+}	Pb^{2+}	V^{5+}
Langmuir	Q_m (mg g^{-1})	37.03	78.74	35.21	90.09
	k_L (L mg^{-1})	0.144	0.055	0.061	0.184
	R^2	0.998	0.921	0.985	0.991
Freundlich	K_F [(mg g^{-1}) (L mg^{-1})$^{1/n}$]	4.42	5.51	6.17	14.11
	n	2.59	1.75	2.79	2.59
	R^2	0.936	0.905	0.953	0.878
Dubinin–Radushkevich	Q_s (mg g^{-1})	18.61	24.98	22.39	38.29
	K_{ad} (mol^2/kJ^2)	2.908	2.161	0.736	0.592
	R^2	0.864	0.848	0.886	0.903
Energy	E (kJ/mol)	0.41	0.48	0.82	0.91

2.5. Mechanism Study

The proposed adsorption mechanism between the selected heavy metal ions and AC-SML is investigated, as graphically shown in Figure 11. Hence, based on the pH study, high efficiency was obtained in the pH ranges of 4–6. Therefore, the possible electrostatic interactions and coordination between Cd^{2+}, Cr^{3+}, Pb^{2+}, and V^{5+} and functional groups (sulfur-metal) of the AC could be the reason for the high removal efficiency at pH 4–6. In addition, the lower removal efficiency at low pH (<4) may be because of the AC-SML's protonation of active sites as well as the competition of H$^+$ for adsorption, while at high pH > 7, the selected metal cations are present as hydroxide form (M(OH)$_x$)$^-$), which is the prominent repulsion force interactive between the adsorbates and negative surface charge of adsorbent (AC-SML$^-$).

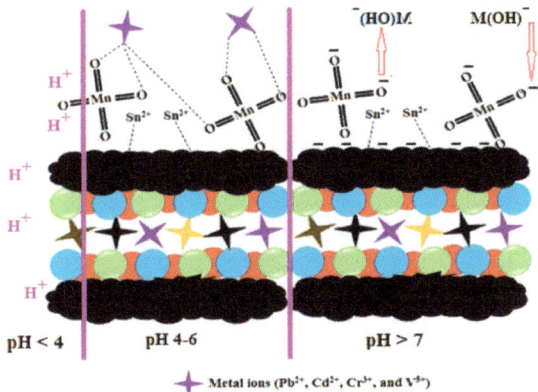

Figure 11. Schematic of the proposed mechanism for heavy metal ion removal by AC-SML.

2.6. Regeneration

To study the regeneration, 80 mg of the AC-SML was positioned into a test tube, including selected Cd^{2+}, Cr^{3+}, Pb^{2+}, and V^{5+}. After each adsorption process, the adsorbates were desorbed using 5 mL (HCl, 0.1 M) after being shaken for 30 min. Then, launched heavy metal ions were measured with ICP-OES. Then, the AC-SML was washed with excess distilled water and reused for a further experimental process for repeated five adsorption–desorption cycles. The adsorption efficiency was obtained 85%, 81%, 87%, 80% for Cd^{2+}, Cr^{3+}, Pb^{2+} and V^{5+}, respectively. This procedure indicated that the removal effectivity of selected heavy metal ions does not reduce notably for up to five cycles. Thus, it can be claimed that the AC-SML nanocomposite can be regenerated for five adsorption–desorption cycles.

2.7. Comparison

Table 3 indicates that the freshly synthesized AC-SML nanocomposite could be noteworthy for removing Cd^{2+}, Cr^{3+}, Pb^{2+}, and V^{5+} ions. Compared to other adsorbents, the AC-SML nanocomposite showed high adsorption capacity at pH 4. The adsorption process duration was relatively fast. The AC-SML's improved adsorption capacity was due to the doping of the sulfur metal layer oxides on the AC, enhancing interactions with heavy metal ions and resulting in significant removal.

Table 3. Comparison of AC-SML with other published adsorbents for heavy metal sorption.

Adsorbent	Heavy Metals	pH	Time (min)	Qe (mg/g)	Ref.
AC-SML	Pb^{2+}, Cd^{2+}, Cr^{3+}, V^{5+}	4	120	37–90	This study
AC-metal phosphate layered	Pb^{2+}, Cd^{2+}, Co^{2+}, V^{5+}, Ni^{2+}	6	140	20–140	[2]
AC-silica	Pb^{2+}, Cd^{2+}, Ni^{2+}	7	200	61–400	[22]
Magnetic sporopollenin-polyaniline	Pb^{2+}	6	90	163.93	[5]
Magnetic graphene oxide-TiLa	Pb^{2+}	5	120	112	[4]
Alginate@MgS	Pb^{2+}	4	140	84.74	[1]

3. Material and Methods

3.1. Material

Tin (II) chloride ($SnCl_2$), manganese (II) chloride tetrahydrate ($MnCl_2 \cdot 4H_2O$), chromium (III) nitrate nonahydrate ($Cr(NO_3)_3 \cdot 9H_2O$), cadmium chloride hemipentahydrate ($CdCl_2 \cdot 2.5H_2O$), lead nitrate ($Pb(NO_3)_2$), ammonium metavanadate (NH_4VO_3), hydrochloric acid (HCl, 37%), sodium hydroxide (NaOH), and sodium chloride (NaCl) were purchased from Merck Chemicals.

3.2. Adsorbent Synthesis

3.2.1. Pretreatment and Synthesis AC

The nanosized pomegranate peel-activated carbon was prepared according to the previous study [9]. Briefly, the raw pomegranate peel (PG) was washed and dried in the laboratory under ambient temperature and dark place. The dried and clean pomegranate peel was ground into a fine powder and sieved properly (<150 μm). Then, lignans were extracted by combining 20 g of pomegranate peel powder with 6 g of NaOH in 200 mL of distilled water and stirring for 24 h. After being diluted and filtered through filter paper, the mixture was rinsed with excess distilled water and finally dried in an oven at a temperature of 85 °C. Next, the uniformly sized powder was carbonized in a furnace at 400 °C for 3 h (under an N_2 atmosphere) to produce biochar. Using a mass ratio of 5:1 (AC/NaOH), 120 mL of distilled water, and vigorously stirring at 120 °C for 2 h, we thoroughly combined the produced carbon with the NaOH solution. Following filtration of the solution, the isolated product was dried in an oven at 120 °C for 10 h. It was transferred to the furnace to finish the activation process and heated to 800 °C for 2 h. After being neutralized in 0.01 M HCl and excess distilled water, AC was dried in an oven at 90 °C for 24 h [2].

3.2.2. Synthesis of SML-AC

Subsequently, 1 g S_2, 0.1 g $SnCl_2$, and 0.1 g $MnCl_2$ were added to a solution containing 2 g of the dried AC in 150 mL of distilled water, and the combination was agitated at 45 °C for 3 h (pH 9). Afterward, the solid was separated from the mixture and heated in the furnace for 2 h at 650 °C. The product (SML-AC) was then rinsed with extra distilled water and dried in an oven at 85 °C for 18 h [23–25] (Figure 12).

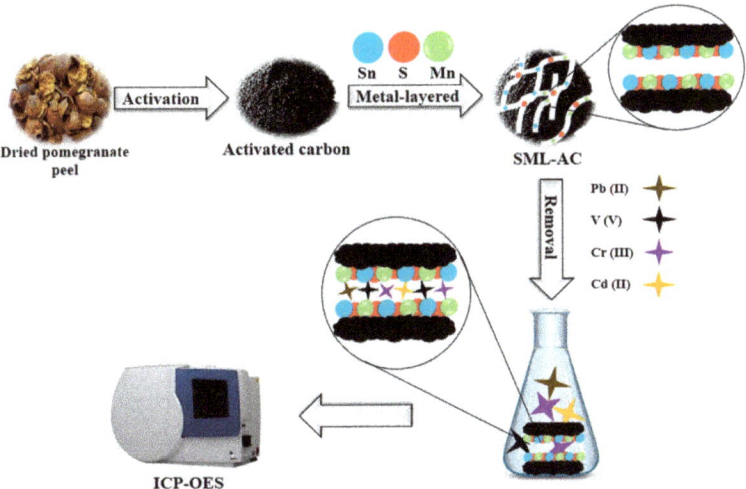

Figure 12. A schematic for SML-AC synthesis and removal procedure.

3.3. Adsorbent Characterization

A Bruker Equinox 55 spectrometer was used to conduct Fourier-transform infrared spectroscopy (*FTIR*) in the 400–4000 cm^{-1} wavenumber region to identify the functional groups of the newly synthesized material. The surface morphology and elemental structure of the AC and SML-AC were examined using a field emission scanning electron microscopy (FESEM) TESCAN MIRA3 equipped Energy Dispersive X-ray Analysis (EDX). Brunauer–Emmett–Teller (BET) determined the surface area and pore size.

3.4. Removal Procedure

First, 30 mL of the sample solution that contained 30 mg·L^{-1} of the target analytes mixed with 40 mg of the SML-AC adsorbent. After 30 min of orbital shaking on a shaker, the mixture was then centrifuged to separate the SML-AC from the mixture (4000 rpm, 6 min). Lastly, a syringe filter made of cellulose with a mesh size of 0.2 was used to filter 5 mL of the supernatant, which included the metal ions in residual concentration. This sample was immediately subjected to an inductively coupled plasma optical emission spectroscopy (ICP-OES) analysis. In the adsorption trials, multiple critical parameters such as pH (ranging from 2 to 7), SML-AC quantity (ranging from 10 to 150 mg), and contact duration (ranging from 5 to 240 min) were explored and improved to produce acceptable results from the adsorbent. Required parameters were calculated below [1,26].

$$\text{Removal efficiency } (R\%) = \left(\frac{C_1 - C_e}{C_1}\right) \times 100 \tag{8}$$

$$\text{Adsorption capacity } (q_e) = \left(\frac{V}{m}\right) \times (C_1 - C_e) \tag{9}$$

C_1 and C_e are initial and residual concentrations of ions (mg·L^{-1}), respectively. V is the initial sample volume (mL), q_e is the equilibrium adsorption capacity (mg·g^{-1}), and W represents the adsorbent dosage (mg).

4. Conclusions

This study developed an efficient nano-sized adsorbent for removing heavy metal ions from aqueous solutions using a metal–sulfur-layered oxide immobilized on pomegranate peel-derived activated carbon. The nanocomposite was prepared and used for removing selected heavy metal ions (Cd^{2+}, Cr^{3+}, Pb^{2+}, and V^{5+}) from the water sample at pH 4. The equilibrium isotherm models were used to explain the adsorption capacity, which obtained 37.03, 78.74, 35.21, and 90.09 mg·g^{-1} for Cd^{2+}, Cr^{3+}, Pb^{2+}, and V^{5+}, respectively. The Langmuir and Freundlich model and free Energy findings were carried out in the selected heavy metal ions adsorption process following monolayer and multilayer sorption patterns underneath the physical adsorption process. The kinetic rate was limited with pseudo-second order followed by electron sharing and electrostatic interactions. Hence, with a high removal efficiency of >90% for all selected heavy metal ions, the AC-SML nanocomposite is an efficient adsorbent for removing Cd^{2+}, Cr^{3+}, Pb^{2+}, and V^{5+} from aqueous media.

Author Contributions: Conceptualization, S.R. and E.P.; methodology, B.H.J. and E.P; software, M.R.; validation, E.P., N.V.D. and M.R.; formal analysis, E.P.; investigation, E.P.; resources, N.D.; data curation, N.V.D.; writing—original draft preparation, E.P.; writing—review and editing, E.P.; visualization, N.D.; supervision, B.H.J.; project administration, S.R.; funding acquisition, S.R. All authors have read and agreed to the published version of the manuscript.

Funding: This research received no external funding.

Institutional Review Board Statement: Not applicable.

Informed Consent Statement: Not applicable.

Data Availability Statement: Not applicable.

Acknowledgments: The authors would like to thank the University of Hafr Al Batin for the facilitation and financial support.

Conflicts of Interest: The authors declare no conflict of interest.

Sample Availability: Samples of the adsorbent are available from the authors.

References

1. Bidhendi, M.E.; Parandi, E.; Mahmoudi-Meymand, M.; Sereshti, H.; Nodeh, H.R.; Joo, S.-W.; Vasseghian, Y.; Khatir, N.M.; Rezania, S. Removal of lead ions from wastewater using magnesium sulfide nanoparticles caged alginate microbeads. *Environ. Res.* **2022**, *216*, 114416. [CrossRef] [PubMed]
2. Esmaeili Bidhendi, M.; Abedynia, S.; Mirzaei, S.S.; Gabris, M.A.; Rashidi Nodeh, H.; Sereshti, H. Efficient removal of heavy metal ions from the water of oil-rich regions using layered metal-phosphate incorporated activated carbon nanocomposite. *Water Environ. J.* **2020**, *34*, 893–905. [CrossRef]
3. Nafi, A.W.; Taseidifar, M. Removal of hazardous ions from aqueous solutions: Current methods, with a focus on green ion flotation. *J. Environ. Manag.* **2022**, *319*, 115666. [CrossRef] [PubMed]
4. Mosleh, N.; Ahranjani, P.J.; Parandi, E.; Nodeh, H.R.; Nawrot, N.; Rezania, S.; Sathishkumar, P. Titanium lanthanum three oxides decorated magnetic graphene oxide for adsorption of lead ions from aqueous media. *Environ. Res.* **2022**, *214*, 113831. [CrossRef]
5. Mosleh, N.; Najmi, M.; Parandi, E.; Nodeh, H.R.; Vasseghian, Y.; Rezania, S. Magnetic sporopollenin supported polyaniline developed for removal of lead ions from wastewater: Kinetic, isotherm and thermodynamic studies. *Chemosphere* **2022**, *300*, 134461. [CrossRef] [PubMed]
6. Vonnie, J.M.; Li, C.S.; Erna, K.H.; Yin, K.W.; Felicia, W.X.L.; Aqilah, M.N.N.; Rovina, K. Development of Eggshell-Based Orange Peel Activated Carbon Film for Synergetic Adsorption of Cadmium (II) Ion. *Nanomaterials* **2022**, *12*, 2750. [CrossRef] [PubMed]
7. Lilhare, S.; Mathew, S.B.; Singh, A.K.; Carabineiro, S.A. Aloe Vera Functionalized Magnetic Nanoparticles Entrapped Ca Alginate Beads as Novel Adsorbents for Cu (II) Removal from Aqueous Solutions. *Nanomaterials* **2022**, *12*, 2947. [CrossRef]
8. Parandi, E.; Pero, M.; Kiani, H. Phase change and crystallization behavior of water in biological systems and innovative freezing processes and methods for evaluating crystallization. *Discov. Food* **2022**, *2*, 1–24. [CrossRef]
9. Esmaeili Bidhendi, M.; Poursorkh, Z.; Sereshti, H.; Rashidi Nodeh, H.; Rezania, S.; Afzal Kamboh, M. Nano-size biomass derived from pomegranate peel for enhanced removal of cefixime antibiotic from aqueous media: Kinetic, equilibrium and thermodynamic study. *Int. J. Environ. Res. Public Health* **2020**, *17*, 4223. [CrossRef]
10. Dhahri, R.; Yılmaz, M.; Mechi, L.; Alsukaibi, A.K.D.; Alimi, F.; ben Salem, R.; Moussaoui, Y. Optimization of the Preparation of Activated Carbon from Prickly Pear Seed Cake for the Removal of Lead and Cadmium Ions from Aqueous Solution. *Sustainability* **2022**, *14*, 3245. [CrossRef]
11. Venugopal, B.; Rajamathi, M.J.J.o.c. Layer-by-layer composite of anionic and cationic clays by metathesis. *J. Colloid Interface Sci.* **2011**, *355*, 396–401. [CrossRef] [PubMed]
12. Manos, M.J.; Kanatzidis, M.G. Metal sulfide ion exchangers: Superior sorbents for the capture of toxic and nuclear waste-related metal ions. *Chem. Sci.* **2016**, *7*, 4804–4824. [CrossRef] [PubMed]

13. Sarma, D.; Malliakas, C.D.; Subrahmanyam, K.; Islam, S.M.; Kanatzidis, M.G. $K_{2x}Sn_{4-x}S_{8-x}$ (x = 0.65–1): A new metal sulfide for rapid and selective removal of Cs^+, Sr^{2+} and UO_2^{2+} ions. *Chem. Sci.* **2016**, *7*, 1121–1132. [CrossRef] [PubMed]
14. Yadollahi, M.; Namazi, H. Synthesis and characterization of carboxymethyl cellulose/layered double hydroxide nanocomposites. *J. Nanopart. Res.* **2013**, *15*, 1–9. [CrossRef]
15. Bian, Y.; Bian, Z.; Zhang, J.; Ding, A.; Liu, S.; Zheng, L.; Wang, H. Adsorption of cadmium ions from aqueous solutions by activated carbon with oxygen-containing functional groups. *Chin. J. Chem. Eng.* **2015**, *23*, 1705–1711. [CrossRef]
16. Santos, V.C.G.D.; Salvado, A.d.P.A.; Dragunski, D.C.; Peraro, D.N.C.; Tarley, C.R.T.; Caetano, J. Highly improved chromium (III) uptake capacity in modified sugarcane bagasse using different chemical treatments. *Química Nova* **2012**, *35*, 1606–1611. [CrossRef]
17. Duan, S.; Tang, R.; Xue, Z.; Zhang, X.; Zhao, Y.; Zhang, W.; Zhang, J.; Wang, B.; Zeng, S.; Sun, D.J.C. Effective removal of Pb (II) using magnetic $Co_{0.6}Fe_{2.4}O_4$ micro-particles as the adsorbent: Synthesis and study on the kinetic and thermodynamic behaviors for its adsorption. *Colloids Surf. A Physicochem. Eng. Asp.* **2015**, *469*, 211–223. [CrossRef]
18. Naushad, M. Surfactant assisted nano-composite cation exchanger: Development, characterization and applications for the removal of toxic Pb^{2+} from aqueous medium. *Chem. Eng. J.* **2014**, *235*, 100–108. [CrossRef]
19. Sing, K.S. Reporting physisorption data for gas/solid systems with special reference to the determination of surface area and porosity (Recommendations 1984). *Pure Appl. Chem.* **1985**, *57*, 603–619. [CrossRef]
20. Saleh, T.A. Isotherm, kinetic, and thermodynamic studies on Hg (II) adsorption from aqueous solution by silica-multiwall carbon nanotubes. *Environ. Sci. Pollut. Res.* **2015**, *22*, 16721–16731. [CrossRef]
21. Mona, S.; Kaushik, A. Chromium and cobalt sequestration using exopolysaccharides produced by freshwater cyanobacterium Nostoc linckia. *Ecol. Eng.* **2015**, *82*, 121–125. [CrossRef]
22. Karnib, M.; Kabbani, A.; Holail, H.; Olama, Z. Heavy metals removal using activated carbon, silica and silica activated carbon composite. *Energy Procedia* **2014**, *50*, 113–120. [CrossRef]
23. Shirani, M.; Aslani, A.; Sepahi, S.; Parandi, E.; Motamedi, A.; Jahanmard, E.; Nodeh, H.R.; Akbari-Adergani, B. An efficient 3D adsorbent foam based on graphene oxide/AgO nanoparticles for rapid vortex-assisted floating solid phase extraction of bisphenol A in canned food products. *Anal. Methods* **2022**, *14*, 2623–2630. [CrossRef] [PubMed]
24. Shirani, M.; Parandi, E.; Nodeh, H.R.; Akbari-Adergani, B.; Shahdadi, F. Development of a rapid efficient solid-phase microextraction: An overhead rotating flat surface sorbent based 3-D graphene oxide/lanthanum nanoparticles@ Ni foam for separation and determination of sulfonamides in animal-based food products. *Food Chem.* **2022**, *373*, 131421. [CrossRef]
25. Aghel, B.; Gouran, A.; Parandi, E.; Jumeh, B.H.; Nodeh, H.R. Production of biodiesel from high acidity waste cooking oil using nano GO@ MgO catalyst in a microreactor. *Renew. Energy* **2022**, *200*, 294–302. [CrossRef]
26. Parandi, E.; Safaripour, M.; Abdellattif, M.H.; Saidi, M.; Bozorgian, A.; Nodeh, H.R.; Rezania, S. Biodiesel production from waste cooking oil using a novel biocatalyst of lipase enzyme immobilized magnetic nanocomposite. *Fuel* **2022**, *313*, 123057. [CrossRef]

Article

Uptake of Pharmaceutical Pollutants and Their Metabolites from Soil Fertilized with Manure to Parsley Tissues

Klaudia Stando [1,*], Ewa Korzeniewska [2], Ewa Felis [3,4], Monika Harnisz [2] and Sylwia Bajkacz [1,3,*]

[1] Department of Inorganic, Analytical Chemistry and Electrochemistry, Faculty of Chemistry, Silesian University of Technology, B. Krzywoustego 6 Str., 44-100 Gliwice, Poland
[2] Department of Engineering of Water Protection and Environmental Microbiology, Faculty of Geoengineering, University of Warmia and Mazury in Olsztyn, Prawocheńskiego 1 Str., 10-720 Olsztyn, Poland; ewakmikr@uwm.edu.pl (E.K.); monikah@uwm.edu.pl (M.H.)
[3] Centre for Biotechnology, Silesian University of Technology, B. Krzywoustego 8 Str., 44-100 Gliwice, Poland; ewa.felis@polsl.pl
[4] Environmental Biotechnology Department, Faculty of Power and Environmental Engineering, Silesian University of Technology, Akademicka 2 Str., 44-100 Gliwice, Poland
* Correspondence: klaudia.stando@polsl.pl (K.S.); sylwia.bajkacz@polsl.pl (S.B.)

Abstract: Manure is a major source of soil and plant contamination with veterinary drugs residues. The aim of this study was to evaluate the uptake of 14 veterinary pharmaceuticals by parsley from soil fertilized with manure. Pharmaceutical content was determined in roots and leaves. Liquid chromatography coupled with tandem mass spectrometry was used for targeted analysis. Screening analysis was performed to identify transformation products in the parsley tissues. A solid-liquid extraction procedure was developed combined with solid-phase extraction, providing recoveries of 61.9–97.1% for leaves and 51.7–95.6% for roots. Four analytes were detected in parsley: enrofloxacin, tylosin, sulfamethoxazole, and doxycycline. Enrofloxacin was detected at the highest concentrations (13.4–26.3 ng g^{-1}). Doxycycline accumulated mainly in the roots, tylosin in the leaves, and sulfamethoxazole was found in both tissues. 14 transformation products were identified and their distribution were determined. This study provides important data on the uptake and transformation of pharmaceuticals in plant tissues.

Keywords: pharmaceuticals; parsley; plant metabolism; plant uptake; transformation products; LC-MS/MS analysis

1. Introduction

Animal manure and liquid manure are commonly used for fertilization and reclamation of agricultural land. Due to their high content of organic matter, nitrogen, phosphorus and micronutrients, they are natural alternatives to nitrogen fertilizers [1]. One of the major concerns of using natural fertilizers is the presence of pharmaceutical residues [2]. Both pharmaceuticals and their metabolites are present in the feces of humans and animals undergoing antibiotic therapy [3]. The use of natural fertilizers in the form of manure results in the introduction and dissemination of pharmaceutical contaminants (FCs) in the environment, and may promote the spread of drug resistance in bacteria [4].

Once released into the environment, depending on their physicochemical properties, FCs can accumulate in the soil, contaminate groundwater, and be uptaken, absorbed or immobilized by plants [5]. FCs can be uptaken with water or nutrients via the root pathway and translocated between tissues [6]. There are three mechanisms of nutrient uptake from the soil via the root pathway: root uptake, mass flow, and diffusion [7]. Accumulation of FCs in plant tissues depends on the cellular structure of the plant's tissues and the molecular weight, polarity, lyophilicity, and ionic form of these compounds [8].

After absorption of FCs from the soil, they can metabolize in the plant. Transformation products (TPs) of phase I, II, and III metabolism of various FCs have been detected in plant tissues [9]. Tian et al. [9] observed the formation of clarithromycin and sulfadiazine TPs in leaves and roots of lettuce grown under hydroponic conditions in the presence of FCs. Eight clarithromycin metabolites were identified during phase I of plant metabolism and two sulfadiazine metabolites were formed during phase II of metabolism. Other studies have also shown that FCs from the macrolide, tetracycline and sulfonamide groups were metabolized according to phase I or phase II reactions after plant uptake [9].

Absorption and subsequent bioaccumulation and biotransformation of FCs in edible plants carry a risk of transmission to the human gastrointestinal tract [10,11]. Studies on the pharmacokinetics of sulfamethoxazole and tetracycline were conducted using Chinese cabbage (Napa cabbage) and water spinach grown under hydroponic conditions [12]. The study found that the accumulation capacity of the pharmaceutical depends on various physicochemical properties, for example tetracycline had a higher concentration (77–160 $\mu g\ g^{-1}$) than sulfamethoxazole (18–38 $\mu g\ g^{-1}$). Moreover, both plants showed that the drug accumulated mainly in the roots and to a lesser extent in the green parts, confirming that these compounds were uptaken from the environment. Studies conducted on herbs and grasses treated with penicillin, sulfadiazine and tetracycline confirmed that the roots were most strongly affected by FCs compared to the steams and leaves [13]. Additionally, the presence of FCs in the soil disrupted homeostasis in the plant body, reduced the elemental content of the plant and led to salt stress.

The aim of this study was to evaluate the bioavailability of 14 selected veterinary FCs by parsley plant (*Petroselinum crispum*) from manure-fertilized soil. The conditions for extraction and determination of 14 FCs were developed using high-performance liquid chromatography coupled with tandem mass spectrometer (LC-MS/MS). LC-MS/MS is a superior analytical technique for the determination of trace amounts of FCs in environmental samples [14]. Both the time of flight (TOF) and LTQ-Orbitrap analyzers were preferred due to their high full-scan detection sensitivity, mass accuracy and fast data acquisition. In our research, we used a QTRAP spectrometer, combining the triple quadrupole operating modes with a linear ion trap. This device is suitable for both targeted analyses using the Multiple Response Monitoring Mode (MRM). Additionally QTRAP produces a high amount of fragment ion data which is necessary for retrospective analysis of analyte transformation products. Representatives of FCs from the groups of tetracyclines (tetracycline (TC), oxytetracycline (OTC), doxycycline (DOX)), sulfonamides (sulfamethoxazole (SMX), sulfadiazine (SFD)), fluoroquinolones (ciprofloxacin (CIP), levofloxacin (LVF), enrofloxacin (ENF)), macrolides (clarithromycin (CLR), tylosin (TYL)), metronidazole (MET), trimethoprim (TRI), vancomycin (VAN), and clindamycin (CLD) were selected based on the World Health Organization (WHO) report [15]. The selected FCs include the most commonly used drugs in human and veterinary medicine, however there are few reports of soil contamination in Poland by FCs. In Poland the most commonly applied FCs in pig, bovine and poultry productions are tetracyclines (TC, DOX, OTC), fluoroquinolones (mainly ENF), sulfonamides (mainly SMX and SFD) combined with TRI and macrolides (mainly TYL) [16,17]. The contamination of agricultural soils in northern Poland with residues of selected FCs was also assessed. Selected FCs were present in 21 agricultural soil samples out of 39 tested, and the concentrations of SMX, MET, TRI, TYL, ENF ranged from 3.6 to 57 $\mu g \cdot kg^{-1}$ [18].

A field experiment was conducted in which two types of animal manure (poultry or cattle) were introduced into the soil, in which parsley was sown and grown. The control sample was unfertilized soil, free from FCs' contamination. The experiment was conducted over a four-month growing period. Bioaccumulation of FCs was evaluated by determining their concentrations in parsley leaves and roots collected after the plant vegetation period (targeted analysis). Additionally, FCs' transformation products present in parsley leaves and roots were identified using semi-untargeted and untargeted analysis.

2. Materials and Methods

2.1. Chemicals

Analytical standards TC, OTC, DOX, ENF, LVF, CIP, TYL, TRI, MET, CLR, CLD, SMX, SFD, and VAN were purchased from Sigma-Aldrich (St. Louis, MO, USA). Hypergrade acetonitrile (ACN), methanol (MeOH) and water were purchased from Merck (Darmstadt, Germany). Analytical-grade formic acid (FA), hydrochloric acid (HCl), acetic acid (AcA), sodium hydroxide (NaOH), ethylenediaminetetraacetic acid (EDTA), 25% ammonium hydroxide solution, MeOH, ethanol, ACN and chloroform were purchased from CHEMM-PUR (Piekary Śląskie, Poland). Analytical-grade phosphate dibasic dehydrate (>98%) and citrinic acid monohydrate (>98%) were purchased from Sigma-Aldrich.

OASIS HLB (500 mg, 6 mL), OASIS WCX (60 mg, 3 mL), and OASIS MAX (150 mg, 6 mL) cartridges were purchased from Waters (Eschborn, Germany). BAKERBOND™ Octadecyl (C18) (500 mg, 6 mL) cartridges were purchased from BAKERBOND® (J.T. Baker, Philipsburg, PA, USA).

2.2. Preparation of Standard Solutions

The standard stock solutions of all FCs were prepared in 1.0 mg·mL^{-1} concentration. TC, OTC, DOX, TYL, TRI, SMX, ENF, LVF, MET, CLR, and CLD were diluted in MeOH. CIP was diluted in 1% FA in MeOH, VAN in MeOH:H$_2$O (1:1; v/v) and SFD in acetone. Calibration solutions of FCs in the range of 1.0 to 1200.0 ng·g^{-1} were prepared in MeOH. At the validation stage, a mixture of working solutions with a defined concentration were added to the parsley root and leaf (blank samples). All the solutions were stored in the freezer at $-18\ ^\circ$C and in the dark. The working solutions were stored for a maximum of one week.

2.3. Sample Collection

The field experiment was conducted between June and September 2019. Eight experimental plots of 4 m^2 were prepared. In the poultry manure samples, the FCs' concentration was relatively low, DOX and CIP were found at the concentrations of 330.3 ± 40.1 ng·g^{-1} and 30.0 ± 2.5 ng·g^{-1} respectively. In the bovine manure and soil, none of the studied drugs were detected. A low amount of selected FCs in manure samples was the basis for additional supplementation of the manure with 4 selected FCs: DOX, ENF, SMX, TYL at a concentration of 50 µg·g^{-1} manure. The selected FCs are the most commonly used in the treatment of cattle and poultry from among the 14 veterinary pharmaceuticals studied in this work. Six plots were treated with manure, three of which were with bovine manure (PMBA–bovine manure-supplemented plots) and three with poultry manure (PMPA–poultry manure-supplemented plots), respectively. Another two control plots were not fertilized with manure (CP). Parsley (*Petroselinum crispum*) was selected as the crop because of its fast growth and high ecological tolerance to anthropogenic pollution. Parsley samples (roots and leaves) were collected using the envelope method from five points of each plot into sterile plastic containers [19]. The samples were transported immediately under darkness and cool conditions to the laboratory, where composite samples were prepared. Parsley root was separated from the leaf and washed with double distilled water, then the samples from univariate plots were combined to form composite samples, homogenized, frozen at $-20\ ^\circ$C, and then freeze-dried. Freeze-drying was carried out under 0.035 mbar at $-50\ ^\circ$C ALPHA 1–2 LDplus (CHRIST, Osterode am Harz, Germany). Immediately before extraction, the freeze-dried parsley samples were ground to a powder using an electric grinder MK70 dott (ELDOM, Katowice, Poland).

2.4. Selection of Conditions for FCs' Extraction from Plant Tissues

2.4.1. Extraction of FCs from Parsley Leaves

(a) Examination of Solid-Liquid Extraction (SLE) Procedure

5–15 µg mixture of FCs in 3 mL MeOH was added to 0.5 g of ground freeze-dried parsley leaf. The sample was mixed and the solvent was evaporated naturally in the air.

SLE was performed using solutions of McIlvain buffer (pH = 4), MeOH, ethanol, ACN, and their mixtures (Table S1). The volume of extractant was 10–20 mL for a single extraction; for double extraction, 2 × 10 mL or 2 × 15 mL of solvent was used. Samples were shaken for 30 min or 2 × 30 min at 750 rpm (single extraction/double extraction) using Vibramax 100 (Heidolph, Schwabach, Germany). For each repetition, the samples were centrifuged for 10 min at 8000 rpm, and the supernatants were combined after double extraction. For liquid chromatography coupled with tandem mass spectrometry (LC-MS/MS) analysis, 1.0 mL of extract was collected.

(b) Examination of Solid-Liquid Extraction Combined with Liquid-Liquid Extraction (SLE-LLE) Procedure

500 ng of FCs' standard mixture was added to 0.5 g of freeze-dried parsley leaf, which was allowed to stand for 24 h to adsorb the analytes and evaporate the solvent in the air. The extraction was performed using 10 mL of solvent, and the sample was centrifuged at 8000 rpm for 10 min and the supernatant collected and extracted using 5 mL of chloroform. The extracts were evaporated under a stream of nitrogen and dissolved in 1 mL MeOH:0.1% FA in H_2O (1:1; v/v).

(c) Examination of Solid-Liquid Extraction Combined with Solid-Phase Extraction (SLE-SPE) Procedure

500 ng of FCs mixture in MeOH was added to 0.5 g of ground freeze-dried parsley leaf, and the solvent was evaporated. The analytes were extracted using a mixture of McIlvaine buffer (pH = 4):ACN (1:1; v/v) (single or double extraction). The extracts were adjusted to pH = 3 using concentrated HCl and diluted to 500 mL using distilled water, so that the volume fraction of organic solvent was below 3%.

The utilization of SPE in the purification step of the extracts was studied. Different types of sorbents were examined, including OASIS HLB (500 mg, 6 mL), OASIS MAX (150 mg, 6 mL), BAKERBONDTM Octadecyl (C18) (500 mg, 6 mL) and tandem OASIS WCX (60 mg, 3 mL) as a precolumn. Sorbents were conditioned in the same manner using 6 mL of MeOH, 0.1 M HCl, and distilled water, respectively. Different solvents or mixtures of solvents were used for the elution of analytes: MeOH, 0.1% acetic acid (AcA) in MeOH, 2% FA in ACN/MeOH mixture (8:2 v/v), and 0.1% NH_3 in MeOH, (3–10 mL) (SLE-SPE L1–SLE-SPE L6; Table S2). Eluates were evaporated to dryness under a stream of nitrogen. Immediately before LC-MS/MS analysis, the evaporated residue was dissolved in a mixture of MeOH:0.1% FA in H_2O (1:1; v/v).

2.4.2. Extraction of FCs from Parsley Root

500 ng of a standard in MeOH was added to 0.5 g of parsley root powder and left in the air for 24 h to evaporate the solvent. SLE was then performed using 10 mL of solvent or mixture of solvent, including MeOH, ACN, 0.2 M NaOH, acetone, McIlvain buffer (pH = 4), and 0.1 M EDTA (SLE-SPE R1: McIlvaine buffer (pH = 4):ACN (1:1; v/v); SLE-SPE R2: McIlvaine buffer (pH = 4):MeOH (1:1; v/v); SLE-SPE R3: McIlvaine buffer (pH = 4):MeOH (1:1; v/v), McIlvaine buffer (pH = 4):ACN (1:1; v/v); SLE-SPE R4: McIlvaine buffer (pH = 4):ACN (1:1; v/v), 0.2 M NaOH:acetone (1:1; v/v); SLE-SPE R5: McIlvaine buffer (pH = 4):MeOH (1:1; v/v), 0.2 M NaOH:acetone (1:1; v/v); SLE-SPE R6: ACN: McIlvaine buffer (pH = 4):0.1 M EDTA (2:2:1; $v/v/v$)). Single extraction was carried out for 30 min by shaking at 750 rpm, and then centrifuged for 10 min at 8000 rpm. In the case of multiple SLEs, the extracts were combined. Samples were diluted to 250 mL using distilled water and adjusted to pH = 3 using FA. SPE was performed using OASIS HLB sorbents (500 mg, 6 mL). Conditioning was performed using 6 mL of MeOH, 0.1 M HCl, and distilled water, respectively. The sorbents were vacuum-dried for 20 min, and the analytes eluted with 6 mL of MeOH and 4 mL of 0.1% AcA in MeOH. The samples were evaporated to dryness under a stream of nitrogen and dissolved in 1 mL of MeOH:0.1% FA in H_2O (1:1; v/v).

2.5. Sample Preparation

SLE extraction or two-step SLE combined with SPE were used to extract FCs from plant tissues. The best FCs' extraction parameters were as follows:

2.5.1. Parsley Leaf–Developed Procedure I (Extraction of TC, OTC, DOX, CIP, ENF, LVF, CLD, VAN)

15 mL of McIlvaine buffer (pH = 4):ACN mixture (1:1; v/v) was added to 0.5 g of freeze-dried parsley leaf and extracted for 30 min. The sample was centrifuged at 8000 rpm for 10 min, the supernatant collected, and a second extraction was performed using the same solvent mixture. The supernatants were combined, diluted to 500 mL using distilled water and adjusted to pH = 3 using HCl. In the case of SPE extraction, OASIS HLB sorbent (500 mg, 6 mL) was used and conditioned with 6 mL of MeOH, 0.1 M HCl, and H_2O, respectively. The sample solution was passed through the sorbent at a rate of 3 mL·min^{-1} and then the sorbent was vacuum-dried. Elution was performed using 10 mL of 2% FA in MeOH:ACN (2:8; v/v) and 10 mL of 0.1% NH_3 in MeOH. The residue was evaporated to dryness under a stream of nitrogen, dissolved in 1 mL of 0.1% FA in H_2O:MeOH (1:1; v/v) and centrifuged at 8000 rpm for 5 min. The supernatant was collected and analyzed using LC-MS/MS.

2.5.2. Parsley Leaf–Developed Procedure II (Extraction of MET, SMX, SFD, TRI, TYL, CLR)

0.5 g of freeze-dried leaf sample was extracted twice with 10 mL of MeOH. Each extraction was conducted by shaking the sample at 750 rpm for 30 min, and centrifuged at 8000 rpm for 10 min. The supernatants from the two extractions were combined and evaporated to dryness under a stream of nitrogen. The residue was then dissolved in 1 mL of 0.1% FA in H_2O:MeOH (1:1; v/v) and centrifuged at 8000 rpm for 5 min. The clear supernatant was analyzed using LC-MS/MS.

2.5.3. Parsley Root–Developed Procedure (Extraction of 14 FCs)

10 mL of ACN:McIlvaine buffer (pH = 4):0.1 M EDTA mixture (2:2:1; $v/v/v$) was added to 0.5 g of freeze-dried parsley root and extracted for 30 min at 750 rpm, and centrifuged at 8000 rpm for 10 min. The supernatant was collected and diluted to 250 mL with distilled water and adjusted to pH = 3 using FA. The sample was applied to OASIS HLB sorbent (500 mg, 6 mL), conditioned as in SLE-SPE procedure for leaves (SLE-SPE L1, Section 2.4.1). Elution was performed using 6 mL of MeOH and 4 mL of 0.1% acetic acid in MeOH. The sample was evaporated to dryness under a stream of nitrogen and dissolved in 1 mL of 0.1% FA in H_2O:MeOH (1:1; v/v) immediately before analysis.

2.6. Instrumental and Analytical Conditions

Dionex UltiMate 3000 HPLC system (Dionex Corporation, Sunnyvale, CA, USA) equipped with: rapid separation pump, autosampler, thermostatted column compartment was used for the analysis of FCs. Dionex Chromeleon TM 6.8 software was used to control the chromatography system. The HPLC system was coupled with an AB Sciex Q-Trap® 4000 mass spectrometer (Applied Biosystems/MDS SCIEX, Foster City, CA, USA). Detailed conditions for LC-ESI-MS/MS determination of 14 FCs were discussed in our previous publication [20]. Briefly, ZORBAX SB-C3 column (150 mm × 3.0 mm i.d., 5 µm, Agilent Technologies, Santa Clara, CA, USA) column was used for chromatographic separation. Elution was performed using a gradient mixture of 0.1% FA in water and ACN. Total analysis time was 10 min, the column temperature was maintained at 30 °C and the injection volume was 5 µL. All FCs were analyzed in positive ionization mode. Each sample was analyzed in multiple reaction monitoring mode (MRM) using the two highest precursor ion/product ion transitions.

2.7. Method Validation

The developed SLE-SPE procedure and LC-MS/MS method for determination of 14 FCs in the parsley root and leaf samples was validated. Linearity range, limit of detection (LOD) and quantification (LOQ), recovery, accuracy, precision and matrix effect were studied by analysis of FCs' standards. Calibration curves were obtained by analyzing the calibration solutions of concentration between 1 ng·g^{-1} and 1200 ng·g^{-1}. Quantitative analysis was conducted by calculating the ratios between each analyte's peak area and the peak area of the calibration curve. Regression equations for each analyte were obtained using the linear regression method. The degree of curve model fit was assessed using the determination coefficient (R^2). LOQ was determined as the lowest point on the calibration curve, LOD was calculated as 33% LOQ.

The matrix effects (ME) were evaluated by comparing the peak area of FCs diluted in the blank sample (extract of parsley root or leaf) to the peak area of the analytes diluted with pure solvent. Blank samples of parsley root and leaf were previously analyzed to confirm the absence of any significant peak at MRM transitions. The optimized MS parameters (declustering potential, collision energy, collision cell exit potential) for the selected MRM transitions for each compound were given in our previous publication [20]. Recovery was calculated as the ratio of the measured signal analyte area in the sample after extraction (FCs added before extraction) related to the signal area of the analyte in the matrix solution (FCs added after extraction). The recovery was determined at three concentration levels: low-quality control (LQC = 100 ng·g^{-1}), middle-quality control (MQC = 400 ng·g^{-1}), and high-quality control (HQC = 1000 ng·g^{-1}). Precision and accuracy were determined at the same concentration levels. Accuracy was defined as the relative error (RE), precision was determined to form on the coefficient of variation (CV). Analyses were performed in six replications.

2.8. Identification of Transformation Products of Selected FCs in Plant Tissues

Two approaches were used to identify TPs in plant tissue samples: screening and untargeted analysis. Environmental samples were analyzed by LC-MS/MS in different modes of linear ion trap. The ion source parameters and ionization mode were the same as for targeted analysis (Section 2.6). Intelligent data acquisition (IDA) mode was used for screening analysis, combining pseudo-monitoring multiple reactions (p-MRM) mode of operation with enhanced product ion scanning (EPI). The development of p-MRM method consisted of collecting literature data on TPs of selected FCs and their characteristic MRM transitions, and then creating a list of TPs in MRM mode. According to IDA criteria, if a compound's signal intensity exceeded 500 cps and was in the range of 100–1500 Da, then EPI mode was activated and the full mass spectrum recorded. Confirmation was obtained through comparison of the obtained mass spectra of TPs with those available in the literature [20,21]. The use of p-MRM-IDA-EPI reduced the amount of data obtained in the next step.

For untargeted analysis, full data collection mode (EMS) was used in combination with EPI. IDA criteria were the same as for p-MRM-IDA-EPI. After recording mass spectra by EPI-IDA-EMS for the identified compounds in the sample, the focus was on retrospective analysis of spectra that were not recorded in p-MRM-IDA-EPI. Where possible, the data obtained were confirmed using databases or literature.

3. Results and Discussion

3.1. Development of the Extraction Procedure for the Isolation of FCs from Plant Tissues

3.1.1. Parsley Leaves–Development of SLE Procedure

McIlvain buffer (pH = 4) and organic solvents, including MeOH, ethanol, ACN, and acetone (Table S1), were used to extract FCs from leaf samples [6,10]. Table S3 shows analyte recoveries obtained using SLE for parsley leaves. The shaking time and intensity were chosen based on the stability data of the determined compounds (unpublished material).

Notably, the utilization of only organic solvents provides inefficient extraction of selected FCs from the parsley matrix. The following extraction efficiencies were obtained: 30.1–80.2% for MeOH, 18.8–87.5% for MeOH:ACN (1:1; v/v), whereas the lowest efficiency was obtained for ACN with only 2.9–63.2%. According to Table S3, it was possible to perform SLE extraction of 7 analytes (SMX, SFD, CLR, MET, TRI, TYL, ENF) using only MeOH, with recoveries above 60%. Performing double extraction with MeOH resulted in higher recoveries of ENF, TYL, TRI, MET, and SFD (above 80%). In the case of SLE of tetracyclines, sulfonamides, and fluoroquinolones from lettuce, carrot, tomato, and walnut leaf samples, a mixture of ACN:acetone (1:1; v/v, pH = 3) can be used as the extractant [22,23]. According to reports, the procedure of double extraction of SLE using MeOH and 5% FA in MeOH without a sample purification step gives the selected extraction of 59 FCs from samples of 8 edible plant species [24]. The McIlvaine buffer (pH = 4) (SLE L4) was suitable for the extraction of all analytes with a recovery of 31.5% (LVF)–102.8% (VAN). The pH of the extraction buffer is important to improve the solubility of FCs due to their low pKa in the range 1.8–6.6 for all analytes except CLD and CLR [20].

According to the literature, improved extraction of FCs can be achieved by modifying the composition of the McIlvaine buffer solution through addition of an organic solvent [11,25]. In order to recover analytes not extracted by McIlvaine buffer (pH = 4), a mixture with an organic solvent (methanol or ACN) was prepared in a volume ratio of 1:1. McIlvaine buffer (pH = 4) with an organic solvent increased the recovery of fluoroquinolones, sulfonamides and TRI and CLR. After using ACN mixture, higher recoveries were obtained for tetracyclines (TCs: 71.3–91.6%) and sulfonamides (SAs: 76.8–95.9%), compared to MeOH (TCs: 64.2–79.6%, SAs: 66.2–90.4%). It was found that the presence of McIlvain buffer allowed the extraction of analytes with high polarity, and the addition of ACN allowed the extraction of more non-polar compounds [25–27]. Therefore, a mixture of McIlvain buffer (pH = 4):ACN was used in further experiments. The use of double extraction with McIlvain buffer (pH = 4):ACN mixture further increased the recovery (66.0% (CIP)–96.7% (CLR)) (SLE-SPE L7) compared to single extraction (SLE-SPE L5). In the final procedure described in Section 2.4, the solvent volume was increased from 10 mL to 15 mL due to the high solvent absorption of the freeze dried plant material, which significantly reduced the final extractant volume.

The extraction mixture of 0.2 M NaOH with acetone (SLE-SPE L8) for the extraction of fluoroquinolones (FQs) was examined [26]. The recovery of FQs was below 40.1%, and the supernatant solution was turbid and thick. SLE with aqueous NaOH solution is often used to extract plant proteins due to alkaline conditions being easier to cleave H-bonds that stabilize the protein structure [28]. It is known that the extraction of FCs from parsley leaves using 0.2 M NaOH:acetone mixture promotes coextraction of plant proteins followed by denaturation [29]. Therefore, a suitable procedure suitable for soil matrices, would not translate to plant tissue matrices. A mixture of MeOH:EtOH:H$_2$O was also studied, as well as a mixture where water was replaced with McIlvaine buffer (pH = 4) (SLE-SPE L9) [30]. The recoveries for both mixtures were lower than McIlvaine buffer (pH = 4):ACN mixture (1:1; v/v).

3.1.2. Parsley Leaves–Combining SLE-LLE Procedures

Leaves have a complex organic matrix, consisting of compounds such as lignin, cellulose, proteins, flavonoids, tannins and plant pigments that can affect the efficiency of the extraction process [31]. The main problem with SLE for the preparation of leaf samples is co-extraction from the plant pigment matrix. Reports commonly employ chloroform to extract chlorophyll from plants and its mixtures [32]. Chloroform has low polarity compared to the other organic solvents used, with dipole moments of chloroform (1.04 D), ACN (3.92 D), methanol (1.70 D), and water (1.85 D), hence, it could be used to purify plant extracts containing FCs. Figure S1 compares the obtained results after purification of plant extracts, with chloroform in aqueous (aq.f) and organic (org.f) fractions.

The use of chloroform at the LLE stage for purification of SLE extracts did not improve extraction efficiency. It was observed that the recovery of tetracyclines and CLD in the aqueous fraction increased (37.6–83.43%) compared to pure MeOH. However, 23.0–63.3% of some analytes (LVF, ENF, TYL, TRI, MET, CLR) were also soluble in the organic fraction. The partial solubility of the aforementioned six FCs in chloroform was due to the formation of H-bonds between the analytes and chloroform. According to the Lewis theory, chloroform is an electron pair acceptor, and the $-NH_2$, $-NHR_2$, $-NR_3$, $-OH$, $=O$ groups present in the drug structure act as an electron pair donor [33]. The combined SLE-LLE method significantly degraded the recovery of SAs and VANs. Therefore, the LLE step was abstained for the purification of the plant matrix.

3.1.3. Parsley Leaves–Combining SLE-SPE Procedures

The use of SPE for the purification of parsley leaf extracts was studied. SPE sorbents of different natures were tested: anion-exchangeable (MAX), cation-exchangeable (WCX), hydrophilic-lipophilic balance (HLB) and silica gel modified with octadecylsilane groups (C18). Sorbent selection consisted of two criteria: effective retention of FCs and removal/retention of matrix components. The obtained results are shown in Figure S2. Previous studies have examined the tandem combination of two different sorbents, where one sorbent was used for retention of matrix constituents and the other for retention of analytes. For example, the tandem combination of cation exchange sorbents (SCX and MCX) with HLB has been successfully applied to extract FCs from plant tissues [34,35]. In our study, we tested the tandem combination of WCX and C18 sorbent (for retention of matrix components) with HLB sorbent (for retention of analytes). The results showed that 10 of 14 analytes were retained on WCX sorbent and the recoveries were in the range of 8.1% (SMX)–49.7% (CLR). C18 sorbent retained 12 of 14 analytes with recoveries ranging from 1.9% (TC)–41.1% (SMX). Due to high analyte loss, WCX and C18 as pre-columns in tandem SPE were not used. The employment of strong anion exchange sorbents (SAX/MAX) and/or HLB for the extraction of FCs from solid samples has been extensively reported [21,32] and promotes the separation of analytes in good recoveries. In our study, we also tested the feasibility of MAX and HLB sorbents. The recovery of FCs with MAX sorbent was in the range of 1.1% (MET)–70.1% (CLR), with recoveries below 35% for 9 of 14 analytes. The best recovery was obtained for the HLB sorbent, where all FCs, except metronidazole, had recovery in the range of 44.4% (CLR)–106.8% (OTC). Therefore, the HLB sorbent was selected for further study.

In the next step, SLE extraction was combined with SPE and parameters such as solvent composition for SLE, sample volume after dilution (so that the volume percentage of organic solvent did not exceed 3%), volume and solvent composition for SPE elution were selected. The selection of solvents for SLE is summarized in Table S3. The comparison of FCs' extraction efficiency in parsley leaf SLE-SPE procedures (SLE-SPE L1–SLE-SPE L6) is shown in Figure 1, and their parameters are summarized in Table S2. First, one-step (SLE-SPE L1) and two-step (SLE-SPE L2) SLE extraction were performed, followed by SPE extraction under the same conditions. Analyte recoveries for the single extraction were lower (13.0–45.9%) than for the two-step extraction (14.2–67.8%). Commonly, multiple solvent SLE is used for FCs' extraction from real samples because, in multi-stage extraction, the distribution coefficient is established at each stage, which makes extraction more effective [36,37]. Additionally, the utilization of solvents of different composition at each step as in L2, allows the extraction of FCs with different polarities [6,9,34]. Hence, in SLE-SPE L3 procedure, the volume of elution solution was increased from 10 mL to 20 mL, which increased the recovery of CIP, LVF, ENF, TYL, CLR, and CLD. In another modification (SLE-SPE L4), 0.1% AcA in MeOH and 0.1% NH_3 in MeOH were used for elution. The application of two-step elution under extreme pH conditions promoted the extraction of a wider group of FCs with different pKa values. Reports have shown that the most common elution of analytes from HLB sorbent after passing the plant matrix extract is carried out under inert conditions–MeOH [6,9], although our study indicated that sequential elution with solvents of different pH gave a higher recovery of FCs.

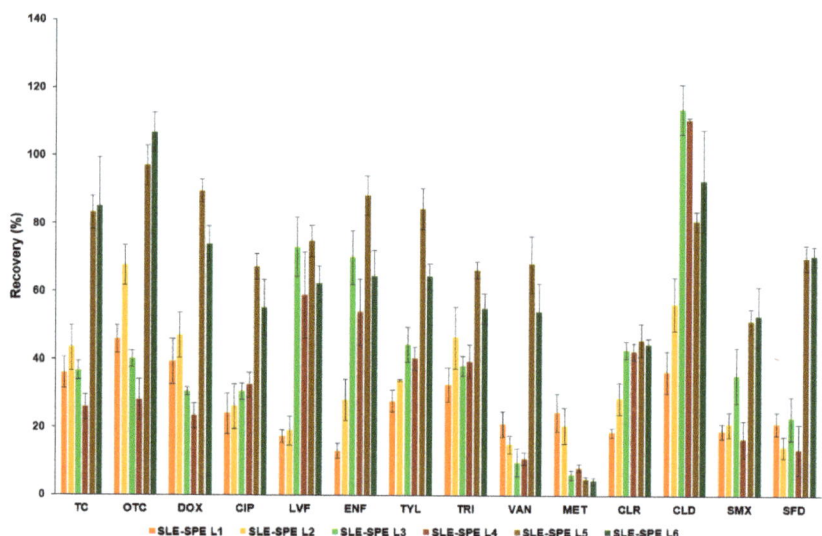

Figure 1. Comparison of the efficiency of FCs' extraction process following SLE-SPE procedure from parsley leaf.

Finally, we modified the SLE step by increasing the volume of solvent in a single extraction from 10 mL to 15 mL and eliminated the extraction with McIlvain buffer (pH = 4) with MeOH, replacing it with ACN in both extraction steps (SLE-SPE L5, SLE-SPE L6). In addition, 2% FA in a mixture of MeOH:ACN (SLE-SPE L5) and 0.1% AcA in MeOH (SLE-SPE L6) were tested in SPE step as low pH eluent, 0.1% NH_3 in MeOH was used for the second elution in both cases. Both procedures gave recoveries above 50% for all analytes except MET (4.3–4.7%) and CLR (44.4–45.8%). Application of SLE-SPE L5 and SLE-SPE L6 procedures allowed efficient extraction of 8 FCs: TC, OTC, DOX, CIP, ENF, LVF, VAN, CLD (Figure 1). The SLE-SPE L5 procedure gave better recoveries of the aforementioned compounds compared to SLE-SPE L6, respectively: 68.1% (VAN)–97.1% (OTC), 54.1% (VAN)–106.8% (OTC), hence, it was used as one of the sample preparation procedures for parsley leaves. The efficiency of SLE-SPE L5 extraction was low for the other 6 analytes (SMX, SFD, MET, TRI, TYL, CLR), so it was necessary to use a procedure based on double SLE extraction with MeOH for their separation (SLE L1, without using SPE; recoveries 61.9–89.9%).

3.1.4. Parsley Root–Development of SLE-SPE Procedure

SLE-SPE extraction of five selected FCs from parsley root has already been studied by our research team [38]. In the present study, the SLE-SPE procedure was also used to extract 14 selected analytes, and the SLE step was modified. Figure 2 compares the efficiency of FCs' separation from parsley root using SLE-SPE procedures. In all procedures (SLE-SPE R1, SLE-SPE R6), SPE extraction conditions were the same (Section 2.4.2). The majority of known SLE-SPE procedures have been validated for soil matrices, but less frequently for plant samples. Compared to parsley leaf, parsley root contains much fewer coeluting matrix compounds (essential oils, flavonoids) [39], which affect the extraction efficiency of FCs. Mixtures of McIlvain buffer (pH = 4) with organic solvents (ACN, MeOH) were used for the SLE of soil matrices [25,40]. For extraction of FCs from plant material (roots, leaves, seeds), the same SLE-SPE-based procedures are commonly applied, which differ in analyte recoveries depending on the studied tissue [9,23]. The selection of SLE-SPE extraction conditions from parsley root was similarly performed as for the leaf samples (Section 3.1.1). Performing a single extraction with a mixture of McIlvaine (pH = 4):ACN buffer (SLE-SPE R1) and McIlvaine (pH = 4):MeOH buffer (SLE-SPE R2), gave FCs recoveries of 23.4% (SMX)–78.2% (CIP) and 20.5% (ENF)–93.3% (DOX), respectively. Single SLE-SPE R1 and SLE-SPE R2 extractions were ineffective in the separation of SAs and FQs from the root matrix. Two-step SLE and combination of the mixtures in a single procedure was examined to increase recovery (SLE-SPE R3). In SLE-SPE R3, a significantly higher recovery of CLR (84.4%) and a slight increase in recovery of SMX (33.8%) and SFD (40.1%) were observed. Improved extraction efficiency of fluoroquinolones was examined using extraction under alkaline conditions (SLE-SPE R4, SLE-SPE R5) as the second SLE step, which worked well for soil matrices [26]. The use of a mixture of 0.2 M NaOH with acetone improved the recovery of fluoroquinolones (CIP: 69.8–108.9%, LVF: 37–69.8%, ENF: 57.5–88.3%), but decreased significantly the recovery of TRI, MET, and CLD, hence, this approach was not examined further. Finally, a one-step extraction with a mixture of ACN:McIlvaine buffer (pH = 4):0.1 M EDTA (2:2:1; $v/v/v$) (SLE-SPE R6) was used, which gave good recoveries for all analytes 51.7–95.6% and the highest repeatability (CV: 2.5–9.6%). Procedures using 0.1 M EDTA in mixtures of different compositions have been reported for soil matrices and manure [25], but have not been applied to the separation of FCs from plant roots.

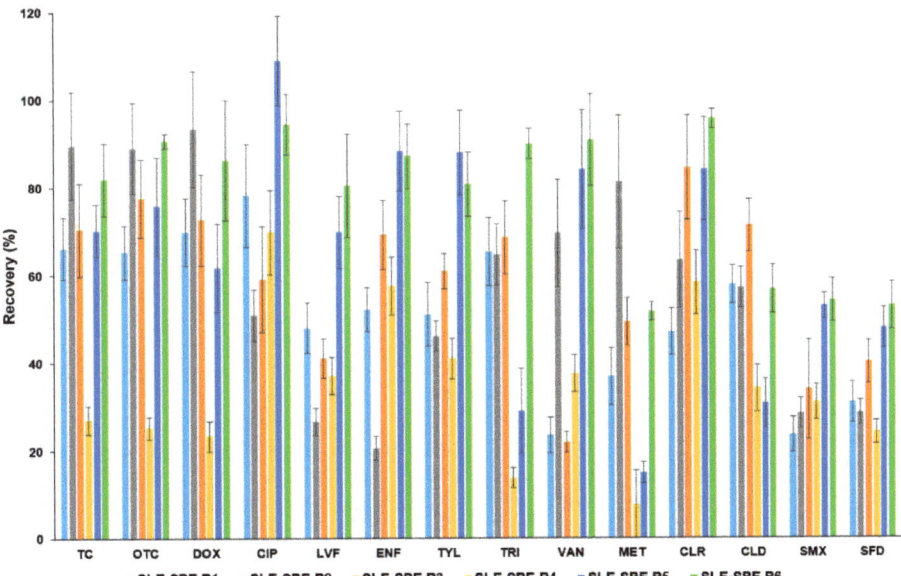

Figure 2. Comparison of efficiencies using SLE-SPE procedures for the separation of analytes from parsley root: SLE-SPE R1: McIlvaine buffer (pH = 4):ACN (1:1; v/v); SPE-SPE R2: McIlvaine buffer (pH = 4):MeOH (1:1; v/v); SLE-SPE R3: McIlvaine buffer (pH = 4):MeOH (1:1; v/v), McIlvaine buffer (pH = 4):ACN (1:1; v/v); SLE-SPE R4: McIlvaine buffer (pH = 4):ACN (1:1; v/v), 0.2 M NaOH:acetone (1:1; v/v); SLE-SPE R5: McIlvaine buffer (pH = 4):MeOH (1:1; v/v), 0.2 M NaOH:acetone (1:1; v/v); SLE-SPE R6: ACN: McIlvaine buffer (pH = 4):0.1 M EDTA (2:2:1; $v/v/v$).

3.2. Method Validation

Tables S4 and S5 present the validation parameters of the developed LC-MS/MS method for parsley root and leaf matrix, respectively. Validation was performed using FCs-free sample extracts prepared according to the procedure described in Section 2.4, enriched with the appropriate amount of analytes. Analyses were performed in six independent replicates. Calibration curves were linear in the range of 1–1200 ng·g^{-1} for all analytes except VAN and TC, for which the range was 5–1200 ng·g^{-1}. The coefficient of fit for the curves (R^2) was in the range of 0.9858–0.9988, indicating a good fit of the curves. The sensitivity of the method was determined by LOD and LOQ, which were 0.3–1.6 ng·g^{-1} and 1.0–5.0 ng·g^{-1}, respectively.

Accuracy and precision were determined at three concentration levels (100 ng·g^{-1}, 400 ng·g^{-1}, 1000 ng·g^{-1}). The precision determined as RE (%) was less than 9.60% for the extracts from parsley root and 7.90% from parsley leaf extracts. The precision of the method determined by CV was in the range of 0.40–7.56% for the root matrix and 0.49–11.42% for the parsley leaf matrix. The recoveries determined for the parsley root matrix were in the range of 51.7% (MET)–95.6% (CLR) in HQL samples. In the case of the parsley leaf sample, the recoveries of 6 analytes (SMX, SFD, MET, TRI, TYL, CLR) in double MeOH extraction procedure were in the range of 61.9% (CLR)–89.9% (TYL) in HQL samples. In the parsley leaf SLE-SPE procedure, the recoveries of the remaining 8 analytes (TC, OTC, DOX, CIP, ENF, LVF, VAN, CLD) were in the range of 67.3% (CIP)–97.1% (OTC) in HQL samples. The recoveries obtained were satisfactory and similar to those previously reported for plant matrices [36,41]. The selectivity of the method was obtained by comparing the chromatograms recorded in MRM mode for the analytes after extraction of the plant material and the blank. No interference of matrix components was observed in the chromatograms, with retention times corresponding to the analytes.

The matrix effect was also examined for both extracts from parsley root and leaf samples (Tables S4 and S5). The matrix effect was influenced by both matrix type and sample preparation [42]. It was observed that the matrix effect significantly affected the signal intensity of analytes SFD (signal attenuation, ME = −11.35%) and CLD (signal enhancement, ME = 11.31%) for parsley root matrix. For the other analytes, the matrix effect was negligible (ME < 7.81). Similarly for parsley leaf, where ME was in the range of −6.74–7.96%.

3.3. Investigation of FCs Accumulation in Plant Tissues

Four FCs were detected in plant tissue samples collected from plots enriched with poultry manure (PMPA) and cattle manure (BMPA): DOX, TYL, SMX and ENF (Table 1). ENF was detected at the highest concentration, ranging from 23.9–29.3 ng·g^{-1} in parsley leaves, and detected at 13.4–25.3 ng·g^{-1} in parsley roots (Table 1). For example, in a report, three crop species (soybean, corn, and bean) were examined, where root absorption was the main route of ENF uptake by the plant, and its content was in the range of 1.68 ng g^{-1}–26.17 µg·g^{-1} [43]. Parsley samples showed slightly higher ENF content in parsley leaves than parsley roots. The different tendency in accumulation was probably related to morphophysiological differences between plant species, in which the type and distribution of tissues and the presence of apoplastic barriers limited the transport of selected ionic forms of pharmaceuticals in the plant [44]. SMX was detected in trace amounts (<6.83 ng·g^{-1}) in all parsley samples grown in manure-enriched soil (PMPA, BMPA). According to literature data, SMX is present in the soil environment in anionic form, which causes repulsion from the root cell apoplast and reduces its uptake from the soil [45]. Another reason for low uptake may be the rapid degradation rate of SMX in the soil under environmental conditions [45]. TYL was only detected in parsley leaves, but its concentration was below LOQ (<1.0 ng·g^{-1}). In 11 vegetables grown in TYL-enriched soil, 0.2–2.4 ng·g^{-1} was detected, and in most cases the result was below LOQ [3]. According to the literature, TYL in water, soil or manure environment is biodegradable up to 30 days [46]. Our results for parsley samples confirmed the low absorption of TYL by the plant.

Table 1. Concentrations of selected pharmaceuticals in plant tissues.

	Concentration (ng·g^{-1} Dry Weight) (SD)					
	PMPA		BMPA		CP	
Tissues	leaf	root	leaf	root	leaf	root
TC	nd	nd	nd	nd	nd	nd
OTC	nd	nd	nd	nd	nd	nd
DOX	2.1 (0.7)	14.0 (0.5)	3.0 (0.1)	13.6 (4.4)	nd	nd
CIP	nd	nd	nd	nd	nd	nd
ENF	23.9 (3.0)	13.4 (0.5)	29.3 (3.9)	25.3 (3.7)	nd	nd
LVF	nd	nd	nd	nd	nd	nd
MET	nd	nd	nd	nd	nd	nd
CLR	nd	nd	nd	nd	nd	nd
CLD	nd	nd	nd	nd	nd	nd
SMX	2.3 (0.1)	<LOQ	2.3 (0.2)	6.8 (2.2)	nd	nd
SFD	nd	nd	nd	nd	nd	nd
TYL	<LOQ	nd	<LOQ	nd	nd	nd
TRI	nd	nd	nd	nd	nd	nd
VAN	nd	nd	nd	nd	nd	nd

PMPA-soil plots supplemented with poultry manure; BMPA-soil plots supplemented with bovine manure; CP-control plots without supplementation; nd–not detected.

In the case of DOX, higher concentrations were detected in the roots (13.62–14.02 ng·g^{-1}) than in the green part of the plant (2.06–3.08 ng·g^{-1}). DOX in soil is resistant to degradation processes and has a half-life of 533 days [47]. Additionally, it does not show toxic properties to plants and adult earthworms [48]. However, studies conducted by Litskas et al. [49] concurrently confirm that DOX has an inhibitory effect on earthworm reproduction in soil, where the abundance of juvenile organisms was significantly lower in soil enriched with this FC. DOX is mobile in the environment and detectable in soil solutions [48]. It is the least studied pharmaceutical in the whole group of tetracyclines, where studies have shown that DOX was not detected in lettuce [36], while it was present in radish and pak choi [50] when grown in soil supplemented with DOX. The range of concentrations of pharmaceuticals determined in our study (1.0–29.3 ng g^{-1}) was comparable to those reported, where 1.6–2.8 ng·g^{-1} ENF, 0.5–1.5 ng·g^{-1}, TYL and SMX 0.05–6.8 ng·g^{-1} were detected [10].

Once absorbed by edible plants, FCs are carried back into the food chain, potentially endangering human health. FCs' residues in edible plants induce antibiotic resistance development and promote the transfer of antibiotic-resistant bacteria to humans [51,52]. In a study conducted in Germany for 1001 food plants for 5 bacterial strains, the drug resistance genes DOX, ENF, SMX and TYL were detected. Relatively high resistance rates *E. faecalis* were also observed for doxycycline (23%) and tylosin (10%) [53]. Another concern about FCs in food is the development of severe allergic reactions. The constant consumption of even low amounts of FCs in food can reduce fertility, be carcinogenic and cause obesity [54].

3.4. Identification of Transformation Products of Selected FCs in Plant Tissues

The photolysis of selected FCs is limited to the top layer of soil where light reaches (7–8 mm), and its rate is two orders lower than for the process carried out in water [55,56]. Direct photodegradation is not considered the main pathway for FCs' degradation in soils. FCs are efficiently transformed and degraded by bacterial strains and fungi present in soils [56,57]. Thus, it was assumed that both TPs formed in the soil and adsorbed by the plant and those formed by FCs' metabolism may be present in the tissues of harvested plants.

10 ENF TPs and two DOX TPs were detected in plant samples grown in manure-fertilized soils. SMX216 was detected in parsley root samples from PMPA and BMPA plots, and SMX158 was only detected in parsley root from PMPA plot. Table 2 shows the

distribution of TPs according to the plot where parsley was grown. Structure identification of TPs was performed in two steps: first, screening analysis using pseudo-MRM mode; then structures were confirmed by recording mass spectra using EMS-IDA-EPI. Table 3 shows the structures of the identified TPs and their characteristic p-MRM transitions. The obtained results were confirmed based on literature data.

Table 2. Identified transformation products in parsley leaf and root samples.

Lp.	TPs	PMPA		BMPA		CP	
		Leaf	Root	Leaf	Root	Leaf	Root
1	ENF263	+	+	+	+	−	−
2	ENF330	−	+	−	+	−	−
3	ENF390	−	+	−	+	−	−
4	ENF282	+	−	+	−	−	−
5	ENF342	+	−	−	+	−	−
6	ENF376	+	−	+	−	−	−
7	ENF334	+	−	+	−	−	−
8	ENF306	+	−	+	−	−	−
9	ENF223	−	+	−	+	−	−
10	ENF291	−	+	−	+	−	−
11	DOX274	−	+	−	+	−	−
12	DOX399	−	+	−	+	−	−
13	SMX158	−	+	−	−	−	−
14	SMX216	−	+	−	+	−	−

According to the obtained results, different tissue distribution of ENF TPs was observed, four ENFs were detected exclusively in leaves (ENF282, ENF376, ENF334, ENF306), four exclusively in roots (ENF330, ENF390, ENF223, ENF291), and two (ENF263, ENF342) in both leaf and root. Two TPs formed by attachment of a hydroxyl group to a quinolone ring (ENF376) or piperazine ring (ENF390) were identified. In ENF390, the ethyl chain at the nitrogen atom on the piperazine ring was additionally oxidized. ENF328 was formed by defluorination of ENF, cleavage of the ethyl group from the piperazine ring, and subsequent oxidation. Hydroxylation and oxidation are typical reactions in biotic transformations of FCs [58]. ENF376 has been described as a ENF metabolite produced by basidiomes [59] and detected in chicken tissues after ENF administration [60]. To date, ENF376, ENF390 and ENF328 have only been identified after ENF photodegradation processes [61,62]. The four ENF TPs were formed by piperazine ring opening. ENF306 was formed by dissociation of the ethyl group from the piperazine ring and opening with loss of C_2H_4. For ENF334, loss of C_2H_4 from the piperazine ring was also observed, but a carbonyl group was attached at the nitrogen atom instead of an ethyl group. ENF330 was formed by dissociation of fluorine and ethyl group and opening of the piperazine ring followed by oxidation. ENF291 was formed by cleavage of the piperazine ring and loss of the secondary amine nitrogen, followed by oxidation of the methylamine group. Therefore, the transformation of FQs present in soil was mainly by oxidation of the piperazine side chain, although the fluoroquinolone ring remained intact [55]. ENF306, ENF330, and ENF291 are photodegradation products of ENF or its metabolite CIP [62,63], and ENF334 is a metabolite isolated from pharmaceutical slime [64]. Hence, it was possible that these four TPs were formed in the environment and uptaken by the plant. Three TPs were formed by detachment of the main groups that form ENF. ENF342 was formed by dissociation of fluorine from the quinolone ring. Fluorine disconnection occurred by photodegradation [62] and microbial degradation [20]. ENF263 was formed by C-N bond disruption and piperazine ring disconnection. ENF223 was formed by dissociation of the cyclopropyl ring from the ENF263 molecule. According to the literature, ENF263 and ENF223 are formed by biodegradation carried out by microorganisms [58,64]. DOX accumulated mostly in the parsley roots, its two TPs, DOX399 and DOX274, were detected. DOX399 was formed via C-C bond cleavage and dissociation of the amide group from the DOX ring. The transformation pathway leading to the formation of DOX274 from

DOX was studied in photodegradation and involved C-C bond breaking, C-N bond breaking and dehydration reactions sequentially [65]. The transformation pathways of DOX under environmental conditions have not been reported in detail. DOX TPs were detected in the photodegradation process [65], but it was disregarded that they were formed by microbial activity or plant metabolism. Two SMX TPs were also detected in parsley roots. SMX158 was formed via S-N bond cleavage and detachment of the isoxazole ring. SMX216 was formed via isoxazole ring opening and dissociation of the -C_2H_4 group. Reports have described SMX158 TPs in the degradation of SMX under ionizing radiation, which may suggest that it was formed by SMX oxidation [66] and microorganisms [67]. However, SMX216 was identified as a metabolic product of sulfate-reducing bacteria, hence, it was observed more frequently in oxygen-deficient environments in its free form [68]. As noted, transformation of pharmaceuticals in the environment can occur due to biotic (microbial activity) and abiotic (photolysis) factors [43]. The response of the plant organism to FCs' transformation products and their tissue distribution depends on the plant species [43,44]. The metabolism of FCs is generally considered to be the detoxification mechanism of the plant; however, there is concern that some metabolites may have the ability to acquire or potentiate bactericidal activity [9,44].

Table 3. Compilation of the determined pharmaceuticals and their identified transformation products.

Analytes	Formula	[M + H]$^+$ (m/z)	Fragmentation Ions (m/z)	Structure	Ref.
			Pharmaceuticals		
DOX	$C_{22}H_{24}N_2O_8$	445.2	428.2 154.1		
ENF	$C_{19}H_{22}FN_3O_3$	360.7	316.2 245.1		[20]
TYL	$C_{46}H_{77}NO_{17}$	916.1	174.2 775.5		
SMX	$C_{10}H_{11}N_3O_3S$	253.9	92.0 108.0		

Table 3. Cont.

Analytes	Formula	[M + H]⁺ (m/z)	Fragmentation Ions (m/z)	Structure	Ref.
			Transformation products		
ENF263	$C_{13}H_{11}FN_2O_3$	263.0	245.0 204.0		[46,49,69–71]
ENF330	$C_{17}H_{19}N_3O_4$	330.0	312.0 284.0		[62]
ENF390	$C_{19}H_{20}FN_3O_5$	390.0	372.0		[61]
ENF328	$C_{17}H_{17}N_3O_4$	328.0	310.0 300.0		[62]
ENF342	$C_{19}H_{23}N_3O_3$	342.0	324.0 301.3 297.0		[62]
ENF376	$C_{19}H_{22}FN_3O_4$	376.1	358.1 340.1		[59,62,72]
ENF334	$C_{16}H_{16}FN_3O_4$	334.0	316.0 217.0		[63,64]

Table 3. Cont.

Analytes	Formula	[M + H]⁺ (m/z)	Fragmentation Ions (m/z)	Structure	Ref.
ENF306	$C_{15}H_{16}FN_3O_3$	306.9	288.0 271.0 145.0		[73,74]
ENF223	$C_{10}H_7FN_2O_3$	223.0	207.0 190.0		[64]
ENF291	$C_{14}H_{11}FN_2O_4$	291.0	273.0 245.0 217.0		[63]
DOX274	$C_{16}H_{18}O_4$	275.0	275.0 257.0 247.0 229.0		[65], this work
DOX399	$C_{21}H_{21}NO_7$	400.0	400.0 283.0 265.0 211.0		[75], this work
SMX158	$C_6H_7NO_2S$	158.0	140.0 92.0		[66]
SMX216	$C_7H_9N_3O_3S$	216.0	156.0 202.2		[68]

4. Conclusions

In this study, efficient methods were developed for the extraction of 14 common FCs' contaminants from parsley root and leaf samples. The LC-MS/MS technique was used for determination. Depending on the matrix and compound properties, double SLE with MeOH (leaf tissue: TYL, TRI, MET, CLR, SMX, SFD) or combined SLE-SPE (leaf tissue: TC, OTC, DOX, CIP, ENF, LVF, CLD, VAN; root tissue: all FCs) was used. Analyte recoveries from leaf and root tissues were at satisfactory levels (51.7–97.1% for HQL samples), and LOQs ranged from 1–5 ng·g^{-1}.

Veterinary drug residues introduced with animal manure to the soil were adsorbed by plants to varying degrees. DOX, TYL, SMX and ENF were detected in plant samples. DOX bioaccumulated mainly in parsley roots, whereas higher concentrations of ENF were detected in parsley leaves. TYL was not detected in plant roots, and concentrations in leaves were below LOQ, indicating its poor bioaccumulation in the plant. SMX was detected in all plants grown in manure-amended soil. Screening analysis identified 14 transformation products of ENF, SMX, and DOX. ENF TPs were formed through hydroxylation, oxidation, and piperazine ring opening reactions. In the case of SMX and DOX, TPs were formed by C-C and S-N bond-breaking.

Due to the widespread practice of zoonotic fertilizer utilization, FCs can be transferred to food crops and subsequently enter the human food chain. Although the concentrations of detected FCs in parsley were at low levels (<29.26 ng·g^{-1}), their ability to accumulate and further metabolize in living tissues is of concern. Further studies are needed to determine the extent of the risks from the use of zoonotic fertilizers.

Supplementary Materials: The following supporting information can be downloaded at: https://www.mdpi.com/article/10.3390/molecules27144378/s1, Figure S1: Comparison of the efficiency of analyte extraction following SLE procedure combined with LLE purification stage (matrix: parsley leaves: LLE solvent: chloroform). Figure S2: Comparison of sorbents used for purification of extracts in SLE-SPE procedure of parsley leaf. Eluent composition for MAX, C18, HLB: 0.1% AcA in MeOH (10 mL), 0.1% NH$_3$ in MeOH (10 mL), for WCX:MeOH (3 mL), 2% FA in MeOH/ACN (2:8; *v/v*) (3 mL). Table S1: Variable parameters in tested SLE procedures. Table S2: Variable parameters in tested SLE-SPE procedures. Table S3: The application of different solvents to SLE procedure for parsley leaf samples. Table S4: The analytical method parameters and extraction recovery of antibiotics in parsley root samples at three different concentrations (*n* = 6). Table S5: The analytical method parameters and extraction recovery of antibiotics in parsley leaf samples at three different concentrations (*n* = 6).

Author Contributions: Conceptualization, E.K. and S.B.; formal analysis, K.S.; funding acquisition, E.K.; methodology, K.S. and S.B.; project administration, E.K.; supervision, E.K. and S.B.; validation, K.S.; visualization, K.S. and S.B.; writing—original draft, K.S.; writing—review & editing, E.K., E.F., M.H. and S.B. All authors have read and agreed to the published version of the manuscript.

Funding: This study was supported by grant No. 2017/27/B/NZ9/00267 from the National Science Centre (Poland).

Institutional Review Board Statement: Not applicable.

Informed Consent Statement: Not applicable.

Data Availability Statement: All data are included in the article.

Conflicts of Interest: The authors declare no conflict of interest.

Sample Availability: Samples of the compounds are not available from the authors.

References

1. Perkasa, A.Y.; Gunawan, E.; Dewi, S.A.; Zulfa, U. The Testing of Chicken Manure Fertilizer Doses to Plant Physiology Components and Bioactive Compound of Dewa Leaf. *Procedia Environ. Sci.* **2016**, *33*, 54–62. [CrossRef]
2. Sarker, Y.A.; Rashid, S.Z.; Sachi, S.; Ferdous, J.; das Chowdhury, B.L.; Tarannum, S.S.; Sikder, M.H. Exposure Pathways and Ecological Risk Assessment of Common Veterinary Antibiotics in the Environment through Poultry Litter in Bangladesh. *J. Environ. Sci. Health Part B Pestic. Food Contam. Agric. Wastes* **2020**, *55*, 1061–1068. [CrossRef] [PubMed]
3. Kang, D.H.; Gupta, S.; Rosen, C.; Fritz, V.; Singh, A.; Chander, Y.; Murray, H.; Rohwer, C. Antibiotic Uptake by Vegetable Crops from Manure-Applied Soils. *J. Agric. Food Chem.* **2013**, *61*, 9992–10001. [CrossRef] [PubMed]
4. Chen, Z.; Jiang, X. Microbiological Safety of Chicken Litter or Chicken Litter-Based Organic Fertilizers: A Review. *Agriculture* **2014**, *4*, 1–29. [CrossRef]
5. Bochkareva, I.; Maymanova, E. Poultry Farms As a Source of Environmental Pollution. *Interexpo GEO-Sib.* **2019**, *4*, 106–111. [CrossRef]

6. Pan, M.; Wong, C.K.C.; Chu, L.M. Distribution of Antibiotics in Wastewater-Irrigated Soils and Their Accumulation in Vegetable Crops in the Pearl River Delta, Southern China. *J. Agric. Food Chem.* **2014**, *62*, 11062–11069. [CrossRef]
7. Brundrett, M.C.; Ferguson, B.J.; Gressshoff, P.M.; Mathesius, U.; Munns, R.; Rasmussen, A.; Ryan, M.H.; Schmidt, S.; Watt, M. Plants in Action. Available online: https://rseco.org/index.html (accessed on 5 July 2022).
8. Kodešová, R.; Klement, A.; Golovko, O.; Fér, M.; Nikodem, A.; Kočárek, M.; Grabic, R. Root Uptake of Atenolol, Sulfamethoxazole and Carbamazepine, and Their Transformation in Three Soils and Four Plants. *Environ. Sci. Pollut. Res.* **2019**, *26*, 9876–9891. [CrossRef]
9. Tian, R.; Zhang, R.; Uddin, M.; Qiao, X.; Chen, J.; Gu, G. Uptake and Metabolism of Clarithromycin and Sulfadiazine in Lettuce. *Environ. Pollut.* **2019**, *247*, 1134–1142. [CrossRef]
10. Pan, M.; Chu, L.M. Fate of Antibiotics in Soil and Their Uptake by Edible Crops. *Sci. Total Environ.* **2017**, *599–600*, 500–512. [CrossRef]
11. Chung, H.S.; Lee, Y.J.; Rahman, M.M.; Abd El-Aty, A.M.; Lee, H.S.; Kabir, M.H.; Kim, S.W.; Park, B.J.; Kim, J.E.; Hacımüftüoğlu, F.; et al. Uptake of the Veterinary Antibiotics Chlortetracycline, Enrofloxacin, and Sulphathiazole from Soil by Radish. *Sci. Total Environ.* **2017**, *605–606*, 322–331. [CrossRef]
12. Chen, H.R.; Rairat, T.; Loh, S.H.; Wu, Y.C.; Vickroy, T.W.; Chou, C.C. Assessment of Veterinary Drugs in Plants Using Pharmacokinetic Approaches: The Absorption, Distribution and Elimination of Tetracycline and Sulfamethoxazole in Ephemeral Vegetables. *PLoS ONE* **2017**, *12*, e0183087. [CrossRef] [PubMed]
13. Minden, V.; Schnetger, B.; Pufal, G.; Leonhardt, S.D. Antibiotic-Induced Effects on Scaling Relationships and on Plant Element Contents in Herbs and Grasses. *Ecol. Evol.* **2018**, *8*, 6699–6713. [CrossRef] [PubMed]
14. Kim, C.; Ryu, H.D.; Chung, E.G.; Kim, Y.; Lee, J.K. A Review of Analytical Procedures for the Simultaneous Determination of Medically Important Veterinary Antibiotics in Environmental Water: Sample Preparation, Liquid Chromatography, and Mass Spectrometry. *J. Environ. Manag.* **2018**, *217*, 629–645. [CrossRef] [PubMed]
15. World Health Organization. *WHO Report on Surveillance of Antibiotic Consumption 2016–2018 Early Implementation*; World Health Organization: Switzerland, Geneva, 2018; ISBN 9789241514880.
16. Patyra, E.; Kwiatek, K.; Nebot, C.; Gavilán, R.E. Quantification of Veterinary Antibiotics in Pig and Poultry Feces and Liquid Manure as a Non-Invasive Method to Monitor Antibiotic Usage in Livestock by Liquid Chromatography Mass-Spectrometry. *Molecules* **2020**, *25*, 3265. [CrossRef] [PubMed]
17. Osiński, Z.; Patyra, E.; Kwiatek, K. HPLC-FLD-Based Method for the Detection of Sulfonamides in Organic Fertilizers Collected from Poland. *Molecules* **2022**, *27*, 2031. [CrossRef]
18. Łukaszewicz, P.; Kumirska, J.; Białk-Bielińska, A.; Dołżonek, J.; Stepnowski, P. Assessment of Soils Contamination with Veterinary Antibiotic Residues in Northern Poland Using Developed MAE-SPE-LC/MS/MS Methods. *Environ. Sci. Pollut. Res.* **2017**, *24*, 21233–21247. [CrossRef]
19. Barnekow, U.; Feseno, S.; Kashparov, V.; Kis-Bendek, G.; Matissof, G.; Onda, Y.; Sanzharova, N.; Tarjan, S.; Tyler, A.; Varga, B. *Guidelines on Soil and Vegetation Sampling for Radiological Monitoring*; International Atomic Energy Agency: Vienna, Austria, 2019.
20. Kokoszka, K.; Zieliński, W.; Korzeniewska, E.; Felis, E.; Harnisz, M.; Bajkacz, S. Suspect Screening of Antimicrobial Agents Transformation Products in Environmental Samples Development of LC-QTrap Method Running in Pseudo MRM Transitions. *Sci. Total Environ.* **2022**, *808*, 152114. [CrossRef]
21. Kokoszka, K.; Wilk, J.; Felis, E.; Bajkacz, S. Application of UHPLC-MS/MS Method to Study Occurrence and Fate of Sulfonamide Antibiotics and Their Transformation Products in Surface Water in Highly Urbanized Areas. *Chemosphere* **2021**, *283*, 131189. [CrossRef]
22. Pan, M.; Chu, L.M. Transfer of Antibiotics from Wastewater or Animal Manure to Soil and Edible Crops. *Environ. Pollut.* **2017**, *231*, 829–836. [CrossRef]
23. Zhao, F.; Yang, L.; Chen, L.; Li, S.; Sun, L. Bioaccumulation of Antibiotics in Crops under Long-Term Manure Application: Occurrence, Biomass Response and Human Exposure. *Chemosphere* **2019**, *219*, 882–895. [CrossRef]
24. ben Mordechay, E.; Mordehay, V.; Tarchitzky, J.; Chefetz, B. Pharmaceuticals in Edible Crops Irrigated with Reclaimed Wastewater: Evidence from a Large Survey in Israel. *J. Hazard. Mater.* **2021**, *416*, 126184. [CrossRef]
25. Ho, Y.B.; Zakaria, M.P.; Latif, P.A.; Saari, N. Simultaneous Determination of Veterinary Antibiotics and Hormone in Broiler Manure, Soil and Manure Compost by Liquid Chromatography-Tandem Mass Spectrometry. *J. Chromatogr. A* **2012**, *1262*, 160–168. [CrossRef] [PubMed]
26. Bian, K.; Liu, Y.H.; Wang, Z.N.; Zhou, T.; Song, X.Q.; Zhang, F.Y.; He, L.M. Determination of Multi-Class Antimicrobial Residues in Soil by Liquid Chromatography-Tandem Mass Spectrometry. *RSC Adv.* **2015**, *5*, 27584–27593. [CrossRef]
27. Łukaszewicz, P.; Białk-Bielińska, A.; Dołżonek, J.; Kumirska, J.; Caban, M.; Stepnowski, P. A New Approach for the Extraction of Tetracyclines from Soil Matrices: Application of the Microwave-Extraction Technique. *Anal. Bioanal. Chem.* **2018**, *410*, 1697–1707. [CrossRef]
28. Cui, Q.; Ni, x.; Zeng, L.; Tu, Z.; Li, J.; Sun, K.; Chen, X.; Li, X. Optimization of Protein Extraction and Decoloration Conditions for Tea Residues. *Hortic. Plant J.* **2017**, *3*, 172–176. [CrossRef]
29. Sari, Y.W.; Mulder, W.J.; Sanders, J.P.M.; Bruins, M.E. Towards Plant Protein Refinery: Review on Protein Extraction Using Alkali and Potential Enzymatic Assistance. *Biotechnol. J.* **2015**, *10*, 1138–1157. [CrossRef] [PubMed]

30. Bajkacz, S.; Rusin, K.; Wolny, A.; Adamek, J.; Erfurt, K.; Chrobok, A. Highly Efficient Extraction Procedures Based on Natural Deep Eutectic Solvents or Ionic Liquids for Determination of 20-Hydroxyecdysone in Spinach. *Molecules* **2020**, *25*, 4736. [CrossRef] [PubMed]
31. Sariyildiz, T.; Anderson, J.M. Variation in the Chemical Composition of Green Leaves and Leaf Litters from Three Deciduous Tree Species Growing on Different Soil Types. *For. Ecol. Manag.* **2005**, *210*, 303–319. [CrossRef]
32. Wood, L.W. Chloroform-Methano Extraction of Chlorophyll α. *J. Fish. Aqunt. Sci* **1985**, *42*, 2–7.
33. Siek, T.J. Effective Use of Organic Solvents to Remove Drugs from Biologic Specimens. *Clin. Toxicol.* **1978**, *13*, 205–230. [CrossRef]
34. Li, W.; Dai, X.; Pu, E.; Bian, H.; Chen, Z.; Zhang, X.; Guo, Z.; Li, P.; Li, H.; Yong, Y.; et al. HLB-MCX-Based Solid-Phase Extraction Combined with Liquid Chromatography-Tandem Mass Spectrometry for the Simultaneous Determination of Four Agricultural Antibiotics (Kasugamycin, Validamycin A, Ningnanmycin, and Polyoxin B) Residues in Plant-Origin Foods. *J. Agric. Food Chem.* **2020**, *68*, 14025–14037. [CrossRef] [PubMed]
35. Wang, C.; Li, H.; Wang, N.; Li, H.; Fang, L.; Dong, Z.; Du, H.; Guan, S.; Zhu, Q.; Chen, Z.; et al. Simultaneous Analysis of Kasugamycin and Validamycin-A in Fruits and Vegetables Using Liquid Chromatography-Tandem Mass Spectrometry and Consecutive Solid-Phase Extraction. *Anal. Methods* **2017**, *9*, 634–642. [CrossRef]
36. Albero, B.; Tadeo, J.L.; del Mar Delgado, M.; Miguel, E.; Pérez, R.A. Analysis of Multiclass Antibiotics in Lettuce by Liquid Chromatography-Tandem Mass Spectrometry to Monitor Their Plant Uptake. *Molecules* **2019**, *24*, 4066. [CrossRef] [PubMed]
37. Wu, X.; Conkle, J.L.; Gan, J. Multi-Residue Determination of Pharmaceutical and Personal Care Products in Vegetables. *J. Chromatogr. A* **2012**, *1254*, 78–86. [CrossRef]
38. Kokoszka, K.; Kobus, A.; Bajkacz, S. Optimization of a Method for Extraction and Determination of Residues of Selected Antimicrobials in Soil and Plant Samples Using HPLC-UV-MS/MS. *Int. J. Environ. Res. Public Health* **2021**, *18*, 1159. [CrossRef]
39. Punoševac, M.; Radović, J.; Leković, A.; Kundaković-Vasović, T. A Review of Botanical Characteristics, Chemical Composition, Pharmacological Activity and Use of Parsley. *Arh. Farm.* **2021**, *71*, 177–196. [CrossRef]
40. Hou, J.; Wan, W.; Mao, D.; Wang, C.; Mu, Q.; Qin, S.; Luo, Y. Occurrence and Distribution of Sulfonamides, Tetracyclines, Quinolones, Macrolides, and Nitrofurans in Livestock Manure and Amended Soils of Northern China. *Environ. Sci. Pollut. Res.* **2015**, *22*, 4545–4554. [CrossRef]
41. Xie, Z.; Gan, Y.; Tang, J.; Fan, S.; Wu, X.; Li, X.; Cheng, H.; Tang, J. Combined Effects of Environmentally Relevant Concentrations of Diclofenac and Cadmium on Chironomus Riparius Larvae. *Ecotoxicol. Environ. Saf.* **2020**, *202*, 110906. [CrossRef]
42. Zhou, W.; Yang, S.; Wang, P.G. Matrix Effects and Application of Matrix Effect Factor. *Bioanalysis* **2017**, *9*, 1839–1844. [CrossRef]
43. Marques, R.Z.; Wistuba, N.; Brito, J.C.M.; Bernardoni, V.; Rocha, D.C.; Gomes, M.P. Crop Irrigation (Soybean, Bean, and Corn) with Enrofloxacin-Contaminated Water Leads to Yield Reductions and Antibiotic Accumulation. *Ecotoxicol. Environ. Saf.* **2021**, *216*, 112193. [CrossRef]
44. Rocha, D.C.; da Silva Rocha, C.; Tavares, D.S.; de Morais Calado, S.L.; Gomes, M.P. Veterinary Antibiotics and Plant Physiology: An Overview. *Sci. Total Environ.* **2021**, *767*, 144902. [CrossRef] [PubMed]
45. Malchi, T.; Maor, Y.; Tadmor, G.; Shenker, M.; Chefetz, B. Irrigation of Root Vegetables with Treated Wastewater: Evaluating Uptake of Pharmaceuticals and the Associated Human Health Risks. *Environ. Sci. Technol.* **2014**, *48*, 9325–9333. [CrossRef] [PubMed]
46. Tasho, R.P.; Cho, J.Y. Veterinary Antibiotics in Animal Waste, Its Distribution in Soil and Uptake by Plants: A Review. *Sci. Total Environ.* **2016**, *563–564*, 366–376. [CrossRef] [PubMed]
47. Walters, E.; McClellan, K.; Halden, R.U. Occurrence and Loss over Three Years of 72 Pharmaceuticals and Personal Care Products from Biosolids-Soil Mixtures in Outdoor Mesocosms. *Water Res.* **2010**, *44*, 6011–6020. [CrossRef]
48. Fernández, C.; Alonso, C.; Babín, M.M.; Pro, J.; Carbonell, G.; Tarazona, J.v. Ecotoxicological Assessment of Doxycycline in Aged Pig Manure Using Multispecies Soil Systems. *Sci. Total Environ.* **2004**, *323*, 63–69. [CrossRef]
49. Litskas, V.D.; Karamanlis, X.N.; Prousali, S.P.; Koveos, D.S. The Xenobiotic Doxycycline Affects Nitrogen Transformations in Soil and Impacts Earthworms and Cultivated Plants. *J. Environ. Sci. Health Part A Toxic/Hazard. Subst. Environ. Eng.* **2019**, *54*, 1441–1447. [CrossRef]
50. Wang, J.; Lin, H.; Sun, W.; Xia, Y.; Ma, J.; Fu, J.; Zhang, Z.; Wu, H.; Qian, M. Variations in the Fate and Biological Effects of Sulfamethoxazole, Norfloxacin and Doxycycline in Different Vegetable-Soil Systems Following Manure Application. *J. Hazard. Mater.* **2016**, *304*, 49–57. [CrossRef]
51. Srichamnong, W.; Kalambaheti, N.; Woskie, S.; Kongtip, P.; Sirivarasai, J.; Matthews, K.R. Occurrence of Antibiotic-Resistant Bacteria on Hydroponically Grown Butterhead Lettuce (*Lactuca sativa* Var. *Capitata*). *Food Sci. Nutr.* **2021**, *9*, 1460–1470. [CrossRef]
52. Wang, H.; McEntire, J.C.; Zhang, L.; Li, X.; Doyle, M. The Transfer of Antibiotic Resistance from Food to Humans: Facts, Implications and Future Directions. *Rev. Sci. Tech. Off. Int. Epiz.* **2012**, *31*, 249–260. [CrossRef]
53. Schwaiger, K.; Helmke, K.; Hölzel, C.S.; Bauer, J. Antibiotic Resistance in Bacteria Isolated from Vegetables with Regards to the Marketing Stage (Farm vs. Supermarket). *Int. J. Food Microbiol.* **2011**, *148*, 191–196. [CrossRef]
54. Chen, J.; Ying, G.G.; Deng, W.J. Antibiotic Residues in Food: Extraction, Analysis, and Human Health Concerns. *J. Agric. Food Chem.* **2019**, *67*, 7569–7586. [CrossRef] [PubMed]
55. Sturini, M.; Speltini, A.; Maraschi, F.; Profumo, A.; Pretali, L.; Fasani, E.; Albini, A. Sunlight-Induced Degradation of Soil-Adsorbed Veterinary Antimicrobials Marbofloxacin and Enrofloxacin. *Chemosphere* **2012**, *86*, 130–137. [CrossRef] [PubMed]

56. Zhang, Y.; Xu, J.; Zhong, Z.; Guo, C.; Li, L.; He, Y.; Fan, W.; Chen, Y. Degradation of Sulfonamides Antibiotics in Lake Water and Sediment. *Environ. Sci. Pollut. Res.* **2013**, *20*, 2372–2380. [CrossRef]
57. Rusch, M.; Spielmeyer, A.; Zorn, H.; Hamscher, G. Degradation and Transformation of Fluoroquinolones by Microorganisms with Special Emphasis on Ciprofloxacin. *Appl. Microbiol. Biotechnol.* **2019**, *103*, 6933–6948. [CrossRef] [PubMed]
58. Parshikov, I.A.; Sutherland, J.B. Microbial Transformations of Antimicrobial Quinolones and Related Drugs. *J. Ind. Microbiol. Biotechnol.* **2012**, *39*, 1731–1740. [CrossRef] [PubMed]
59. Karl, W.; Schneider, J.; Wetzstein, H.G. Outlines of an "Exploding" Network of Metabolites Generated from the Fluoroquinolone Enrofloxacin by the Brown Rot Fungus Gloeophyllum Striatum. *Appl. Microbiol. Biotechnol.* **2006**, *71*, 101–113. [CrossRef]
60. Morales-Gutiérrez, F.J.; Barbosa, J.; Barrón, D. Metabolic Study of Enrofloxacin and Metabolic Profile Modifications in Broiler Chicken Tissues after Drug Administration. *Food Chem.* **2015**, *172*, 30–39. [CrossRef]
61. Guo, H.; Ke, T.; Gao, N.; Liu, Y.; Cheng, X. Enhanced Degradation of Aqueous Norfloxacin and Enrofloxacin by UV-Activated Persulfate: Kinetics, Pathways and Deactivation. *Chem. Eng. J.* **2017**, *316*, 471–480. [CrossRef]
62. Li, Y.; Zhang, F.; Liang, X.; Yediler, A. Chemical and Toxicological Evaluation of an Emerging Pollutant (Enrofloxacin) by Catalytic Wet Air Oxidation and Ozonation in Aqueous Solution. *Chemosphere* **2013**, *90*, 284–291. [CrossRef]
63. Paul, T.; Dodd, M.C.; Strathmann, T.J. Photolytic and Photocatalytic Decomposition of Aqueous Ciprofloxacin: Transformation Products and Residual Antibacterial Activity. *Water Res.* **2010**, *44*, 3121–3132. [CrossRef]
64. Pan, L.; Li, J.; Li, C.; Tang, X.; Yu, G.; Wang, Y. Study of Ciprofloxacin Biodegradation by a *Thermus Sp.* Isolated from Pharmaceutical Sludge. *J. Hazard. Mater.* **2018**, *343*, 59–67. [CrossRef] [PubMed]
65. Hong, P.; Li, Y.; He, J.; Saeed, A.; Zhang, K.; Wang, C.; Kong, L.; Liu, J. Rapid Degradation of Aqueous Doxycycline by Surface $CoFe_2O_4/H_2O_2$ System: Behaviors, Mechanisms, Pathways and DFT Calculation. *Appl. Surf. Sci.* **2020**, *526*, 146557. [CrossRef]
66. Kim, H.Y.; Kim, T.H.; Cha, S.M.; Yu, S. Degradation of Sulfamethoxazole by Ionizing Radiation: Identification and Characterization of Radiolytic Products. *Chem. Eng. J.* **2017**, *313*, 556–566. [CrossRef]
67. Wang, J.; Wu, Y.; Zheng, Y.; Liu, J.; Zhao, F. Efficient Degradation of Sulfamethoxazole and the Response of Microbial Communities in Microbial Fuel Cells. *RSC Adv.* **2015**, *5*, 56430–56437. [CrossRef]
68. Jia, Y.; Khanal, S.K.; Zhang, H.; Chen, G.H.; Lu, H. Sulfamethoxazole Degradation in Anaerobic Sulfate-Reducing Bacteria Sludge System. *Water Res.* **2017**, *119*, 12–20. [CrossRef]
69. Herklotz, P.A.; Gurung, P.; vanden Heuvel, B.; Kinney, C.A. Uptake of Human Pharmaceuticals by Plants Grown under Hydroponic Conditions. *Chemosphere* **2010**, *78*, 1416–1421. [CrossRef]
70. Bewick, M.W.M. The Use of Antibiotic Fermentation Wastes as Fertilizers for Tomatoes. *J. Agric. Sci.* **1979**, *92*, 609–674. [CrossRef]
71. Wang, J.; Wang, S. Microbial Degradation of Sulfamethoxazole in the Environment. *Appl. Microbiol. Biotechnol.* **2018**, *102*, 3573–3582. [CrossRef]
72. Junza, A.; Barbosa, S.; Codony, M.R.; Jubert, A.; Barbosa, J.; Barrón, D. Identification of Metabolites and Thermal Transformation Products of Quinolones in Raw Cow's Milk by Liquid Chromatography Coupled to High-Resolution Mass Spectrometry. *J. Agric. Food Chem.* **2014**, *62*, 2008–2021. [CrossRef]
73. Sturini, M.; Speltini, A.; Maraschi, F.; Profumo, A.; Pretali, L.; Fasani, E.; Albini, A. Photochemical Degradation of Marbofloxacin and Enrofloxacin in Natural Waters. *Environ. Sci. Technol.* **2010**, *44*, 4564–4569. [CrossRef]
74. Haddad, T.; Kümmerer, K. Characterization of Photo-Transformation Products of the Antibiotic Drug Ciprofloxacin with Liquid Chromatography-Tandem Mass Spectrometry in Combination with Accurate Mass Determination Using an LTQ-Orbitrap. *Chemosphere* **2014**, *115*, 40–46. [CrossRef] [PubMed]
75. Borghi, A.A.; Silva, M.F.; al Arni, S.; Converti, A.; Palma, M.S.A. Doxycycline Degradation by the Oxidative Fenton Process. *J. Chem.* **2015**, *2015*, 492030. [CrossRef]

MDPI
St. Alban-Anlage 66
4052 Basel
Switzerland
www.mdpi.com

Molecules Editorial Office
E-mail: molecules@mdpi.com
www.mdpi.com/journal/molecules

Disclaimer/Publisher's Note: The statements, opinions and data contained in all publications are solely those of the individual author(s) and contributor(s) and not of MDPI and/or the editor(s). MDPI and/or the editor(s) disclaim responsibility for any injury to people or property resulting from any ideas, methods, instructions or products referred to in the content.

www.ingramcontent.com/pod-product-compliance
Lightning Source LLC
LaVergne TN
LVHW070414100526
838202LV00014B/1457